Tools and Techniques for Social Science Simulation

Ramzi Suleiman · Klaus G. Troitzsch
Nigel Gilbert (Eds.)

Tools and Techniques for Social Science Simulation

With 101 Figures
and 22 Tables

Physica-Verlag
A Springer-Verlag Company

Dr. Ramzi Suleiman
Department of Psychology
University of Haifa
31905 Haifa
Israel

Prof. Dr. Klaus G. Troitzsch
Institut für Sozialwissenschaftliche Informatik
University of Koblenz-Landau
Rheinau 1
56075 Koblenz
Germany

Prof. Nigel Gilbert
Centre for Research on Simulation in the Social Sciences
University of Surrey
Guildford GU2 5XH
United Kingdom

ISBN 978-3-7908-1265-7 ISBN 978-3-642-51744-0 (eBook)
DOI 10.1007/978-3-642-51744-0

CIP-Kurztitelaufnahme der Deutschen Bibliothek
Tools and techniques for social science simulation / Ramzi Suleiman; Klaus G. Troitzsch; Nigel Gilbert (eds.). – Heidelberg: Physica-Verl., 2000

Softcover reprint of the hardcover 1st edition 2000

Cover design: Erich Kirchner, Heidelberg

SPIN 10750445 88/2202-5 4 3 2 1 0 – Printed on acid-free paper

Preface

The use of computer simulations to study social phenomena has grown rapidly during the last few years. Many social scientists from the fields of economics, sociology, psychology and other disciplines now use computer simulations to study a wide range of social phenomena. The availability of powerful personal computers, the development of multidisciplinary approaches and the use of artificial intelligence models have all contributed to this development.

The benefits of using computer simulations in the social sciences are obvious. This holds true for the use of simulations as tools for theory building and for its implementation as a tool for sensitivity analysis and parameter optimization in application-oriented models. In both, simulation provides powerful tools for the study of complex social systems, especially for dynamic and multi-agent social systems in which mathematical tractability is often impossible. The graphical display of simulation output renders it user friendly to many social scientists that lack sufficient familiarity with the language of mathematics.

The present volume aims to contribute in four directions: (1) To examine theoretical and methodological issues related to the application of simulations in the social sciences. By this we wish to promote the objective of designing a unified, user-friendly, simulation toolkit which could be applied to diverse social problems. While no claim is made that this objective has been met, the theoretical issues treated in Part 1 of this volume are a contribution towards this objective. (2) To contribute to the literature on the sensitivity analysis of simulations. This objective, addressed in Part 4, is of particular significance given the notorious sensitivity of simulation results to variations in the initializing conditions and in the parameters that define the dynamics. (3) To discuss various multi-agent based simulations and demonstrate their application to the study of social processes. (4) To apply cellular automata based simulations to the investigation of social evolution processes in various contexts.

The present volume brings together scientists from the fields of economics, sociology and psychology, as well as computer scientists interested in the application of computer simulations in the social sciences. Earlier versions of the book chapters were presented at an international seminar entitled "Social Science Microsimulation: Tools for Modeling, Parameter Optimization, and Sensitivity Analysis". The seminar was held in May 5-9, 1997 at Schloss Dagstuhl, Germany, and was organized by Professor Ulrich Mueller together with the editors of this volume.

We are grateful to the International Conference and Research Center for Computer Science at Schloss Dagstuhl for hosting the conference. We also wish to thank the University of Haifa, the University of Koblenz-Landau and the University of Surrey. Special thanks to the Research Authority, the Institute for Information Processing and Decision Making, and the Jewish-Arab Center at the

University of Haifa for their generous financial support in producing this volume. Finally, we wish to thank Ms. Galit Gilad for her valuable help in formatting this volume and preparing the author and subject indexes.

Haifa, Koblenz, and Guilford, July 1999

Ramzi Suleiman Klaus G. Troitzsch Nigel Gilbert

Contents

I Simulations as Tools for Modelling and Theory Building

1 Models, Processes and Algorithms: Towards a Simulation Toolkit 3
by Nigel Gilbert

2 Questions in the Methodology of Artificial Societies 17
by Jim Doran

3 Computer Assisted Interfacing: On the Use of Computer Simulation for Theory Construction 26
by Georg Müller

4 Interactive Neural Networks as Tools for Modeling Social Systems ... 48
by Christina Stoica

II Multi-Agent Based Simulations

5 SMASS: A Serial Multi-Agent System for Social Simulation 65
by Wolfgang Balzer

6 PECS: A Reference Model for the Simulation of Multi-Agent Systems .. 83
by Christoph Urban

7 Bedrock: Building Multi-Agent Simulation Applications 115
by Matthew J. Rockloff

8 Diversity in Strategies of Partnership Formation 131
by Rosaria Conte

III Game Theory and Cellular Automata

9 Cellular Automata Models as a Modelling Tool 151
by Rainer Hegselmann, Andreas Flache & Volker Möller

10 Evolution of Learning Rules in Space 179
by Oliver Kirchkamp

11 Maintaining Diversity: Simulating the Role of Nonliniarity and Discreteness in Dynamic Social Impact 196
by Bibb Latané & Hiroaki Morio

12 Representation Methods and the Emergence of Inter-Group Cooperation.. 218
by Ilan Fischer & Ramzi Suleiman

IV Sensitivity Analysis

13 Sensitivity Analysis in the Social Sciences: Problems and Prospects 243
by Edmund Chattoe, Nicole J. Saam & Michael Möhring

14 Tests of Dynamic Social Models with Time Related Surveys: an Experimental Approach 274
by Christof Schatz

15 What Is the Use of Gossip? A Sensitivity Analysis of the Spreading of Respectful Reputation 302
Mario Paolucci, Mario Marsero & Rosaria Conte

V Applications

16 Modelling Greening Investors: Economic Development, Opinion Formation, and Technological Change in a Multilevel Simulation Model 317
by Kai-H. Brassel, Ottmar Edenhofer, Michael Möhring & Klaus G. Troitzsch

17 Microsimulating Farm Business Performance 344
by Philip Kokic, Ray Chambers & Steve Beare

18 Ordering Parameters in the Rule Space of Social Systems 351
by Jürgen Klüver, Jörn Schmidt & Ralph Kier

Biographical Sketches 373

Subject Index 380

Author Index 382

Part I

Simulations as Tools for Modelling and Theory Building

1. Models, Processes and Algorithms: Towards A Simulation Toolkit

Nigel Gilbert[1]

This chapter begins the specification of the ideal features of a toolkit for social simulation, starting from a consideration of the standard methodology for simulation research. Several essential components, commonly used in social science simulation research, are identified and it is argued that implementations of these will need to be included in the toolkit. Additional modules, providing graphical output, scheduling, random number generation and parameter editing are also required.

Some existing toolkits are described and it is suggested that, while there are both advantages and costs in taking the toolkit approach to building simulations, the process of specifying the requirements for a toolkit may be a valuable activity in its own right because it can help to clarify simulation methodology.

1.1 Target, Model, Simulation

The use of computer simulation in the social sciences now has a history stretching back nearly twenty years (Troitzsch, 1997), and the sciences have experience over an even longer period. During this time, some basic methodological ideas have evolved. First, there is the idea of modelling itself. We assume that there is some 'real world' phenomenon in which the researcher is interested. This we call the *target* (the terminology was borrowed by Doran and Gilbert, 1994, from Zeigler, 1976). The objective is to create a *model* of this target which is simpler to study than the target itself. We hope that conclusions drawn about the model will also apply to the target because the two are sufficiently similar. However, if it is to be useful, the model always needs to be and, indeed, will be simpler than the target since our modelling abilities are limited. For example, we might model the real market for the wholesale supply of fish with a simpler system where both suppliers and purchasers are represented by computer programs standing in for complex and multi-faceted businesses and their customers (cf. Weisbuch et al., 1997).

---

[1] Centre for Research on Simulation in the Social Sciences, Department of Sociology, University of Surrey, Guildford GU2 5XH, England.
email: n.gilbert@soc.surrey.ac.uk

In the social sciences, the target is always a dynamic entity, changing over time and reacting to its environment. It has both structure and behaviour. This means that the model must also be dynamic. We can represent the model itself as a *specification* — a mathematical equation, a logical statement or a computer program — but to learn something from the specification, we need to examine how the model develops over time. One way of doing this is using analytical methods. This entails deriving the future structure from the specification by reasoning, perhaps using logic or more often by using mathematics. For example, we might have a model of the relationships between a set of macro-economic variables and use algebra to derive the outcome if one of those variables changes over time. With complex models, especially if the specification is non-linear, such analytic reasoning can be very difficult or impossible. In these cases, *simulation* is often the only way. Simulation means 'running' the model forward through (simulated) time and watching what happens. Whether one uses an analytical technique or simulation, the *initial conditions*, that is, the state at which the model starts, are always very important. Often, the dynamics are very different depending on the precise initial conditions used.

1.2 The Lifecycle of Simulation Research

With these terms in mind, we can outline the 'ideal' set of steps in using simulation in the social sciences (cf. Doran, 1997). We start with identifying a 'puzzle', a question whose answer is not known and which it will be the aim of the research to resolve. For example, we might be curious about the reasons for the particular pattern of Puebloan settlements in Mexico from AD 900 to 1300 (Kohler et al., 1996). This leads us to the *definition* of the target for modelling (settlement dynamics in the Mesa Verde region). Normally, some *observations* of the target will be required in order to provide the parameters and initial conditions for our model. For Kohler et al's (1996) work, these were provided by detailed archaeological work by Van West (1994). One can then develop the model, probably in the form of a computer program. The *simulation* itself is performed by executing this program, and the output of the simulation is recorded. So far, the steps involved are fairly obvious, although often not simple to carry out. The remaining steps are less often attended to, yet they are crucial. We need to ensure that the model is correctly implemented and working as intended. This is *verification*, in effect, a 'debugging' step. Unfortunately, with complex simulations this process can be difficult to carry out and, in particular, it is always difficult to know whether one has eradicated the last remaining bug. The difficulty is compounded by the fact that most social science simulations are dependent on pseudo random numbers to simulate the effects of unmeasured variables and random effects (Gilbert, 1996) and so repeated runs can be expected to produce different outcomes. Next, there is *validation*, ensuring that the behaviour of the model does correspond to the behaviour of the target. If settlement patterns in the Mesa Verde are being modelled, the simulation needs to reproduce to some adequate degree the observed pattern of settlements. Unless there is some correspondence, the simulation is unlikely to be a plausible model of the processes which led to the formation of those settlements. Finally, one needs to know how sensitive the model is to slight changes in the parameters and initial conditions: a *sensitivity analysis*.

1.3 The Role of Toolkits

Against the background of this standard simulation methodology, one can ask how the social scientist researcher can best be supported in his or her endeavour. One possibility often suggested is that there should be a 'toolkit' for simulation research, with which the researcher can build appropriate models for a wide variety of targets by assembling pieces from the kit. There are some exemplars of such a toolkit in related areas. For example, statistical analysis is almost always done with a statistical package, of which the best known is SPSS (Statistical Package for the Social Sciences, Norusis, 1990). This provides the researcher with a large number of data analysis methods within a unified data management capability and with a consistent user interface. Once one has learnt how to use SPSS, applying a new statistical procedure requires learning about the theory behind the statistical technique, but does not involve knowing more about the computational side of the analysis. Can one envisage a similar package for simulation? Another interesting example is the family of discrete event simulation packages which are becoming available for the construction of, for example, models of business work flow. There are also packages for modelling using the system dynamics approach, of which STELLA is the best known. These packages provide not only a standardised and verified calculation module which will carry out the simulation, but also graphical interfaces which enable researchers to construct models without doing much conventional programming.

These examples indicate what we should look for in a simulation toolkit. It should provide:

- a set of modules incorporating procedures or algorithms appropriate for social simulation. The second part of this paper will make suggestions about what these might consist of.
- compatible interfaces for all the modules, so that data can flow from one to another.
- a consistent user interface, so that as far as possible what the user learns about one module can be transferred to the other modules
- a set of supporting modules to provide, for example, easy to use facilities for graphical output (graphs, charts, maps, etc.) and data input.

There would be considerable advantages in using a toolkit such as this. If properly designed, it would enable modellers to construct simulations with much less effort, since they would not have to start 'from the ground up' for each new program. The toolkit would also hide many of the implementation details from the modeller, making the task of learning how to simulate much easier. Because the modeller would be using code which had already been verified and tested in many different models, there would be less opportunity for programming errors to creep in. It would be worth studying the properties of the algorithms implemented in the toolkit in depth, because they would be used regularly in many different models. The toolkit thus allows a division of labour: the toolkit builders would be the specialists in the construction, implementation and testing of algorithms, while the users could concentrate on applying the toolkit to social science problems. Finally, the toolkit could become an effective teaching device, helping beginning modellers by guiding them about the decisions which need to be made in constructing a model and restraining them from committing errors.

There are also some disadvantages in the use of a toolkit, of course. The converse of guiding the user is that other, perhaps more innovative ways of using the tools are discouraged. The number and scope of the tools within a toolkit will always be limited and it may be difficult for users with an unconventional problem to achieve what they need. The provision of compatible module interfaces, a graphical user interface and the need to cater for the 'average' model, rather than the particularities of any specific model, mean that it is very hard for toolkits to be as efficient as specially written programs. The loss in efficiency can be substantial, perhaps one or two orders of magnitude, which means that a simulation which could take a minute or two to run using a hand-crafted program might take a couple of hours using a toolkit. The time taken to run a simulation is an important consideration at the stages of verification and sensitivity analysis, when the simulation may have to be executed hundreds or thousands of times with different parameters. Finally, there are the economics to consider. Building and maintaining a toolkit is best achieved by a business which has the continuity and administrative mechanisms required to ensure that support is always available. Maintaining a toolkit fits uneasily into the academic environment, where innovation and change are the main characteristics. However, maintenance and administration have to be paid for, which means that commercial toolkits in other fields are usually expensive ($500 and up) and sometimes very expensive, depending on the size of the market. In contrast, provided that one has programming skills and access to a PC, building one's own model can be done for nothing in one's spare time.

1.4 The Ingredients of a Toolkit for Social Science Simulation

Even if one believes that the advantages of using a toolkit outweigh the costs, it is still necessary to determine whether there are actually any 'modules' which are of sufficiently general application to warrant inclusion in a social science toolkit. With statistical packages, following the quantitative revolution in the United States in the 1950s, there was a clear case for centring them around multiple regression, cross-tabulation and analysis of variance. Hence, the earliest packages featured these techniques and have gradually been extended to include more esoteric procedures. But are there any equivalent basic and essential techniques within the computer simulation field?

The answer to this question is partly dependent on theoretical positions taken with respect to the targets to be simulated and partly dependent on the stock of techniques which social scientists have found to implement their simulations. With respect to the first point, I have argued elsewhere (Gilbert, 1999) that there are a fairly small number of 'abstract social processes' which are important in the social world, in that they keep reappearing in many different settings and lie behind many different social phenomena. For example, there are many circumstances in which people tend to interact preferentially with those who are in some way similar to themselves. Since the interaction usually tends to strengthen the similarity between the interacting partners, this process has an element of positive feedback. In due course the effect is to cluster the people into neighbourhoods, either through physical or social proximity. This kind of process may be at the root of the formation of dialects (e.g. Hutchins and Hazlehurst, 1995; Latané, 1996; Latané and Nowak, 1997), the creation of some kinds of segregation (e.g. Schelling, 1971) and many other familiar institutions. A toolkit needs to be able to model this social process, which has been called 'selection according to similarity' (Kontopoulos, 1993), as well as the others which may be identified by social scientists.

Even without using toolkits, those interested in using computer simulation have in practice tended to use quite a small number of techniques as the basis for their models. It would clearly be desirable for a toolkit to include at least these techniques. Giving the modeller easy access to them will maximise the potential number of users of the toolkit. The following techniques or algorithms seem to be candidates for the list of 'standard' simulation methods which a toolkit should aim to support:

1.4.1 Microsimulation

Microsimulation (Harding, 1990) involves simulating the passage of a large number of 'base' units (usually individuals, households or firms) through time, while applying transfer functions which represent the effects of ageing to each unit. For example, the base units might be individuals and the objective of the simulation might be to predict the effect of state pension arrangements some decades in the future, both in terms of the aggregate cost of pensions and the distribution of wealth among a population. To make the simulation as close as possible to the target — the actual population — data about the base units are typically drawn from a random sample of a national representative survey.

Each unit is 'aged' independently of the others and there is no direct interaction between the units (although there may be some effect of characteristics at the macro i.e. population level, for example, simulated individual behaviour may depend on the overall unemployment rate). As a result, the processing of each unit from one time period to the next is straightforward, but to achieve acceptably accurate results, large numbers of units need to be simulated. Consequently, there is a need for the efficient sequential processing of large numbers of essentially identical cases.

1.4.2 Cellular Automata

Cellular automata (CA) models (for an introduction see Hegselmann, 1996) are also based on the processing of a large number of identical, simple units, but these units are organised spatially, usually in a regular grid, and each unit can interact with and affect its immediate neighbours. As with microsimulation, each unit is advanced through time using an identical transfer function. CA introduce two new features: the need to store and represent units in space, and the need to process units in a way which takes account of the fact that their next state depends on the states of their neighbours, which are also changing. The former imposes a requirement on the storage of data inside the program and also on the kind of user interface which is appropriate for observing the simulation. The latter generally means that the new state of each unit needs to be stored separately from the old state, because each unit's new state is calculated from the 'old' state of its neighbours.

1.4.3 Production Systems

A completely different approach to simulation is taken by those using production systems. A production system consists of a number of rules, each with a condition and an action part, a memory and a rule interpreter. In operation, the rule interpreter scans the rules one by one. Each rule is examined to see whether the conditions it specifies match against the current contents of the memory. If so the rule is 'fired' and the action part of the rule is executed. This action part will usually modify the system's memory, so when the interpreter rescans the rules, a different rule will fire. This process of scanning and firing rules continues indefinitely.

While the basic mechanism is rather simple, there are some complications. There needs to be some procedure for resolving which rule is fired when the conditions of more than one match the memory. It may be that the first matching rule is fired, given some ordering of the rules, or the last one, or the rule which is most 'general'— the modeller has to make a decision about which of these conflict resolution strategies is most appropriate. Another difficulty is that condition-action rules are rather fine-grained and a large number are often required. However, understanding the interactions between large numbers of rules can become very difficult and it is easy to introduce 'bugs'. One solution has been to package the rules into rule-sets, each of which runs as though it was an independent production system, with a minimum of interaction between the rule-sets. One of the most sophisticated production system based models is SOAR, which aims to simulate aspects of the human cognitive system (Laird et al., 1987; Newell, 1990). This has been used as the basis for modelling individual agents in organisations (Ye and Carley, 1995; Carley et al., 1996).

1.4.4 Multi-Agent Systems

Production systems or other techniques can be used to build agents, entities which receive information, assess it in the light of their past experience, decide what action to take, and carry out that action (Wooldridge and Jennings, 1995; Müller, 1996). The decision-making may be relatively simple and *reactive*, or complex, involving planning, inference, learning and sophisticated knowledge representation in so-called *deliberative* agents. If the agents communicate, we have a multi-agent system, the topic of interest to those concerned with Distributed Artificial Intelligence (DAI), a subfield of artificial intelligence "concerned with distributing and coordinating knowledge and actions in multiple agent environments" (O'Hare and Jennings, 1996).

Agents in a multi-agent system (MAS) are most naturally modelled using an object-oriented programming language (Uhrmacher, 1996; 1997). This makes it easy to develop many agents each with the same or similar behaviour and knowledge, and provides appropriate debugging and program management and specification tools. In addition to agents, a MAS needs to include a representation of the environment in which the agents are located. Often, this environment is a spatial grid (as with cellular automata, although agents may move through the environment). The environment can also be an object in the system, although one with different properties from the agents. The agents interact through the environment. Messages from one agent are collected by the environment and then passed on to the recipient agent. There needs to be some form of message encoding and decoding built into the agents (agents cannot see directly 'what is in another agent's head'), using for example the standard protocol designed for agent communication, KQML (Finin and Fritzon, 1994).

MAS are the clearest examples of multi-level systems, although all the techniques mentioned here can be used to build multi-level systems. A multi-level system is one in which the population attributes, aggregated from the attributes of all the agents, influence the actions or characteristics of the individual agents (Troitzsch, 1996). We have already had one example of this when considering the microsimulation of individual behaviour when it depends on the aggregate unemployment rate, and there are many others.

1.4.5 Learning and Adaptive Models

A characteristic of humans which it may be important to model is their ability to learn. Organisations and other institutions also adapt and 'learn' through experience. To model learning and adaptation, the most common techniques used are variations on genetic algorithms (Holland, 1975; Michalewicz, 1996) and artificial neural networks (Swingler, 1996).

Genetic algorithms operate on a population of 'genes', consisting of coded representations of the design of some part of an agent, for example, its strategy for interacting with other agents (an early example is Axelrod, 1987). Two agents are chosen using a procedure which selects proportionately to their "fitness" and their genes are combined using a 'crossover' procedure which takes some of one gene and some of the other. The resulting gene may then be slightly changed by means of a random mutation. The new gene is used to generate a new individual and further offspring are generated from other gene combinations. Once the new agents have been created, the old ones are killed off. In this way, the population evolves while optimising the agents' collective fitness, although each individual agent does not change during its 'life'. There are numerous variations on this theme, including genetic programming (Koza, 1992; Koza, 1994) and cultural algorithms (Reynolds, 1994), differing in the way the parents are selected, the representation used by the genes, and the way that fitness is defined.

While genetic algorithms may be used to adapt a population, artificial neural networks are typically used to enable individual agents to learn. For example, Parisi et al (1995) model agents as neural networks in an environment in which they have to learn to obtain food.

Both genetic algorithms and neural networks learn or adapt by exposure to many instances: many generations of large populations in the case of genetic algorithms, and many training instances in the case of neural networks. The implication is that the underlying implementation needs to be efficient if the adaptation is to proceed reasonably quickly.

This discussion of the basic techniques appropriate for social simulation has indicated that it will be hard to build them all into one toolkit in a way which allows for compatible interfaces and a consistent user interface. This is because the underlying assumptions about the structure and behaviour of the agents to be modelled differ considerably from one technique to the next. For example, most microsimulation models assume a large number of more or less identical, simple agents without any relationship between them. In contrast, cellular automata require the implicit relationship between agents (e.g. proximity) to be modelled, while the typical MAS explicitly models these relationships in terms of an environment. Similarly, while DAI based models will have complex agents, possibly involving planners and parsers and other AI modules, CA agents are typically extremely simple in construction.

Another difficulty this brief survey has revealed is that several of the techniques demand considerable computer resources. We remarked on this specifically with microsimulation and learning and adaptive models. Unfortunately, as we noted earlier, one of the consequences of building a technique into a general toolkit is that efficiency tends to be sacrificed. These points remain as challenges for the toolkit builder.

1.5 Other Ingredients of a Toolkit

While we have concentrated on the simulation modules of a potential toolkit, there are other features which are also essential if it is to be useful. Experience has shown that the following all need attention:

1.5.1 Graphics

The nature of simulation is that it generates very large amounts of diagnostic output. For example, if a simulation is executed for one thousand time units, and at the end of each cycle, 10 values describing the state of each of 50 agents are collected, the researcher will need to comprehend half a million data points for every simulation run. The only way to deal with this volume of data is to view it graphically. However, implementing graphical output is a skilled and time-consuming task unless a library of charts is made available as part of the toolkit. This library needs to include a full range of charts, including varieties of line plots, bar charts, and scatter plots, and these need to be implemented so that they can be dynamically fed with data as the simulation progresses, with the charts automatically rescaling themselves to fit the accumulating data. In addition to the usual plots used by data analysts, raster plots are useful to display the grids used by cellular automata models. These too need to be flexible, with easy control over the appearance of each cell, and to be updateable dynamically.

1.5.2 Scheduling Time

There are two basic methods of stepping through simulated time: either time is modelled moment by moment (in practice, a 'moment' consists of one cycle through the simulation), or the simulation 'jumps' directly to the time when the next 'event' is due. Both methods have their advantages and both ought to be available in a toolkit. A complication arises with the simulation of events which are intended to be carried out at the same time. Using an ordinary computer, parallel execution of two events is impossible, and so must be simulated by making them happen sequentially. However, this can be tricky, especially when the two events affect each other. A typical example of this difficulty occurs with CA models, where the state of one cell depends on the states of its neighbours and the neighbours' states depend on the cell's states. In some situations results of the simulation can depend crucially on the precise updating procedure used to simulate parallel execution of each cell (Hegselmann, 1996). A toolkit must therefore make it possible to control such factors in the scheduling of events.

1.5.3 Random Number Generators

Almost all social science simulations use random number generators, because the models are intrinsically stochastic (e.g. genetic algorithms), because there are 'omitted factors' which are not of interest and are therefore not modelled (e.g. agents' internal cognitive and affectual processes or environmental factors), or because the researcher wants to run simulations which sample parameter values. Moreover, random number generators are often used very intensively, with streams of thousands or millions of numbers being needed. It is therefore essential that the generators are of high quality, that is, produce numbers which are practically indistinguishable from really random numbers.

Often random numbers are required to be taken, not from the usual uniform distribution, but from more complex distributions such as the Normal, Poisson or Bernoulli distributions. Standard techniques exist for creating random numbers which appear to be drawn from these distributions and a toolkit needs to implement these.

1.5.4 Parameter Editors

Most simulations involve at least tens of parameters and initial conditions, and in some cases, there can be thousands of such values. A method of inputting these parameters efficiently and logging their values and their consequences for the outcome of the simulation is essential. Often one wants to repeat a simulation with most parameter values kept constant, and a few altered, for example, when conducting a sensitivity analysis. This is best done by reading the constant values from a file, but entering the changing value through a graphical interface (e.g. a slider). There needs to be a facility to make this easy to set up.

1.6 Existing Toolkits and Programming Environments

The previous sections have amounted to a very informal specification of what a toolkit for social simulation should include. To what extent do existing commercial or publicly available toolkits measure up to these requirements? It is difficult to compile a complete list of all toolkits, and the list would change month by month. Many systems which could be classified as toolkits are in fact little more than somewhat generalised versions of simulation programs written to implement a specific model. Others are mainly intended as proofs by demonstration of particular modelling approaches, rather than useful and portable software. However, there a number of toolkits which are relatively well-known and three of these are very briefly described below, with references to where further information may be found. A longer and more current list of toolkits may be found at http://www.soc.surrey.ac.uk/research/simsoc/languages.html.

1.6.1 MIMOSE

MIMOSE (an abbreviation for MIcro and multilevel MOdelling SoftwarE) is a general model description language and interpreter for building multi-level models (Moehring, 1996). The MIMOSE programming language is object-oriented and functional (and so does not include variables with state). The toolkit includes a programming environment with which the simulation can be initialised and controlled. Facilities are provided for editing input parameters and recording results and for drawing graphs. MIMOSE runs under Unix (Solaris, Linux, and NextStep variants) with X-windows and is available without charge (see http://www.uni-koblenz.de/~sozinf/projecte/mimose/).

1.6.2 Swarm

Swarm is a multi-agent object-oriented toolkit. In Swarm, the basic unit is the 'swarm', a collection of agents executing a schedule of actions. Swarms can be composed of other swarms, enabling several levels of multilevel modelling. Libraries of components for building models are provided which offer reliable random number generators, dynamic graphics, and implementations of genetic algorithms and neural networks. Swarm's base programming language is a little known variant of C called Objective C. It uses X-windows to provide a graphical user interface

Swarm is available without charge for Unix (Linux, Solaris and others). A version to run on Windows PCs is under development at the time of writing. Current information on Swarm can be obtained at - http://www.santafe.edu/projects/swarm/.

1.6.3 LispStat

LispStat is a version of the programming language, Common Lisp, which has been extended by providing functions for statistical data analysis, with an emphasis on providing dynamic graphical methods. LispStat extends standard Lisp arithmetic operations to perform element-wise operations on lists and vectors, and adds a variety of basic statistical and linear algebra functions. A suite of functions provide graphical plots of various kinds, all of which can be extended and customised. An object-oriented programming system is used as the basis for statistical model representations, such as linear and non-linear regression models and generalised linear models. Although LispStat is designed for the development of statistical models, it has many of the features also needed for simulation models, such as its dynamic graphics and statistical functions.

LispStat is available without charge and in implementations for Unix, Macintosh and Windows PCs. Further information can be found at http://stat.umn.edu/~luke/xls/xlsinfo/xlsinfo.html.

1.7 Conclusion

Toolkits could be both a blessing and a curse on social simulation. They will be a blessing to the extent that they open up the possibility of developing simulation models to social scientists who previously did not have the skills or the dedication to build their own models from scratch, in the way that statistical packages have made it possible for social scientists to develop and evaluate their own regression models. They also facilitate replication of models and comparison between models. However, it remains to be seen to what extent any toolkit can entirely relieve the user of programming. For instance, although the Swarm toolkit is unusually comprehensive compared with other toolkits, in order to build a model, the user has to write code in a language, Objective C, which he or she is almost certain to be unfamiliar with.

Even if one remains sceptical about the eventual benefits of constructing a simulation toolkit, there may still be advantages in trying. Simply specifying requirements, as I have tried to do in this chapter, can clarify what simulation involves as a research method and point to ways of doing it better. While the goal of a successful social simulation toolkit may be no more than a dream, the road to the end of the rainbow may be worth travelling.

References

Axelrod, R. (1987): Genetic Algorithm for the Prisoner Dilemma Problem. In: Davis, L. (ed.): Genetic Algorithms and Simulated Annealing. Morgan Kaufmann, San Mateo, CA, 32-41.

Carley, K. M., Prietula, M. J. & Zhiang, L. (1996): Design versus Cognition: the interaction of agent cognition and organizational design on organizational performance. Carnegie Mellon University.

Doran, J. (1997): From computer simulation to artificial societies. Transactions of the Society for Computer Simulation International 14(2), 69-78.

Doran, J. & Gilbert, G. N. (1994): Simulating societies: an introduction. In: Gilbert, G. N. & Doran, J. (eds.): Simulating Societies: the computer simulation of social phenomena. UCL Press, London, 1-18.

Finin, T. & Fritzon, R. (1994): KQML: a language and protocol for knowledge and information exchange. Proceedings of the 13th International Distributed Artificial Intelligence Workshop, Seattle, WA.

Gilbert, G. N. (1996): Simulation as a research strategy. In: Troitzsch, K. G., Müller, U., Gilbert, G. N. & Doran, J. E. (eds.) Social science microsimulation. Springer, Berlin, 448-454.

Gilbert, G. N. (1999): Modeling Sociality: The View from Europe. In: Kohler, T. & Gumerman, G. (eds.): Dynamics in Human and Primate Societies: Agent Based Modeling of Social and Spatial Processes. Oxford University Press, Forthcoming.

Harding, A. (1990): Dynamic microsimulation models: problems and prospects. Discussion Paper 48, Welfare State Programme London School of Economics.

Hegselmann, R. (1996): Understanding social dynamics: the cellular automata approach. In: Troitzsch, K. G., Müller, U., Gilbert, G. N. & Doran, J. E. (eds.) Social science microsimulation. Springer, Berlin, 282-306.

Holland, J. H. (1975): Adaptation in Natural and Artificial Systems. Ann Arbor, University of Michigan Press.

Hutchins, E. & Hazlehurst, B. (1995): How to invent a lexicon: the development of shared symbols in interaction. In: Gilbert, G. N. & Conte, R. (eds.): Artificial Societies. UCL Press, London, 157-189.

Kohler, T. A., Van West, C. R., Carr, E. P. & Langton, C. G. (1996): Agent-based modelling of prehistoric settlement systems in the Northern American Southwest. Third International Conference integrating GIS and environmental modelling National Center for Geographic Information and Analysis, Santa Fe, Santa Barbara.

Kontopoulos, K. M. (1993): The logics of social structure. Cambridge University Press, Cambridge.

Koza, J. R. (1992): Genetic programming. MIT Press,Cambridge, Mass.

Koza, J. R. (1994): Genetic programming 2. MIT Press,Cambridge, Mass.

Laird, J. E., Newell, A. & Rosenbloom, P. S. (1987): Soar: an architecture for general intelligence. Artificial Intelligence 33(1), 1-64.

Latané, B. (1996): Dynamic social impact. In: Hegselmann R., Müller U. & Troitzsch, K. G. (eds.): Modelling and simulation in the social sciences from the philosophy of science point of view. Springer-Verlag, Berlin, 287-310.

Latané, B. & Nowak, A. (1997): Self-Organizing Social Systems: necessary and sufficient conditions for the emergence of clustering, consolidation and continuing diversity. In: Barnett, G. A. & Boster, F. J. (eds.): Progress in communication sciences: persuasion. Ablex, Norwood, NJ, 43-74.

Michalewicz, Z. (1996): Genetic algorithms + data structures = evolution programs, 3rd edition. Springer-Verlag, Berlin.

Moehring, M. (1996): Social science multilevel simulation with MIMOSE. In: Troitzsch, K. G., Müller, U., Gilbert, G. N. & Doran, J. E. (eds.) Social science microsimulation. Springer, Berlin, 123-137.

Müller, J. P. (1996): The design of intelligent agents: a layered approach. LNAI vol. 1177. Springer, Berlin.

Newell, A. (1990): Unified Theories of Cognition. Harvard University Press, Cambridge, MA.

Norusis, M. J. (1990): SPSS/PC+ 4.0 base manual. SPSS Inc.,Chicago.

O'Hare, G. & Jennings, N. (eds.) (1996): Foundations of Distributed Artificial Intelligence. John Wiley and Sons, London.

Parisi, D., Cecconi, F. & Cerini, A. (1995): Kin-directed altruism and attachment behaviour in an evolving population of neural networks. In: Gilbert, N. & Conte, R. (eds.): Artificial Societies. UCL Press, London, 238-251.

Reynolds, R. G. (1994): Learning to cooperate using cultural algorithms. In: Gilbert, N. & Doran, J. (eds.): Simulating societies UCL Press, London, 223-244.

Schelling, T. C. (1971): Dynamic models of segregation. Journal of Mathematical Sociology 1, 143-186.

Swingler, K. (1996): Applying neural networks: a practical guide. Academic, London.

Troitzsch, K. G. (1996): Multilevel simulation. In: Troitzsch, K. G., Müller, U., Gilbert, G. N. & Doran, J. E. (eds.) Social science microsimulation. Springer, Berlin, 107-122.

Troitzsch, K. G. (1997): Social science simulation: origins, prospects, purposes. In: Conte, R., Hegselmann, R. & Terna, P. (eds.): Simulating social phenomena. Springer, Berlin, 41-54.

Uhrmacher, A. (1996): Object-oriented and agent-oriented simulation: implications for social science applications. In: Troitzsch, K. G., Müller, U., Gilbert, G. N. & Doran, J. E. (eds.) Social science microsimulation. Springer, Berlin, 432-445.

Uhrmacher, A. M. (1997): Concepts of object- and agent-oriented simulation. Transactions of the Society for Computer Simulation International 14(2), 59-68.

Van West, C. (1994): Modeling Prehistoric Agricultural productivity in Southwestern Colorado: a GIS approach. Washington State University.

Weisbuch, G., Kirman, A. & Herreiner, D. (1997): Market organisation. In: Conte R., Hegselmann, R. & Terna, P. (eds.): Simulating social phenomena. Springer, Berlin, 221-240.

Wooldridge, M. & Jennings, N. R. (1995): Intelligent agents: theory and practice. Knowledge Engineering Review 10(2), 115-152.

Ye, M. & Carley, K. M. (1995): Radar-Soar: towards an artificial organization composed of intelligent agents. Journal of Mathematical Sociology 20(2-3), 219-246.

Zeigler, B. P. (1976): Theory of modelling and simulation. Wiley, New York.

2. Questions in the Methodology of Artificial Societies

Jim Doran[1]

2.1 Introduction

My aim in this short paper is to examine some methodological problems associated with computer simulation of social systems and especially with the use of so called "artificial societies". I shall proceed by posing and discussing a series of questions intended to provoke consideration of key issues.

Adding "intelligent" agent technology within conventional computer simulation models creates both major opportunities and serious problems. The opportunities are obvious enough: a multiple agent model enables precise study of the connection between micro level cognitive phenomena and macro level social phenomena, with relatively rich representations of each. This has never been possible before. But a problem soon encountered is the difficulty of following through standard computer simulation procedure, for models embodying non-trivial aspects of cognition are impossible to validate in any detailed way (Doran, 1997b). Avoiding this problem leads to the notion of artificial societies and associated methodology, where the emphasis is not on the validity of the model against some specific target system, but rather on the discovery of processual regularities within computer based societies. These regularities state what happens under specified conditions—or, in what ways certain phenomena may originate. Of course, once such insights have been obtained, they then become available for use in a variety of application contexts.

Artificial societies escape the problem of specific validation, but instead encounter problems concerning the soundness and relevance of the structural assumptions they incorporate. These concerns partly reflect our uncertainty as to the soundness or otherwise of our own culturally determined assumptions about human society, and how we may escape them or even become fully aware of them.

[1] Department of Computer Science, University of Essex, Colchester, UK

2.2 The Methodology of Artificial Societies

The central method of artificial societies is to select an abstract social research issue, and then to create an artificial society within a computer system in which that issue may systematically be explored. Building artificial societies is a matter of designing and implementing agents and inter-agent communication, in a shared environment, and existing agent technology provides a wealth of alternatives in support of this process.

On each trial of a society, that is each time we track it through time, we obtain a result of the form:

structural assumptions + parameter settings + initial conditions

==>

a trace of the "history" of the artificial society

Given a set of artificial society histories, with varying initial conditions and parameter settings, we are in a position to detect and make practical use of regularities. It is possible to make predictions and construct explanations of phenomena observed in the society—subject, of course, to the assumptions and settings incorporated into it.

2.2.1 Methodological Problems

There are, however, important methodological problems:

- How should we search the space of parameter settings and initial conditions—by systematic exploration of parameter combinations, or by some form of "Hill Climbing"? This problem is closely related to standard techniques of sensitivity analysis.
- How, other than at an intractable level of detail, can we discover what actually happened? How can we describe what is found? How can we identify those properties that are in some sense "emergent"?
- The problem of "chaotic" instability—very small changes in initial conditions and parameter settings often have major consequences in individual experimental trials. "Averaging out" to find regularities may not be as simple as merely calculating means.

These problems are rather poorly understood, but are not overwhelming. Most past studies cope with them, if in a rather heuristic manner. However, there is a more serious problem.

2.2.2 What is an Agent?

The major methodological difficulty with the use of artificial societies concerns the degree to which the core assumptions we build into them are relevant to reality, have scientific significance, and avoid incorporating our own cultural biases and pre-conceptions. To illustrate this point, consider the concept of an "agent" that lies at the heart of the notion of artificial societies. Social scientists and other non-specialists often seem to read "agent" as meaning "with human-like behaviour and abilities" and probe no further, as if the details of agent definition and construction are either straightforward or immaterial. But, in fact, the notion of an agent is unclear even in the relevant technical literature:

- different functionality sets are taken as characteristic of agents (e.g. reactive agents, whose behaviour is based closely on condition-action rules; deliberative agents which use internal representations; social agents which build representations of others and take part in social relationships with others).
- many different types of agent architecture are considered in the agent literature (e.g. blackboard based, production system based, planner based, layered or hybrid type – see Huhns and Singh, 1998; Jennings and Wooldridge, 1998)
- any given type or architecture of agent is capable of being implemented in detail in many different ways, and with varying levels of effectiveness to the limit of our current technical abilities.

For example, agents may be taken to be little more than independently executable programs, may be taken to be in some sense autonomous and capable of goal setting and complex planning, or may be closely tied to the notion of the human computer interface.

A well-known definition of an agent from the artificial intelligence literature is:

An agent is anything that can be viewed as perceiving its environment through sensors and acting upon that environment through effectors. (Russell and Norvig, 1995, page 31).

But we may reasonably ask what "perceiving" and "acting" mean in this definition. Computationally, this appeal to our intuitions is far from clear. And notice the force of the phrase "viewed as". Anything may be viewed as an agent if the observer so insists (see the detailed discussion of Shoham, 1993). Furthermore, is the foregoing definition intended to exclude spatially dispersed (deliberative) agents, (deliberative) agents composed of other (deliberative) agents, or (deliberative) agents which spatially overlap? If so, there seem to be conditions on our notion of agenthood which are not being made explicit. If not, then we have to be a little careful for our intuitions sit uncomfortably with such possibilities.

A way to address these uncertainties, and perhaps to remove some of them, would be to pose the question:

How may agents be recognised?

In computational terms. This requires that the agent definition itself be made computational. So posed, the question encompasses the idea that agents can "emerge" from a computational context e.g. a (modelled) physical or other environment, and so would force us to remove ambiguities from the definition and to explore without pre-conception the circumstances in which agents might arise.

More generally, what commonly happens in practice is that simplified and somewhat arbitrary elements are included amongst the base assumptions of the artificial societies. But this is troubling. Is there any way to escape from pre-conceptions and let the properties of social processes emerge from the most basic and non-controversial assumptions possible? I shall return to this topic after looking briefly at two particular examples of artificial societies and their use, the first relatively conventional, the second less so.

2.3 Example 1: Trajectories to Social Complexity

Anthropologists have long been interested in how human society first found its way to a more complex form involving hierarchy and centralised decision making (Johnson and Earle, 1987). It is natural to ask;

> *How may experiments with artificial societies be used to study and characterise the set of possible trajectories from simple to complex social organisation?*

The EOS and related projects (reported in detail elsewhere) have gone some way to answering this question. I have suggested that three major social factors seem to mediate emergence trajectories:

- self-interested rationality and its collective consequences (see Doran, Palmer, Gilbert and Mellars, 1994; Doran and Palmer, 1995)
- collective (mis)belief (see Doran, 1998)
- emotions and their group dynamics (see Doran, 1996)

These factors are difficult to pin down precisely when closely examined. This is particularly true of emotions, which attract divergent computational interpretations. What is the architecture of an agent that incorporates (pseudo-)emotions? And it is particularly hard to discover, rather than "design in" from pre-conceptions, the interaction between the factors.

2.3.1 World Histories

One way to remove questionable assumptions, an ambitious one, is to formulate artificial societies right down at the level of world histories, that is, space/time structures which combine a "physical" environment (with agents within it) with an explicit temporal dimension. World histories in this sense may be viewed as sets of variables subject to constraints, world constraints, and handled and studied accordingly. The world constraints will normally be compatible with the laws of physics that we take as established in the real world, but, of course, we might choose to experiment with constraints not compatible with physical laws.

In principle, it is then possible to formulate the task of finding possible world histories meeting specified requirements as a constraint satisfaction problem (Tsang, 1993), and to apply standard search algorithms to find solutions. This is not as impractical as may at first appear, as my second example will illustrate (and see also Doran, 1997a).

2.4 Example 2: Foreknowledge and its Social Impact

It is a reasonable conjecture that a fundamental determinant of societies and their dynamics is their collective set of high level beliefs. Some of the most interesting beliefs are those about time and possible knowledge. For example, according to their chroniclers, the Aztecs believed that knowledge of the future was available to them by way of signs and omens and the carefully study of their own history. They structured their decision making accordingly, and thereby, it is commonly argued, seriously disadvantaged themselves (e.g. Todorov, 1985). Hence a natural question to pose is:

Under what conditions does (mis)belief in the availability of foreknowledge in an artificial society have negative evolutionary impact?

where foreknowledge means direct, unmediated, and recognised awareness (by an agent) of future events or situations, which is fully accurate (though possibly partial). In human terms we are therefore concerned with such phenomena as precognition, oracles, omens, dreams and the many other means by which direct awareness of events in the future may allegedly be obtained. Note that foreknowledge in this sense is not just prediction or an awareness of what (is likely) to happen based on information obtained, say, from another agent. Note also that the question requires establishment of the specific conditions for there to be negative evolutionary impact, not merely whether that impact exists at all.

There is a closely related question which is technically (even) more challenging and thought provoking. Suppose that we assume that foreknowledge is possible, so that belief in it is correct. Then we obtain:

Under what conditions does the availability of foreknowledge in an artificial society have negative evolutionary impact?

To envisage foreknowledge as part and parcel of a society and its context seems to challenge our most basic scientific models, and yet at the same time to be not wholly implausible. Certainly belief in some kind of foreknowledge is ubiquitous. To explore the properties of such artificial societies is therefore not idle.

2.4.1 Implementing Societies with Foreknowledge

To understand how we may create and explore a society and world in which a form of foreknowledge exists and is used, we need to return to the notion of a world history and the world constraints upon it.

The requirement of foreknowledge at certain points in a world history is, in effect, to require that certain additional *foreknowledge constraints* be satisfied at those points. In outline these are that:

- agents' internal representations must agree with the foreknown event
- an agent's actions must correspond (rationally) to its internal representations
- a foreknown event must be compatible with the effects of an agent's actions.

If we are working with agents without internal representations, then the first two of these foreknowledge constraints conflate into:

- an agent's actions must (rationally) be compatible with the foreknown event.

2.4.2 The Hazards Scenario

In order to explore the foregoing ideas, an artificial world has been constructed of which the main features are:

- a two dimensional spatial environment;
- mobile agents—but all action and movement is uncertain;
- immobile "resources" which agents can "harvest";
- immobile "hazards" which can (with uncertainty) "kill" and be killed by agents;
- at any moment agents can (try to) harvest a resource or (try to) kill a hazard;
- agents are simple with bounded "perceptual" and "mental" abilities;
- agents can reproduce by simple local cloning with variation;
- unless it "dies" otherwise, an agent will ultimately die of "old-age";

and, crucially

- some agents are given foreknowledge of when they will die and by which hazard.

Implicitly agents are rational and seek to live long and harvest much. This means that agents which know when and how they are to die are set not to attack hazards, since it is only rational to attack hazards when they pose a threat which can be varied.

2.4.3 A Question and a Paradox

The Hazards Scenario is implemented (in the C programming language, as SCENARIO-3F) by means of an *ad hoc* constraint satisfaction process, which proceeds by creating the world history from early to late, but with systematic partial "lookahead" to establish a "core-history" of foreseeable events in advance of the main history construction process.

This requires frequent *intervention* — that is, manipulation of chance in the artificial world to ensure that all the requisite constraints are satisfied (compare the discussion of backwards causality and "bilking" in Horwich, 1987, page 92 *et seq.*). Note that if an agent is to be killed by a hazard at a certain time, then both agent and hazard must be protected until that time.

The question asked of this world is the following: starting with a population of agents each with a limited temporal span of foreknowledge, and with variation in the temporal span of foreknowledge from parent to child, will the amount of foreknowledge in the population increase or decrease i.e. do agents with foreknowledge survive better than those without?

But how can we study the impact of foreknowledge on life spans when these are necessarily predetermined if agents are to have foreknowledge! Agents may well behave differently because of their foreknowledge of their own death, but that different behaviour, by assumption, cannot change their death date and hence cannot have evolutionary impact! This seems something of a paradox. We can escape from this difficulty by asking not:

What is the life expectancy of agents with and without foreknowledge?

but rather

If certain combinations of agent life expectancies and degrees of foreknowledge are specified, how frequently do histories with those combinations occur in the set of all possible histories compatible with the world and foreknowledge constraints?

where the frequency with which a particular type of world history exists in the set of all possible histories is to be estimated by the difficulty of creating one, as measured by the amount of intervention it requires to do so.

2.4.4 Results

In experiments with the Hazards Scenario, we find that a relatively high level of intervention is needed to construct world histories in which agents who have foreknowledge of their deaths come to predominate. The implication is that, in this scenario, foreknowledge is a liability.

This result depends strongly, not only on the view taken of rational behaviour in an agent, but also on the parameters of the artificial world, notably the degree of "hazardousness" of a typical hazard. If hazards are *not* very dangerous, then relatively little intervention is needed to protect agents against them. In these circumstances, consider world histories in which most agents have foreknowledge and so do not attack hazards. Hazards will tend to exist in substantial numbers. In spite of this, such histories will need relatively little intervention to create and be correspondingly well represented in the set of all possible histories.

2.5 Conclusions

These two examples demonstrate that a very wide range of phenomena may be investigated in artificial societies. We can explore regularities in the relationships between assumptions and emergent outcomes, even when these run quite counter to our own intuitions about the real social world. Of course, the power of investigation that is in principle thus available to us is bounded by sheer computational complexity, and by the current limitations of agent technology.

But if the discovery of abstract social regularities is to be more than the computational equivalent of an investigation in pure mathematics, if it is to have potential practical application, then the base level assumptions that we build into our artificial societies must have some objective validity and interest. At this point my earlier concern with the ambiguous notion of "agent" is relevant. For agent based simulation is built around this deeply uncertain notion. We can go at least some way towards avoiding building into our artificial societies our own unrecognised cultural assumptions by pitching the assumptions that we do make at the lowest possible level, even below the level of agent, and letting the interesting phenomena emerge. We may, for example, generate world histories and ask not: "What are the agents doing?" but "What agents are there to be found and what are they doing?"

A further development, flowing from the second of the two foregoing examples, is that there seems to be a fruitful change of perspective from a temporally unfolding simulation to a world history created in whatever way may be computationally convenient. This change of perspective encourages us to think less about generative probabilities (how far this or that event is or is not now likely) and more about the properties of the set of all world histories compatible with specified constraints.

References

Doran, J. (1996): Simulating Societies Using Distributed Artificial Intelligence. In: Troitzsch, K. G., Müller, U., Gilbert, G. N. & Doran, J. E. (eds.): Social Science Simulation. Springer-Verlag, Berlin, Heidelberg, New York, 381-393.

Doran, J. (1997a): Foreknowledge in Artificial Societies. In: Conte, R., Hegselmann, R. & Terna, P. (eds.): Simulating Social Phenomena. Lecture Notes in Economics and Mathematical Systems 456. Springer, Berlin, 457-469.

Doran, J. (1997b): From Computer Simulation to Artificial Societies. SCS Trans Vol 14(2).

Doran, J. (1998): Simulating Collective Misbelief. Journal of Artificial Societies and Social Simulation 1(1) (http://www.soc.surrey.ac.uk/JASSS/1/1/3.html).

Doran, J., Palmer, M., Gilbert, N. & Mellars, P. (1994): The EOS Project: modelling Upper Palaeolithic Social Change. In: Doran, J. & Gilbert, N. (eds.): Simulating Societies: the computer simulation of social phenomena. UCL Press, London, 195-221.

Doran, J. & Palmer, M. (1995): The EOS Project: integrating two models of Palaeolithic Social Change. In: Gilbert, N. & Conte, R. (eds.): Artificial Societies: the computer simulation of social life. UCL Press, London, 103-125.

Horwich, P. (1987): Asymmetries in Time: Problems in the Philosophy of Science. A Bradford Book \ The MIT Press, Cambridge, Mass. & London.

Huhns, M. N. & Singh, M. P. (eds.) (1998): Readings in Agents. Morgan Kaufmann.

Jennings, N. R. & Wooldridge, M. J. (eds.) (1998): Agent Technology: Foundations, Applications and Markets. Springer, Berlin.

Johnson, A. W. & Earle T K (1987): The Evolution of Human Societies. Stanford University Press, Stanford CA.

Russell, S. & Norvig, P. (1995): Artificial Intelligence: a Modern Approach. Prentice Hall.

Shoham, Y. (1993): Agent-Oriented Programming. AI Journal 60, 51-92.

Todorov, T. (1985): The Conquest of America. Colophon Books, Harper & Row, New York.

Tsang, E. (1993): Foundations of Constraint Satisfaction. Academic Press, London.

3. Computer-Assisted Interfacing: On the Use of Computer Simulation for Theory Construction

Georg Müller[1]

3.1 Introduction

A new social theory generally claims to give non-trivial insights into the functioning of a social process. If this claim holds, the different stages and outcomes of this process are logical consequences of the theory in question, which thus explains or predicts the phenomena of interest. It is obvious that this kind of interfacing is crucial not only for the practical use of a theory but also for the process of theory-construction: It enables the architects of a new theory to cross-check whether its implications are compatible with the already existing body of theoretical knowledge.

In order to endow a new theory with this kind of interfacing facilities it has to be composed of a set of highly interconnected theoretical propositions which all use a precisely defined common terminology. Since theories formulated in natural languages rarely meet these standards, there is a long sociological tradition of using mathematical languages for theory construction. H. Simon's formalization of Homan's theory of group dynamics (Simon, 1957, pp. 99-114), Hamblin, Jacobsen, and Miller's theory of social change (Hamblin et al., 1973), or J. Coleman's mathematics of collective action (Coleman, 1973) are just a few examples for this intellectual tradition. However, in spite of the mathematical rigor of many of these formalizations they are often not very useful for drawing conclusions in a deductive way: If they are applied to real world problems, their formal complexity in terms of variables and equations soon increases beyond the capabilities of deductive reasoning[2].

Fortunately, the growing interest in formalized social theories was paralleled by technical progress in computer science which enabled the scientific community to simulate complex social processes. Consequently, social scientists began to translate their theories into simulation models in order to use computers for making inferences about the different stages and outcomes of social processes. Forrester's model of the urban dynamics of American cities was one of the early and more prominent scientific endeavors in this field (Forrester, 1969). However, this model also pointed to the limits of simulation with regard to interfacing: The theoretical conclusions drawn from simulation models often refer to very specific sets of model equations or model parameters and consequently their universal validity is

---

[1] Department of Social Work and Social Policy, University of Fribourg, Switzerland

[2] See e.g. the GLOBUS world model of Bremer (1987) or the Urban Dynamics model of Forrester (1969).

often endangered by one or several of the following deficiencies:

1) In order to reduce the number of unknown parameters of a model its structure is often simplified:[3] The use of linear instead of monotonic relations between model variables is one of these often used simplifications which endanger the universal validity of the theoretical conclusions drawn from simulation models.
2) Many simulations are based on estimates of parameter values which refer to specific populations or historical situations.[4] Thus, inferences from such models generally do not cover the universe of actually existing and virtually possible societies.
3) Often the technique of experimentation with different values of model parameters is used in order to explore the behavior of a model.[5] Inferences drawn from such experiments are only "proven" for a very limited set of experimental parameter values and hence do not allow us to draw general conclusions about the implications of a model.

Theoretical inferences claim to be generally valid for all actually existing and virtually possible societies. The previously mentioned shortcomings of the actual practice of simulation-assisted interfacing have thus to be considered as serious problems for computer-based theory construction. They have led the author of this paper to attempt to develop a new methodology for computer-assisted interfacing (cf. Müller, 1996) which will be explained and refined in the following sections.

3.2 A New Methodology for Computer-Assisted Inferencing

3.2.1 Inferencing by Falsification

One of the real advantages of computer simulation is its ability to the handle models with numerically indeterminate parameters such as unknown initial conditions or coefficients of equations: Most simulation models allow to substitute these formal parameters by hypothetical numerical values of virtual societies and to calculate within a relatively short time the evolution and outcome of the corresponding social process. This ease of parameter handling is useful for making general inferences for two reasons:

1) It allows to study the consequences of a model without using empirically based parameter estimates which generally make conclusions time- and context-dependent.
2) It enables us to study models with many parameters, thus reflecting the full complexity of the verbally formulated theories behind them.

In spite of the ease of parameter handling in computer simulation, it is more difficult to get general conclusions about a model than it seems at a first glance. One is indeed tempted to believe that a few simulation experiments with arbitrarily

[3] See e.g. arms race model of Richardson (1960).

[4] See e.g. Urban Dynamics model of Forrester (1969).

[5] See e.g. simulation methodology of Hanneman (1988).

chosen parameter values would allow us to draw general conclusions about the behavior of a model and the consequences of the theory behind it. R. Hanneman (1988, 1995a, 1995b) e.g. has advocated this type of quasi-experimentation as a third way of knowledge production between traditional statistical analysis and pure mathematical reasoning. However, if general validity of a hypothetical inference means that it has to be true for all virtual actors and functional relations between variables, "playing around" with parameter values is not sufficient to confirm a hypothetical inference: In order to confirm a general theoretical proposition one would have to check it for an infinity of parameter combinations which would have to exhaust the parameter space of virtual societies. Since this is practically not possible we have no chance to *confirm* by simulation a hypothesis which we believe to follow from a given theory.

The impossibility to confirm hypothetical inferences by computer simulation resembles the problem of confirming a general theoretical proposition with observational data. As K. Popper (1959: 70) stated a long time ago, there are never enough empirical data in order to confirm a hypothesis for all possible social situations. Instead, K. Popper has suggested to attempt to *falsify* a general hypothesis by systematically searching for disconfirming counter-examples. This is much easier than trying to confirm a theory for an infinity of situations since one single counter-example is sufficient to discard the hypothesis in question. However, if a falsification-attempt fails, the tested hypothesis can only be provisionally accepted: There is always a small risk that the hypothesis will be falsified by future investigations.

Due to the isomorphic relation between the data space of empirical analyses and the parameter space of computer simulation it seems to be natural to use the principle of falsification also for computer-aided interfacing: Instead of trying to *confirm* a general proposition hypothetically following from a theory, we propose to attempt to *falsify* it by searching for parameter sets which disconfirm the general character of the tested theoretical proposition. If a serious and systematic search reveals no falsifying counter-examples, the tested proposition is *provisionally* accepted as a logical consequence of the model and the theory behind it. Otherwise, the detection of a falsifying counter-example compels us to deny an inferential relation between the model and the tested proposition. As with empirical tests of theories, it is easier to find one falsifying counter-example than to confirm an inference by trying to check an infinite number of parameter combinations.

If e.g. long-term stability of a social process is one of the hypothetical inferences of a model, the proposed principle of falsification suggests us to search for initial conditions or structural parameters for which the mentioned process becomes *unstable.* If this search succeeds, the tested hypothetical inference must be discarded; if it fails, we are allowed to provisionally accept the hypothetical inference as universally valid consequence of the simulated model.

For logical purists the proposed interfacing procedure is probably not acceptable since their deductive way of thinking is not prepared to admit incomplete searches which result in the provisional truth of a theoretical proposition. However, "provisional truth" is nothing new for the empirical sciences: Their whole theoretical knowledge is permanently "endangered" by new discoveries which may be in contradiction with these theories. Hence, from a pragmatic point of view it seems to be common practice to work with theories which have a logical status that is not one-hundred percent clear.

3.2.2 Optimization as a Heuristic for the Search of Falsifying Parameter Values

One of the key features of the proposed method of falsification is the rigor of the search for falsifying counter-examples. Theoretically, a learned guess may help to find a falsifying example. However, with regard to the many dimensions of the parameter space and the risk of failing to notice a counter-example, non-falsification means nothing if the search has been performed by hand. Hence, it is suggested to use for this task an automated computer-based artificial intelligence method.

One of the most flexible methods of artificial intelligence is constrained numerical optimization. Its goal is to find the minimum or maximum of a real-valued target function

$$\phi: (p_1, p_2, \ldots) \in \mathbb{R}^n \rightarrow \mathbb{R} \qquad (1)$$

by varying the parameters $p_1, p_2, \ldots$ within the constraints

$$L_i \leq p_i \leq U_i, \; i = 1, 2, \ldots, n \qquad (2)$$

Since constrained optimization has successfully been applied to many different fields, the author of this article has proposed to use it as a heuristic for the search for a falsifying parameter set (Müller, 1996).

In order to be able to use optimization techniques for parameter search, the target function

$$\phi: (p_1, p_2, \ldots) \in \mathbb{R}^n \rightarrow \mathbb{R} \qquad (1)$$

has to be interpreted as a *correspondence-function* that measures the correspondence between a hypothetical inference H and the results of the simulation run with the model parameters $p_1, p_2, \ldots$ Consequently, the search for a falsifying parameter set corresponds to the *minimization* of the function ϕ by varying p_1, $p_2, \ldots$ within the meaningful limits

$$L_i \leq p_i \leq U_i, \; i = 1, 2, \ldots, n \qquad (2)$$

of the parameter space of a virtual society. The limits L_i and U_i are on the one hand given by the definitions of the parameters p_i which exclude e.g. negative values, on the other hand they can be fixed on the grounds of the observed maximal and minimal values.

By definition, the minimization of ϕ yields a set of *worst-case* parameters $\pi_1, \pi_2, \ldots$ for which the correspondence $\phi(\pi_1, \pi_2, \ldots)$ between the simulation-experiment for $\pi_1, \pi_2, \ldots$ and the hypothetical inference H is as *small* as possible. Whether these worst-case parameters $\pi_1, \pi_2 \ldots$ falsify the hypothesis that proposition H is an inference of the model depends on a *correspondence-threshold* θ: If

$$\phi(\pi_1, \pi_2, \ldots) < \theta \qquad (3)$$

the correspondence is unacceptably low and the simulation run for the worst-case parameters $\pi_1, \pi_2, \ldots$ falsifies the general validity of the proposition H which consequently cannot be considered as an inference of the model. Otherwise, the worst-case parameters $\pi_1, \pi_2, \ldots$ do not falsify the general validity of this proposition which thus must at least provisionally be accepted as an inference of the model.

There are many different methods for defining the value of the correspondence-threshold θ: It can e.g. be derived from a *statistical test,* from a *zero-tolerance interpretation* of the proposition H, from the planned *practical use* of the proposition H, etc. In practice, θ depends on the type of model and the type of inference being studied.

3.2.3 The Different Steps of Computer-Assisted Interfacing

On the grounds of the previous methodological discussion, the test whether a proposition H is an inference of a model or not can be summarized by the following seven-step procedure:

1) Define a simulation model M. Make it as independent of real world data and as close to the original theory as possible by introducing indeterminate parameters.
2) Formulate the general proposition H whose inferential relation to model M has to be checked.
3) Define on the grounds of the criteria mentioned in the preceding chapter 2.2 the search space of the meaningful parameter combinations $(p_1, p_2, \ldots)$ by fixing for every parameter p_i the upper and lower limits U_i and L_i such that
   $$L_i \leq p_i \leq U_i, \; i = 1, 2, \ldots, n \qquad (2)$$
4) Define the correspondence function
   $$\phi\colon \; (p_1, p_2, \ldots) \in \mathrm{IR}^n \to \mathrm{IR} \qquad (1)$$
   which measures the correspondence between the general proposition H and the results of the simulation run for the parameters $p_1, p_2, \ldots$ The implementation of this function for different types of propositions H will be discussed in chapters 3.2.4 to 3.2.6.
5) Determine the correspondence tolerance θ on the grounds of the criteria mentioned in chapter 2.2.
6) Minimize the correspondence $\phi(p_1, p_2, \ldots)$ by varying the parameters p_1, $p_2, \ldots$ within the limits defined by inequality (2) and determine this way the worst-case parameters $\pi_1, \pi_2, \ldots$
7) Discard proposition H if for the worst-case parameter set
   $$\phi(\pi_1, \pi_2, \ldots) < \theta \qquad (4)$$
   If inequality (4) does not hold, provisionally accept proposition H as a generally valid implication of the tested theory.

In the following, the implementation of this seven-step procedure will be shown for three important types of inferences commonly used in social science literature.

3.2.4 Inferences About Differences Between Two Categories of Actors

The literature of empirical social science is full of examples where two mutually exclusive categories of social actors are compared with regard to their different characteristics: Men and women with regard to violent behavior, participants of experiments and control groups with regard to learning progress, foreigners and nationals with regard to their unemployment rates, etc. In spite of the rather descriptive nature of such comparisons they are part of our empirical knowledge about society. New theories should be compatible with this knowledge and consequently inferences about differences between categories of actors matter for theory construction.

Categorical comparisons between two groups of social actors A and B with regard to a property X typically read as

$$X(a) > X(b) \quad \text{for all actors } a \varepsilon A \text{ and } b \varepsilon B \tag{5}$$

Hence, it is near at hand to define for this type of inference H the correspondence function as a difference

$$\phi(p_1, p_2, \ldots) = X(a) - X(b) \tag{6}$$

where the properties

$$X(a) = f(p_1, p_2, \ldots) \tag{7}$$

and

$$X(b) = g(P_1, P_2, \ldots) \tag{8}$$

of the actors α and β are considered to be functions of the parameters to be varied. Thus the minimization of $\phi(P_1, P_2, \ldots)$ implies the construction of two artificial actors a and b which belong to the groups A and B, respectively, and for which

$$\phi(\pi_1, \pi_2, \ldots) = X(\alpha) - X(\beta) \tag{9}$$

is the worst-case correspondence between H and the model.

For non-stochastic models the natural correspondence-threshold θ is obviously 0. Consequently, the proposition H is only accepted as an universally valid implication of the model if

$$\phi(\pi_1, \pi_2, \ldots) = X(\alpha) - X(\beta) > \theta = 0 \tag{10}$$

3.2.5 Inferences About Correlations Between Variables

In theoretical social science literature one often finds another type of proposition which refers to correlations between theoretical variables and which generally reads as follows:

The higher the variable X, the higher (or: lower) the variable Y. (11)

Computer simulation models produce two kinds of *artificial* data which can be used in order to confirm or disconfirm this kind of correlational proposition: *Longitudinal* data, where the focus of interest is on the evolution of two variables $X(t)_{t \varepsilon T}$ and $Y(t)_{t \varepsilon T}$ for the different *time-points t* of a sample T, and *cross-sectional* data, where the relation between two variables $X(a)_{a \varepsilon A}$ and $Y(a)_{a \varepsilon A}$ is studied for different artificial *actors a* of a sample A. For both cases it seems to be natural to use in the correspondence-function ϕ one of the traditional statistical correlation measures *corr* in order to formalize the proposition represented by formula (11). Consequently, ϕ is defined as

$$\phi(p_1, p_2, \ldots) = s * \mathrm{corr}(X(a), Y(a)), \quad a \in A \tag{12}$$

respectively

$$\phi(p_1, p_2, \ldots) = s * \mathrm{corr}(X(t), Y(t)), \quad t \in T \tag{13}$$

where for both (12) and (13)

$$s = \begin{cases} +1 \text{ if the tested hypothesis H postulates a } \textit{positive} \text{ correlation} \\ -1 \text{ if the tested hypothesis H postulates a } \textit{negative} \text{ correlation} \end{cases} \tag{14}$$

For the calculation of ϕ the Spearman- and the Pearson-correlation are both acceptable (Blalock, 1979: 396 ff., 434 ff.). The Spearman rank correlation is numerically more difficult to calculate; it is, however, more congenial with the idea of a monotonic nonlinear relation between the variables X and Y.

For both, the Spearman- and the Pearson-coefficients there are *critical values c* that are normally used in empirical statistical analyses in order to decide whether a positive or negative correlation is statistically significant or not. Hence, it seems to be logical to use one of these critical values c as correspondence-threshold θ. More precisely, we propose to set

$$\theta = c \tag{15}$$

Consequently, if $\phi(p_1, p_2, \ldots)$ is minimized and the worst-case set of parameters yields

$$\phi(\pi_1, \pi_2, \ldots) < \theta \tag{16}$$

there is a counter-example for which the correlation corr is zero or has the wrong sign such that H cannot be an universally valid inference of the analyzed model and the theory behind it.

3.2.6 Inferences About Causal Relations Between Events

A third type of important social science hypothesis refers to causal relations between events. The hypotheses that frustration is followed by aggression or that socio-economic modernization stabilizes the population growth are two examples of this type of causal proposition. They can both be summarized by the formula:

If event A, then event B. (17)

By definition, events like A and B are accompanied by changes of variables X and Y which we assume to take place during two *subsequent* time intervals $[t_\alpha, t_0]$ and $[t_0, t_\omega]$ (see figure 1). The chain of events (17) may thus be formalized as follows:

$$X(t_0) - X(t_\alpha) \begin{matrix} > \\ = \\ < \end{matrix} D_X \rightarrow Y(t_\omega) - Y(t_0) \begin{matrix} > \\ = \\ < \end{matrix} D_Y \tag{18}$$

where D_X and D_Y are levels or thresholds of changes. The sample-hypothesis that socio-economic modernization stabilizes the population growth of a nation may thus be formalized as:

$$\text{modernization}(t_0) - \text{modernization}(t_\alpha) > D_X = 0 \rightarrow \tag{19}$$

$$\text{population\_growth}(t_\omega) - \text{population\_growth}(t_0) = D_Y = -\,\text{population\_growth}(t_0)$$

and consequently

$$\text{modernization}(t_0) - \text{modernization}(t_\alpha) > 0 \rightarrow \text{population\_growth}(t_\omega) = 0 \quad (20)$$

An obvious method for testing an hypothesis H about a causal relation between two events A and B is thus to *minimize* the correspondence-function

$$\phi(p_1, p_2, \ldots) = s * | Y(t_\omega) - Y(t_0) - D_y | \quad (21)$$

under the constraint of equation (18), i.e.

$$X(t_0) - X(t_\alpha) \begin{matrix} > \\ = \\ < \end{matrix} D_X \quad (22)$$

where

$$s = \begin{cases} + \text{signum}(Y(t_\omega) - Y(t_0) - D_y) & \text{if H postulates } Y(t_\omega) - Y(t_0) > D_y \\ -1 & \text{if H postulates } Y(t_\omega) - Y(t_0) = D_y \\ - \text{signum}(Y(t_\omega) - Y(t_0) - D_y) & \text{if H postulates } Y(t_\omega) - Y(t_0) < D_y \end{cases} \quad (23)$$

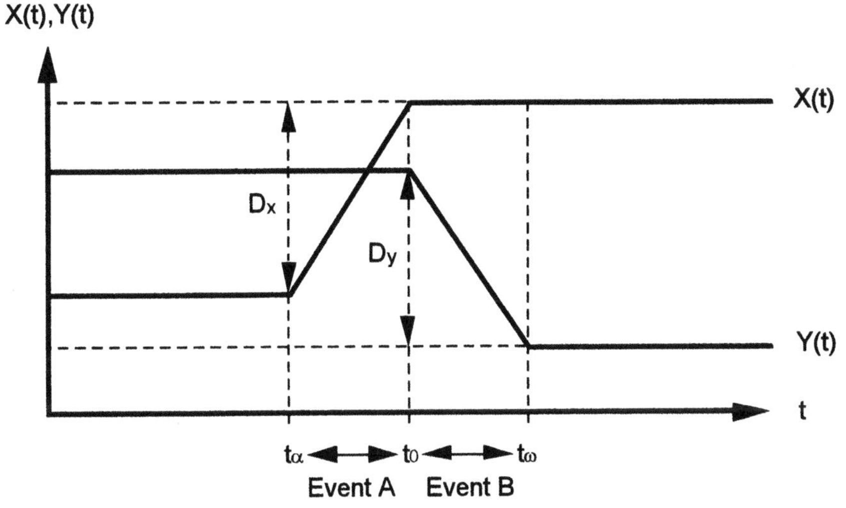

Figure 1: Changes of X(t) and Y(t) as correlates of causally related events A and B.

Whether this attempt to find a falsifying counter-example fails or not, depends on the structure of the proposition H. If H postulates that

$$Y(t_\omega) - Y(t_0) > D_y \quad (24)$$

or

$$Y(t_\omega) - Y(t_0) < D_y \quad (25)$$

any *violation* of

$$\phi(\pi_1, \pi_2, \ldots) = s * | Y(t_\omega) - Y(t_0) - D_y | > \theta = 0 \quad (26)$$

(see (21)) entails a violation of the inequalities (24) or (25) and thus questions the general validity of hypothesis H. If H states that

$$Y(t_\omega) - Y(t_0) = D_y \quad (27)$$

the situation is more difficult, especially if it refers to a converging process

$$Y(t_\omega) \rightarrow Y(t_0)+D_Y \quad \text{for } t_\omega \rightarrow \infty \tag{28}$$

Since equation (28) is by the logic of the convergence process for finite t_ω hardly ever precisely fulfilled, the correspondence-threshold θ has to be different from zero. Unfortunately, the time series $Y(t)_{t=1,2,...}$ is often autocorrelated such that statistical test criteria (e.g. t-tests) are of little use for fixing θ. A viable solution to this problem is to set θ to a small *negative* value which fits the practical use of the inference H. If the correspondence-function ϕ fulfills

$$\phi(\pi_1, \pi_2, \ldots) \geq \theta \tag{29}$$

for the time-point t_ω and if this inequality (29) also holds for a set of subsequent time-points, then the hypothesis H is a generally valid inference of the model.

3.3 Computer-Assisted Interfacing in Practice

3.3.1 Some Problems of the Proposed Method

The optimization of the parameters of simulation models often means intensive numerical computing: The optimization algorithm which we will use for this purpose typically needs 5 to 20 iterations in order to find a worst-case set of parameters that minimizes the respective correspondence function. Since for each of these iterations there are not only the next set of improved parameters but also all gradients of the correspondence function which have to be calculated, each iteration entails several simulation runs. Hence, the resulting overall number of arithmetical operations is considerable and entails two major risks:

a) The power of the processor of a standard PC may be insufficient to do all these calculations within due time.
b) The rounding errors of the many arithmetical operations may accumulate up to the point where the results of the minimization process are wrong or meaningless.

Even if computers were sufficiently fast and reliable there would be a third kind of problem caused by the logic of the gradient-based search process of most minimization algorithms: This process sometimes leads only to a local instead of a global minimum and thus prevents the detection of a counter-example which falsifies the hypothesis under discussion.

In order to assess the significance of these risks in the following sections, the proposed methodology will be applied to three practical examples describing the effects of a dyadic international arms race. The examples refer to the three types of inferences mentioned in chapters 3.2.4 to 3.2.6 and thus will at the same time illustrate our methodology.

3.3.2 The Richardson Arms Race Model

In order to show the practical aspects of our method of inferencing we will use it to draw conclusions from the so-called Richardson arms race model (Richardson,

1960; Taagepera et al., 1975; Nicholson, 1989:147 ff.). It is a relatively old model, however, its simplicity allows to deduce theoretical propositions by analytical means and to compare these propositions with the results of computer-assisted interfacing. Hence, the primary concern of the following sections is the demonstration of our method of computer assisted interfacing and not the production of new theoretical results by means of this method.

As the name of the Richardson arms race model suggests it describes the armaments $X(t)_{t=1,2,...}$ and $Y(t)_{t=1,2,...}$ of two hostile nations by means of two difference equations:

$$\Delta X(t) = X(t+1) - X(t) = a_X * Y(t) - c_X * X(t) + g_X \quad (30)$$

$$\Delta Y(t) = Y(t+1) - Y(t) = a_y * X(t) - c_y * Y(t) + g_y \quad (31)$$

where:

$X(t), Y(t)$: Armaments of the two nations considered.

$\Delta X(t), \Delta Y(t)$: Growth of the armaments of the two nations.

g_X, g_y : Aggressiveness of the geopolitical ambitions of the two nations, where $g_X > 0$ and $g_y > 0$.

a_X, a_y : Weights of the threat of the rival armament, where $a_X > 0$ and $a_y > 0$.

c_X, c_y : Weights of the tax load of the own armament, where $c_X > 0$ and $c_y > 0$.

Equations (30) and (31) are obviously symmetrical. They both assume that the growth of the own armament depends on:

a) The *armament of the rival nation:* The higher it is, the higher the need for additional armament of the own military in order to defend the own country.
b) The *own armament:* The higher it is, the higher the tax load resulting from military spending and the lower the chance that additional armament is accepted by the national population.
c) The *geopolitical ambitions* of a nation: The more aggressive they are, the higher the need for additional armament of the own military in order to realize these ambitions.

Equations (30) and (31) postulate that the influence of these variables on the growth rates of armaments is linear and not monotonic. In order to keep the model analytically treatable, this simplification is *not* corrected by the introduction of additional parameters.

3.3.3 The Minimization of Correspondence-Functions with SPSS

For the implementation of the proposed method of computer-assisted interfacing a program-module is needed that performs the derivative-free optimization of parameters with constraints. Many program libraries such as NAG[6], IMSL[7], or MATLAB[8] indeed offer such modules which have, however, to be integrated in a user-written simulation-program. This means that the user needs knowledge of a

---

[6] See e.g. Phillips (1986).

[7] See e.g. IMSL (1991).

[8] See e.g. Paert-Enander et al. (1996).

module-specific programming language such as e.g. FORTRAN or MATLAB. Since social scientists are generally much more familiar with statistical packages such as e.g. SPSS than with the languages mentioned before, we propose for optimization the use of an appropriate SPSS-module that can be embedded in an SPSS-simulation-program.

At first, SPSS seems to be the wrong software product for this purpose: SPSS is designed for statistical analysis and not for optimization or model programming (Norusis, 1993, passim). However, if SPSS is used with dummy data for *only one single case,* the case-driven structure of an SPSS-user-program becomes linear instead of circular. In this way, it is possible to program simulation models in a FORTRAN-like manner by using COMPUTE, LOOP, and similar SPSS-commands (see annex 3.6.1). Moreover, the constrained nonlinear regression module *CNLR* of the *SPSS Advanced Statistics Package* (Norusis, 1995: 209 ff.) becomes for this one-case situation a minimization module that optimizes any loss- or correspondence-function (see annex 3.6.1).

CNLR is based on a sequential quadratic programming algorithm originally designed for the mathematical NPSOL-package (SPSS Inc., 1992 : 34 ff.) and needs, as most other minimization algorithms, user-supplied start values for the parameters to be optimized. These values are iteratively improved by the optimization algorithm until the corresponding function value of the user-defined loss-function cannot be lowered any more (see annex 3.6.2). Thus, for the purposes discussed here, the final iteration of CNLR generally yields the minimum value of the correspondence-function ϕ.

3.3.4 Example 1: The Correlation Between the Levels of Armaments in a Dyadic Arms Race

Richardson's model of arms races is based on the idea of a mutual reference of two virtually hostile nations and thus suggests a positive correlation between the respective trajectories of armaments $X(t)_{t=1,2,...}$ and $Y(t)_{t=1,2,...}$:
The higher at a given time-point t the armament X(t) of the first nation, the higher the armament Y(t) of its opponent.
Although this *correlational hypothesis* seems to be very plausible, its general validity is of course not assured. In the following it will be checked with our methodology of computer-assisted interfacing, as outlined in section 3.2.5.

As one of the first steps after the construction of the simulation model (see annex 3.6.1), section 3.2.3 suggests to define the dimensions and constraints of the parameter space. For an endogenous arms race model the dimensions of this space are given by those model variables which have to be externally supplied in order to start a simulation run. Their lower boundaries L_i represent a small positive value near zero which corresponds to the lower limit of their definition intervals; their upper boundaries U_i are empirically plausible maximum values that have to be set in order to prevent numerical problems during optimization. More precisely, the parameter space used for finding a falsifying counter-example is defined by

$$X(0), Y(0), a_x, a_y, c_x, c_y, g_x, g_y \geq 0.001 \qquad (32)$$

$$a_x, a_y, c_x, c_y, g_x, g_y \le 0.5 \quad (33)$$

$$X(0), Y(0) \le 1.0 \quad (34)$$

In section 3.2.5 (see equation (13)) it was recommended to define a correspondence-function

$$\phi(p_1, p_2, \ldots) = 1 * \mathrm{corr}(X(t),Y(t)), \ t \varepsilon T \quad (35)$$

in order to find by a process of minimization a counter-example which falsifies the hypothesis H of a positive correlation between the time series $X(t)_{t=1,2,...}$ and $Y(t)_{t=1,2,...}$. Although H is not a *linear* but rather a *general monotonic* hypothesis, the correlation-function *corr* of equation (35) is not operationalized by the appropriate Spearman rank-correlation r_S but rather by the linear *Pearson correlation r* which allows the computational complexity of ϕ to be kept relatively low. Assuming that a trajectory of 50 time units is long enough in order to eventually falsify H, equation (35) thus becomes

$$\phi(a_x,a_y,c_x,c_y,g_x,g_y,X(0),Y(0)) = r\,(X(t),Y(t)), \ t \varepsilon T = \{1,2,\ldots,50\} \quad (36)$$

For the *initial* parameter estimates as mentioned in figure 2, the correspondence-function (36) is equal to the positive maximum value

$$\phi(a_x,a_y,c_x,c_y,g_x,g_y,X(0),Y(0)) = 1 \quad (37)$$

Hence, the corresponding trajectory of armaments $(X(t),Y(t))_{t=1,2,..}$ is represented by an ascending linear "pearl-string curve" in the upper phase diagram of figure 2. If ϕ is minimized on the basis of the start values mentioned before, the situation changes completely: Run on a Power Macintosh 4400/160 with a clock rate of 160 megacycles, the SPSS-software (see annex 4.6.1) reaches after 9 seconds or 11 optimizing iterations the absolute minimum

$$\phi(a_x,a_y,c_x,c_y,g_x,g_y,X(0),Y(0)) = r(X(t),Y(t)) = -1.00, \quad (38)$$

where $t \varepsilon T = \{1,2, \ldots,50\}$

of the correspondence-function (36). The respective set of worst-case parameters appears at the bottom of the second phase diagram of figure 2 and entails an armament trajectory $(X(t),Y(t))_{t=1,2,..}$ with a perfectly negative slope (see "pearl-string line" in lower phase diagram of figure 2).

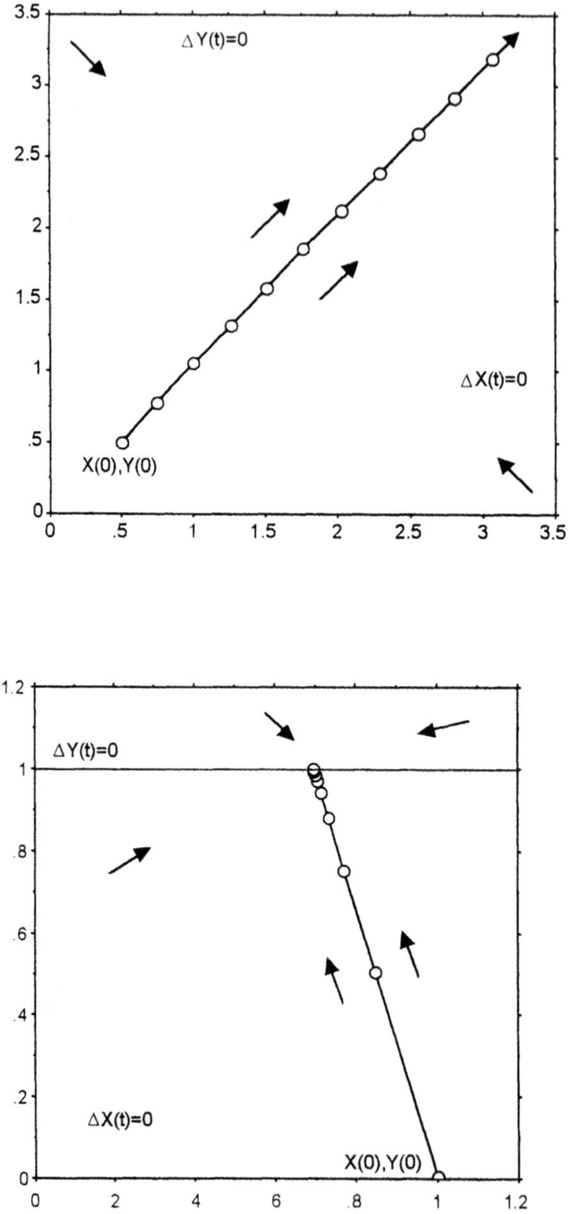

a_x=0.11, a_y=0.00, c_x=0.15, c_y=0.50, g_x=0.00, g_y=0.50,
X(0)=1.00, Y(0)=0.00

Figure 2: Phase diagrams and trajectories of the armaments X(t) and Y(t) for the initial (above) and the optimised (below) parameter values.

Since a one-sided statistical test of the null hypothesis

H_0: $r(X(t),Y(t)) \geq 0, \; t \in T = \{1,2,\ldots,50\}$ (39)

yields for the significance level α=0.01 a confidence limit of -0.322 (van der Waerden, 1965: 350), the correspondence-tolerance is set to

$\theta = -0.322$ (40)

and consequently

$\phi(a_X,a_Y,c_X,c_Y,g_X,g_Y,X(0),Y(0)) =$

$r(X(t)_{t=1,2,\ldots}, Y(t)_{t=1,2,\ldots}) = -1.00 < -0.322 = \theta$ (41)

Thus, even allowing for the autocorrelation of the time series variables $X(t)_{t=1,2,\ldots}$ and $Y(t)_{t=1,2,\ldots}$, hypothesis H_0, postulating a positive *linear* correlation between these variables, is surely invalid. Thus, we have found a linear trajectory as a counter-example which falsifies of course also the more general hypothesis H postulating a positive *monotonic* relation between $X(t)_{t=1,2,\ldots}$ and $Y(t)_{t=1,2,\ldots}$. Hypothesis H is consequently *not* an universally valid inference of the Richardson model.

3.3.5 Example 2: The Consequences of an Unilateral Armament Moratorium for a Dyadic Arms Race

One of the big advantages of computer simulation is the possibility of studying the consequences of virtual experiments which are too expensive, complicated, or unethical to be realized in practice. In the following, we will use such a "Gedanken Experiment" in order to study the application of computer-assisted interfacing to the analysis of *causal relations* between events (see chapter 3.2.6): We will analyze the consequences of an unilateral armament moratorium where the parameters a_X, c_X and g_X and consequently also the armament growth $\Delta X(t)$ of one of the nations are permanently reduced to level zero. It is hypothesized that Richardson's arms race model also entails for the second nation a decrease of the growth of armaments $\Delta Y(t)$ to level zero. More precisely we have to test whether

Setting at t_0 $a_X = c_X = g_X = 0 \rightarrow \Delta Y(t_\omega) - \Delta Y(t_0) = -\Delta Y(t_0)$ for $t_\omega >> \infty$ (42)

or equivalently

Setting at t_0 $a_X = c_X = g_X = 0 \rightarrow \Delta Y(t_\omega) = 0$ for $t_\omega >> \infty$ (43)

For this particular situation the Richardson-model consists of two simplified model equations:

$\Delta X(t) = 0*Y(t) - 0*X(t) + 0 = 0$ (44)

$\Delta Y(t) = a_Y*X(t) - c_Y*Y(t) + g_Y$ (45)

Conventional mathematical analysis of these equations shows that hypothesis (43) is universally valid: From the equations (44) and (45) follows that Y(t) will grow or shrink until

$Y(t) = (a_Y*X(t) + g_Y) / c_Y = (a_Y*X(t_0) + g_Y) / c_Y$ (46)

Table 1: Searching for parameters which falsify hypothesis (43).

Initial values						Optimized values					
a_y	c_y	g_y	$X(t_0)$	$Y(t_0)$	$[t_0,t_\omega]$	a_y	c_y	g_y	$X(t_0)$	$Y(t_0)$	$\phi(a_y,c_y,\ldots)$
.10	.10	.50	0.10	1.00	[0,500]	.50	.01	.50	1.00	0.01	$-6.57*10^{-3}$
.10	.20	.10	1.00	1.00	[0,500]	.50	.01	.50	1.00	0.01	$-6.57*10^{-3}$
.20	.10	.20	1.00	0.10	[0,500]	.50	.01	.50	1.00	0.01	$-6.57*10^{-3}$
.20	.20	.10	1.00	1.00	[0,500]	.50	.01	.50	1.00	0.01	$-6.57*10^{-3}$
.50	.20	.10	0.10	0.10	[0,500]	.50	.20	.10	0.10	0.10	$-1.39*10^{-16}$

General constraints of the parameter search: $X(t_0), Y(t_0), a_y, c_y, g_y \geq 0.01$;
$a_y, c_y, g_y \leq 0.5$; $X(t_0) \leq 1.0$; $Y(t_0) \leq 1.0$.
Special constraints of hypothesis (43): $a_X = c_X = g_X = 0$.

and consequently (see equation (45))

$$\Delta Y(t) = 0 \tag{47}$$

In order to test the armament-stop hypothesis (43) by computer-assisted interfacing we have to minimize the correspondence-function:

$$\phi(a_y,c_y,g_y,X(t_0),Y(t_0)) = -\,|\,\Delta Y(t_\omega)\,| \tag{48}$$

under the constraint

$$a_X = c_X = g_X = 0 \tag{49}$$

Table 1 contains the results of this minimization process for different initial values of parameters. If we set

$$\theta = -0.01 \tag{50}$$

and thus consider a growth rate of 0.01 as sufficiently low for the practical purposes of stabilizing an arms race, table 1 confirms the hypothesis at least for t_ω=500.

However, table 1 also points to some of the numerical difficulties of this optimization process: In certain situations such as for the initial values on the last line of table 1, the minimization algorithm is unable to locate the global minimum of ϕ and eventually even hides the existence of a falsifying worst-case. From an epistemological point of view, this deficiency is only a minor problem since a confirmation of a hypothesis is always a provisional decision which can be corrected by future search for falsifying evidence. From a practical point of view, the risk of overlooking a falsifying example is relatively small if the optimization is repeated with several sets of different initial parameter values. If the different optimizations lead to different minima, the focus of analysis is on the worst-case set of parameters associated with the lowest detected minimum, i.e. in our example, on the optimized parameters on the first line of table 1.

Hypothesis (43) postulates zero-growth of $\Delta Y(t_\omega)$ not only for t_ω=500 but also for all subsequent time-points t_ω>500. Hence, in order to check the hypothesis (43), the correspondence-function

$$\phi(a_y,c_y,g_y,X(t_0),Y(t_0)) = -\,|\,\Delta Y(t_\omega)\,| \tag{48}$$

has also been evaluated for some values of t_ω beyond t_ω=500. As figure 3 shows, this test reveals no falsifying evidence for the correspondence-tolerance θ=-0.01. Thus, the hypothesis that an unilateral armament moratorium also entails zero-growth of the armament of the opponent can provisionally be accepted.

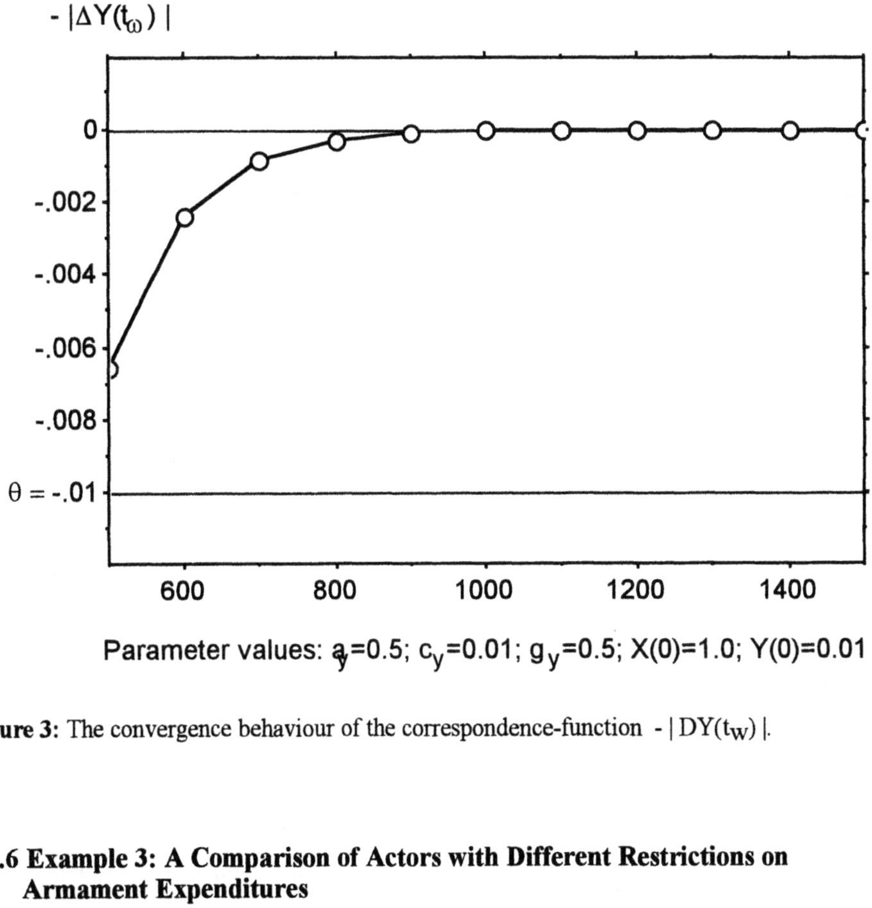

Figure 3: The convergence behaviour of the correspondence-function - | DY(t_W) |.

3.3.6 Example 3: A Comparison of Actors with Different Restrictions on Armament Expenditures

In an abstract dyadic arms race of a Richardson-model, the rival actors primarily differ with regard to their model parameters a_X, c_X, g_X and a_y, c_y, g_y. In the following, the focus of analysis will be on the implications of these differences for the behavior of the two rival actors. More precisely, we shall concentrate on the effects of the budgetary restrictions c_X and c_y which are hypothesized to influence the short-term growth of armaments in the following way:

$c_y > c_X \rightarrow X(t_0+1)-X(t_0) = \Delta X(t_0) > \Delta Y(t_0) = Y(t_0+1)-Y(t_0)$, other things being equal. (51)

The ceteris paribus clause at the end of hypothesis (51) means equality of parameters and initial conditions. Consequently, we have to study the behavior of the model

$$\Delta X(t_0) = a_X * X(t_0) - c_X * X(t_0) + g_X \quad (52)$$

$$\Delta Y(t_0) = a_X * X(t_0) - c_y * X(t_0) + g_X \quad (53)$$

By building the difference

$$\Delta X(t_0) - \Delta Y(t_0) = (a_X * X(t_0) - c_X * X(t_0) + g_X) - (a_X * X(t_0) - c_y * X(t_0) + g_X) = = (c_y - c_X) * X(t_0) \quad (54)$$

between the equations (52) and (53), it immediately becomes clear that for positive $X(t_0)$ the difference $c_y > c_X$ entails $\Delta X(t_0) > \Delta Y(t_0)$, as postulated in hypothesis (51).

Although hypothesis (51) is from the analytical point of view a trivial consequence of the Richardson model, it will nevertheless be tested by means of our methodology of interfacing: This way, we will give a practical illustration of the use of our methodology for making inferences about *differences between categories of actors* (see section 3.2.4) and show a test case of a minimization problem with two indeterminate superfluous parameters a_X and g_X.

Table 2: The search for parameters which falsify hypothesis (51).

Initial values					Optimized values					
a_X	c_X	c_y	g_X	$X(t_0)$	a_X	c_X	c_y	g_y	$X(t_0)$	$\phi(a_X, c_X, \ldots)$
.10	.10	.50	.10	1.00	.10	.30	.31	.10	0.01	$1.00*10^{-4}$
.40	.10	.50	.10	1.00	.40	.30	.31	.10	0.01	$1.00*10^{-4}$
.40	.10	.50	.40	1.00	.40	.30	.31	.40	0.01	$1.00*10^{-4}$
.40	.20	.30	.40	0.50	.40	.25	.26	.40	0.01	$1.00*10^{-4}$
.40	.30	.40	.40	0.50	.40	.35	.36	.40	0.01	$1.00*10^{-4}$

General constraints of the parameter search: $a_X, c_X, c_y, g_X \geq 0.01$;
$a_X, c_X, c_y, g_X \leq 0.5$; $0.01 \leq X(t_0) \leq 1.0$.
Special constraint of hypothesis (51): $c_y \geq c_X + 0.01$.

According to section 3.2.4 we have to take the difference (54) in order to define a correspondence-function

$$\phi(a_X, c_X, c_y, g_X, X(t_0)) = \Delta X(t_0) - \Delta Y(t_0) = (c_y - c_X) * X(t_0) \quad (55)$$

for testing the hypothesis (51). The results of the minimization of function (55) are compiled in table 2. For different sets of initial values they are relatively heterogeneous. In despite of this heterogeneity, the results of the parameter optimization seem to be quite reasonable: Correctly, the optimization algorithm perceives no advantage in changing the superfluous variables a_X and g_X such that the initial and the final values of these parameters are always the same. The variation of the *optimized* values of c_y and c_X reflects the fact that not the levels c_y and c_X but only the difference $(c_y - c_X)$ influences the correspondence function ϕ as defined by equation (55). This difference is for all sets of initial values correctly reduced to the minimal value

$$c_Y - c_X = 0.01 \tag{56}$$

as permitted by the numerical optimization constraint mentioned at the end of table 2. Finally, under this constraint the results in table 2 never violate the inequality

$$\phi(a_X,c_X,c_Y,g_X,X(t_0)) = \Delta X(t_0) - \Delta Y(t_0) > 0 = \theta \tag{57}$$

This means that hypothesis (51) cannot be falsified, even if the difference $(c_Y\text{-}c_X)$ is very small and the correspondence-tolerance θ is set to zero.

3.4 Conclusions

The practical examples presented in sections 3.3.4 to 3.3.6 show that our method of computer-assisted interfacing is less problematic than it might be on the grounds of the numerical algorithms behind it: Fears that the power of a standard PC may be insufficient for parameter optimization have turned out to be unjustified since we were generally able to get a worst-case set of parameters within less than 15 seconds computation time. By repeated optimization with different initial parameter values we have managed to identify and discard local minima which are known to be a risk for successful numerical optimization. Finally, for long-term simulation (see 3.3.5) we have encountered some rounding-problems which were probably caused by the insufficient numerical accuracy of the internal variables of SPSS. However, since other programming languages such as e.g. FORTRAN offer the possibility to control rounding problems by the use of variables with extra-long mantissas (double precision variables), there are alternatives for those cases where the numerical problems of the SPSS-software become insurmountable.

To sum up, our interfacing methodology seems to work, at least for linear models with a relatively small number of parameters. If linearity is given up and if the number of parameters of the model increases, the numerical optimization probably becomes more difficult. Thus, in the future the limits of the proposed method will also have to be explored in depth. However, if in these future tests non-linear models with more parameters are also manageable with this method, it would open an important new perspective for the formalization of social science theories: Instead of using simplified linear models in order to represent fuzzy monotonic relations between variables of verbally formulated theories, we could work with nonlinear model structures whose shape could be changed by the variation of parameters. By the minimization of an appropriate correspondence-function ϕ, a whole family of monotonic functional relations between model variables could be analyzed with regard to their logical implications. It is obvious that in this way computer simulation could make a much more important contribution to the exploration of verbally formulated theories than it has done up to now.

References

Blalock, H. M. (1979): Social Statistics. McGraw-Hill Inc., New York.

Bremer, S. A. (ed.) (1987): The GLOBUS Model. Computer Simulation of Worldwide Political and Economic Developments. Campus Verlag and Westview Press, Frankfurt a.M. and Boulder.

Coleman, J. S. (1973): The Mathematics of Collective Action. Aldine, Chicago.

Forrester, J. W. (1969): Urban Dynamics. M.I.T. Press, Cambridge (Mass.).

Hamblin, R. L., Jacobsen, R. B. & Miller, J. L. L. (1973): A Mathematical Theory of Social Change. John Wiley & Sons, New York.

Hanneman, R. A. (1988): Computer-Assisted Theory Building. Sage Publications, Newbury Park.

Hanneman, R. A. (1995a): Simulation Modeling and Theoretical Analysis in Sociology. Sociological Perspectives 38, 457-462.

Hanneman, R. A., Collins, R. & Mordt, G. (1995b): Discovering Theory Dynamics by Computer Simulation: Experiments on State Legitimacy and Imperialist Capitalism. Sociological Methodology 25, 1-46.

IMSL (1991): User's Manual. IMSL Inc., Houston.

Müller, G. (1996): Exploring and Testing Theories: On the Role of Parameter Optimization in Social Science Computer Simulation. In: Troitzsch, K. G., Müller, U., Gilbert, N. and Doran, J. (eds.): Social Science Microsimulation. Springer, Berlin, 66-77.

Nicholson, M. (1989): Formal Theories in International Relations. Cambridge University Press, Cambridge.

Norusis, M. J. and SPSS Inc. (1993): SPSS Base System User's Guide. SPSS Inc., Chicago.

Norusis, M. J. and SPSS Inc. (1995): SPSS Advanced Statistics User's Guide. SPSS Inc., Chicago.

Paert-Enander, E., Isaksson, P., Merlin, B & Sjoeberg, A. (1996): The MATLAB Handbook. Addison-Wesley, Harlow.

Phillips, J. (1986): The NAG Library: A Beginner's Guide. Clarendon-Press, Oxford.

Popper, K. R. (1959): The Logic of Scientific Discovery. Hutchinson Publishers, London.

Richardson, L. F. (1960): Arms and Insecurity. Boxwood Press, Pittsburgh.

Simon, H. A. (1957): Models of Man. John Wiley & Sons, New York.

SPSS Inc. (1992): SPSS Statistical Algorithms. SPSS Inc., Chicago.

Taagepera, R., Shiffer, G.M., Perkins, R. T. & Wagner, D. L. (1975): Soviet-American and Israeli-Arab Arms Races and the Richardson Model. General Systems 20, 151-158.

Van der Waerden, B. L. (1965): Mathematische Statistik. Springer-Verlag, Berlin.

3.6 Annex: A Sample Program for Checking Inferences About Correlations Between Variables

3.6.1 The SPSS-6 Program

```
* Defining a one-case-situation.
data list free / Tend AVOID.
begin data.
50 1
end data.
*.
*.
* Setting the initial values of the parameters.
model program Ax=0.1, Ay=0.1, Cx=0.1, Cy=0.1, Gx=0.25, Gy=0.25,
INITx=0.5, INITy=0.5.
*.
* Initialization of the auxiliary variables of the correspondence-function.
compute N=0.
compute Sy=0.
compute Sx=0.
compute S2y=0.
compute S2x=0.
compute CROSS=0.
*.
*.
* Begin of the simulation-loop.
loop T=1 to Tend.
* Initialization of the model.
if (T=1) Ysim=INITy.
if (T=1) Xsim=INITx.
*.
* The model equations.
compute Xsim=Ax*Ysim + (1-Cx)*Xsim + Gx.
compute Ysim=Ay*Xsim + (1-Cy)*Ysim + Gy.
*.
* Calculation of the auxiliary variables of the correspondence-function.
compute N=N+1.
compute Sy=Sy + Ysim.
compute Sx=Sx + Xsim.
compute S2y=S2y + Ysim*Ysim.
compute S2x=S2x + Xsim*Xsim.
compute CROSS=CROSS + Xsim*Ysim.
end loop.
* End of the simulation loop.
*.
```

```
*.
* Calculation of the correspondence-function CORR.
compute COV=CROSS - Sy*Sx/N .
compute CORR=COV / sqrt(abs(S2y-Sy*Sy/N) * abs(S2x-Sx*Sx/N)).
*.
*.
* Minimization of the correspondence-function CORR for given boundaries.
cnlr AVOID / pred=CORR / loss=CORR / save pred /
criteria iter=100 fprecision=1.e-10 /
bounds 0.5>Ax>0.001; 0.5>Ay>0.001; 0.5>Cx>0.001; 0.5>Cy>0.001;
0.5>Gx>0.001; 0.5>Gy>0.001;1>INITy>0.001; 1>INITx>0.001.
```

3.6.2 Selected Results of the Previous SPSS-6 Program Calculated with a Power Macintosh 4400/160

All the derivatives will be calculated numerically.

There are 11 cases. There is enough memory for them all.

Iteration	Loss funct[9]	AX GX	AY GY	CX INITX	CY INITY
0.1	.9999900379	.100000000	.100000000	.100000000	.100000000
		.250000000	.250000000	.500000000	.500000000
1.6	.7969156080	.142609006	.001000000	.042671013	.209132176
		.155606379	.344412195	.518865722	.481119418
2.4	.7969154905	.142724238	.001006412	.042657577	.209225960
		.155556530	.344462361	.518937864	.481047862
3.2	.7732332895	.500000000	.003441247	.169254169	.500000000
		.001000000	.500000000	1.00000000	.001000000
4.8	-.999999997	.105740256	.002924988	.154571531	.500000000
		.001000000	.500000000	1.00000000	.001000000
5.14	-.999999998	.105739986	.002925005	.154571229	.500000000
		.001001286	.499998714	1.00000000	.001000000
6.10	-.999999999	.105739870	.002925038	.154568647	.499995807
		.001001286	.499990323	.999983201	.001000000
7.5	-.999999999	.105739796	.002925036	.154568544	.499995589
		.001001286	.499989826	.999982206	.001000006
8.2	-.999999999	.105739787	.002925037	.154568536	.499995568
		.001001291	.499989826	.999982206	.001000006
9.4	-1.00000000	.105739523	.002925824	.154569001	.499996774
		.001001154	.499989430	.999982415	.001000006
10.2[10]	-1.00000000	.105739439	.002926145	.154569064	.499996930
		.001001098	.499989351	.999982454	.001000006

---

9 Loss function = correspondence-function = Pearson correlation coefficient.

10 Final iteration yielding the worst-case set of parameters.

Run stopped after 11 major iterations.

Cannot improve on the current point.

Preceding task required 9.00 seconds elapsed.

4. Interactive Neural Networks as Tools for Modeling Social Systems

Christina Stoica[1]

In recent years numerous computer programs have been developed which are used in the social sciences, including expert systems that are employed in social research (Ohly, 1994), particularly for modelling social processes (e.g. Pfaffenberger, 1988; Benfer et al., 1991) and theories of social science (e.g. Manhart, 1989; Klüver, 1995). Additionally, differential equations (Troitzsch, 1990; Bossel, 1992) and distributed artificial intelligence (DAI) are now often used to construct "Artificial Societies" (e.g. Gilbert and Conte, 1995) and to simulate social processes.

Nevertheless, artificial neural networks are still rarely used in social science simulations. Yet there are many network topologies developed for special problems in the natural sciences, and an overview of these networks can pick out with ease those that are suitable for modelling social science problems (e.g. Halpin, 1998; Stoica, 1999). In our research group at Essen we now work primarily with what is known as the "Interactive Network" (IN), which I will describe in this paper in more detail. IN are well known in linguistics (e.g. Waltz and Pollack, 1985; Kintsch, 1988; Thagard, 1988,1992), and they are especially suited for problems like analysing the behaviour of speakers and the content of meaning referring to some statements respectively for modelling logical connections between individual statements and/or concepts. The IN is especially suitable to model these sorts of problems because of its peculiar functional manner (which I will explain below). We at Essen use the IN for modelling complex social systems to analyse their behaviour (Stoica, 1999).

Here, I wish to demonstrate some examples of simulations with interactive networks, in order to show which kind of problems can be simulated with this sort of technique. In general interactive networks are well suited for simulations of top-down models, i.e. social areas that are globally observed, in which the analysis attempts to grasp the whole behaviour of the system. In contrast to top-down models there are bottom-up approaches for simulating processes on the micro-level (Klüver, 1999; Stoica, 1999).

---

[1] Center for Research in Higher education at the university of Essen

A significant problem derives from the fact that social systems are adaptive. That is to say, they may change their rules or enlarge themselves according to the demands of their environment. To model these characteristics of adaptive social systems one has to use hybrid techniques, i.e. the coupling of two or more different programs.

We use different hybrid systems to model adaptive processes of social systems, for example cellular automata (CAs) coupled with genetic algorithms (GAs) or interactive networks (INs) coupled with GAs (see below). The different behaviour of the different hybrid systems allows us to analyse all aspects of adaptive systems, for example: learning processes, social differentiation, and the self modelling of social systems (Klüver et al. 1997).

There are different definitions of "hybrid systems."[2] We at Essen prefer the terms "vertically coupled systems" and "horizontally coupled systems" (Klüver et al., 1997; Stoica, 1999). Vertically coupled systems are combinations of programs in which one program (defined as a metasystem) operates on the rules and/or parameters of the second (the base system). Possible examples of base systems include cellular automata (CA), knowledge-based systems (KS) and interactive networks (IN). The metasystem that operates on the base systems is usually the genetic algorithm (GA). In horizontally coupled systems (HCS) an independent system transfers its results to a second system that performs another part of the task. In this sort of arrangement both systems operate on the same logical level. An example of an HCS is THEOPRO (Klüver, 1995) which is comprised of a knowledge based system coupled with an interactive network. The knowledge based system constructs the net, consisting of concepts of a theory of society, and a weight matrix. The neural net takes this information and performs simulations according to the theory.[3]

Higher order hybrids combine hybrids of these two types. With SOZAIN (Klüver et al. 1997) we have developed a system that contains a CA coupled horizontally with an IN, and different GAs operating on both base systems as metasystems. The main idea behind this arrangement is to simulate the concept of "self-modelling", an important capacity of complex systems. "Self-modelling" here refers to the fact that complex systems such as societies are able to construct models of themselves by mapping their structures into some of their parts. In our case, the characteristics of the system performed by the hybrid CA is transformed into the IN. The IN simulates the actual (global) state of the CA, and the hybrid IN tests possibilities of changing specific structures before returning the obtained

---

[2] See Goonatilake and Khebbal (1995)

[3] In contrast to this procedure, Gallant (1993) presents the idea to let a neural network generate rules for an expert system which would also be a horizontally coupled system.

results to the CA. The hybrid CA proceeds on the basis of this new information, and so on.

By using these techniques it is possible to model self-referential processes of social systems, which are both extremely important and as yet scarcely understood. We have only begun to analyse these very difficult properties of social systems, but our first results are quite promising.

Returning now to the IN, I will present some details about this network and how we have used it.

4.1 Interactive Neural Networks

The Interactive Networks (IN) belong to the broad category of feedback or recurrent networks, but with the specification that they are not trained.[4] The distinctive quality of this network is the spreading of activation from externally activated components to all other components. The activation of each component (unit) depends on its weight-values. These values are fixed according to the criteria given by the problem and are represented in a square matrix. The values can either be positive or negative, and depend on whether there is an excitatory or inhibitory connection between the components.

The weights of the connections between the units do not change during the processing of the net. Because of this, INs are highly suitable for testing logical hypotheses and for simulating processes in which the singular parts are connected recursively. When using an IN, the construction of a model for simulations always goes as follows: The relevant parts of a social "system" are defined as the units (the neurons) of the net. The relations are defined as delaying or strengthening the units, in other words, as the meaning of the inhibitory or excitatory connections.

The simulation starts by giving an external activation input, i.e. a positive numerical value, to one or more of the units. The net runs a number of steps that the user has fixed, and if the network reaches a steady state, activation values of all units define the result of the simulation. The IN uses the standard activation rule:

$$A_i = \sum_j A_j * w_{ij}$$

where A_i is the activation value of the receiving unit i, A_j is the activation value of a sending unit j and w_{ij} is the weight value of the relation between the units j and i.

---

[4] For more information about the IN see Kruse et al. (1991)

We use the IN principally to investigate two rather different fundamental questions. First, we investigate prescriptive models. By means of the simulation we try to examine how specific developments of a social system might have taken place. The analysis of the model relates to the possible (sufficient or necessary) conditions that might have produced a particular result (e.g. by the model of Eder shown below).

Second, we study open simulations. These models concentrate on the analysis of future events and developments. Here the case is reversed. The conditions and the rules that determine the operation of the systems are known and given as input. In this sort of case we want to find out how the system develops, especially when it is able to vary its own rules and relations between the system units. For the reasons I mentioned earlier, hybrid techniques are necessary here. A simple IN would not be sufficient.

The following example belongs to the first type of problem.

4.2 The Genesis of State Organised Societies

On the basis of ethnological material Eder (1976) developed the thesis that the transition of tribal societies into State-organised advanced civilisations was caused by processes of social learning. These advances in learning led in turn to new social roles and consequently to reorganisation. The following assumptions are relevant for the simulation (Eder, 1976; Klüver, Stoica, 1995):

(1) The transition from tribal to advanced societies occurs only if there were an *outside pressure* (Eder, 1976, 62ff), i.e. an external pressure of selection, caused either by ecological and/or demographic (overpopulation) problems.

(2) The evolutionarily required learning processes take place because at first new (*mythical*) *religions* push away the *animistic religions*. As a result of this new view of life *new* principles of *law* come into being, based on individualistic guilt and punishment, crowding out the *old law* in the process. From this it follows that the political role of the *chief* is displaced by that of *judge/ruler*. To institutionalise these new roles it is necessary that socially unequal class structures form.

(3) Allegedly this process of transition takes place only in societies which already show a mild form of *social inequality* (Eder, 1976, 85ff). In this regard Eder distinguishes between tribe societies and chiefdom societies. Unfortunately, he is not very explicit here.

The theses 1 and 3 are necessary conditions for the sketched evolution process described in thesis 2. The marked terms are the components (units) for the interactive network (IN).

Figure 1 contains the main components of the model:

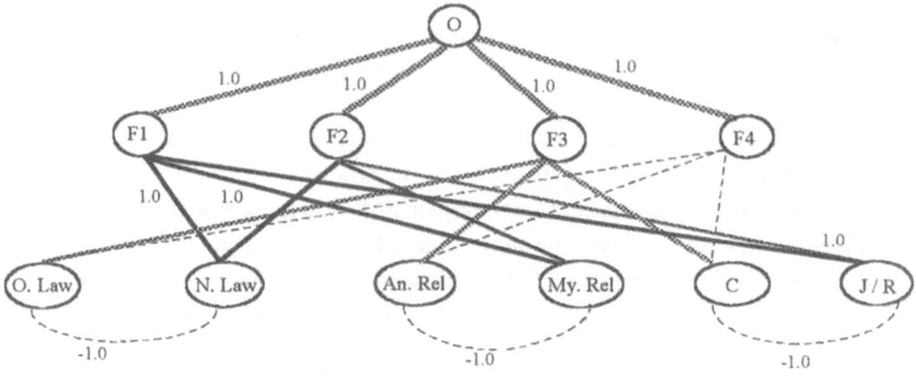

Figure 1: Characteristics of the model based on the model of Eder

Interpretation:

The network consists of 11 components. Unit *O* symbolises the *outside pressure* of selection. The units *F1 to F4* represent four *social segments* (here *families*) which demonstrate a hierarchical difference, and thus *social inequality*. The social structure is directly influenced by the unit "outside pressure". During the stage of transition the units *old law* (*O.Law*) and *new law* (*N.Law*), *animistic* (*An.Rel*) and *mythic religions* (*My.Rel*) and the roles of the *chief* (*Chief*) and *judge/ruler* (*J/R*) all coexist. Finally, we assign the role of the chief to the old law principles and animistic religion and the new role of the judge/ruler to the new law and mythic religions.

Some additional explanatory remarks are necessary:

Eder's first assumption requires an outside pressure for the process of transition to begin. Therefore the model contains the unit "outside pressure", which has a strictly technical, rather than sociological, meaning. INs need the external activation of one or more units as was described in chapter 1.

The four family units symbolise the fact that the society to be modelled is a tribe society. That means that it is only differentiated into social segments, specifically *families*. Of course different numbers of families could have been chosen.

The other units are the representation of Eder's second assumption, that law, religion and roles of rulership exist in different forms at the same time. So our model is both an operationalization and standardisation of Eder's comparatively informal text.

Eder's third assumption is represented in the model by our defining asymmetric relations between the families by according values in the weight matrix (see below).

Another assumption has to be made. Only the segments F1 and F2 are interested in new social structures, whereas the segments F3 and F4 are bounded to the old principles. The reason for this assumption lies in the hypothesis that in segmentary societies only the socially higher segments profit by the introduction of new social rules and roles. As a supporting example, consider the case of the transition from polytheistic religions to Christendom in the early Germanic societies, which demonstrates the fact that this was an innovation by (and *for the benefit of*) the chiefs and kings. In our model we represent this insight by stressing the hierarchical structures supporting the families F1 and F2, in order to make their dominance clear. As a result, the families F1 and F2 have the power to carry through their interests within the new social structures. In contrast, families F3 and F4 are not in a position to manipulate the social innovations made necessary by the outside pressure. In our modelling of this assumption it was necessary to enlarge the theory itself and not only to "translate" it into a formal model. This is not unusual. The business of model constructing often forces one to fill gaps in an original theory which was stated only informally.

At this point an obvious objection could be raised to our model. Its units represent distinctly different types of social categories. So it would seem to be problematic to integrate such different categories within a single model. But this problem vanishes if we understand the *sociological* meaning of such categories like family, law or social rules as nothing other than sets of rules of social interactions. For example, *sociologically* a religion may be understood to comprise of the rules of actions which followers of the religion in question take as basis for their interactions with peoples from the same religion or with outsiders (i.e. infidels). On this sort of understanding the different units in our model all represent the same type of social categories: Sets of social rules (Giddens, 1988),[5] and may be grouped together with no problems.

Our weight matrix was constructed according to Eder's theses (1) - (3). We begin with the unit outside pressure (O), whose connections to the units F1 to F4

[5] To clarify: within a family the rules of interaction are fixed, meaning the way of communication or acting with other members of this family. We can observe if someone belongs to this family or not, if the behavior corresponds to the specific rules of this family. Of course there are also rules concerning the communication with members of other families and so on. This example could be apply to the other units respectively.

are equally strong, with the weight value +1.0. Hence it follows that these components all must react immediately to outside pressure. The hierarchical structure between the families is represented with different values referring to the connections, e.g.: a comparatively small social distance exists from F1 to F2 (weight value 0.30), a greater to F3 (0.60) and a still great social distance to the family F4 (0.90). Accordingly, high weight values (+1.0) extend from F1 to the new social structures, represented with the units new law and mythic religions. In contrast, there are connections represented with the weight value about 0.70 from F4 to the old principles.

The construction of the weight matrix extending from the units F1 to F4 was based on the following principle: the hierarchical structures are represented in the different weight values to the other components and vice versa (in other words we are working with asymmetric connections)[6]. The roles of the chief in relation to the judge/ruler are only connected immediately with the old in relation to new structures; in other words, thesis (2) entails that the role of the chief with respect to the role of judge is the consequence or result of the religious *weltanschauung*. In this case there is no direct connection from each family to these units. Rather, these units have an influence on the families. This is represented with the weight values from the units chief and judge/ruler to the correspondent family (see columns 10 and 11 in the weight matrix below). Finally, are inhibitoric connections (weight value -1.0) between the units representing the old structures and new structures (see weight matrix below).

With these relation in place we can start the simulation with the IN. The result of the simulation must decide whether the weight values fixed by me are valid with regard to the assumptions described above. To start the simulation with the IN it is necessary to decide which unit gets an external activation. In this case it is evident that the unit *outside pressure* is to be externally activated, according to the theses. Figure 2 displays the result, including the weight matrix:

[6] The matrix is to be read in the following way: the weight w_{ij} means the value of the connection from unit j to unit i.

Matrix of the connections

	[1]	[2]	[3]	[4]	[5]	[6]	[7]	[8]	[9]	[10]	[11]
[1] Outer pressure	0.00	0.00	0.00	0.00	0.00	0.00	0.00	0.00	0.00	0.00	0.00
[2] Family 1	**1.00**	0.00	0.20	0.10	0.05	0.00	**1.00**	0.00	**1.00**	0.00	**1.00**
[3] Family 2	**1.00**	0.30	0.00	0.20	0.10	0.00	**0.70**	0.00	**0.70**	0.00	**0.70**
[4] Family 3	**1.00**	0.60	0.30	0.00	0.20	0.70	0.20	0.70	0.20	0.40	0.20
[5] Family 4	**1.00**	0.90	0.60	0.30	0.00	**0.60**	0.00	**0.60**	0.00	**0.60**	0.00
[6] Old Law	0.00	0.00	0.00	0.50	**0.60**	0.00	-1.00	1.00	-1.00	**1.00**	-1.00
[7] New Low	0.00	**1.00**	**0.70**	0.30	0.00	-1.00	0.00	-1.00	1.00	0.00	**1.00**
[8] Animististic Religion	0.00	0.00	0.00	0.70	**0.60**	0.00	-1.00	0.00	-1.00	**1.00**	-1.00
[9] Mythic Religion	0.00	**1.00**	**0.70**	0.30	0.00	-1.00	0.00	-1.00	0.00	0.00	**1.00**
[10] Chief	0.00	0.00	0.00	0.00	0.00	1.00	0.00	1.00	0.00	0.00	**-1.00**
[11] Judge / Ruler	0.00	0.00	0.00	0.00	0.00	0.00	1.00	0.00	1.00	**-1.00**	0.00

Time step: 50	External Input	Activation
[1] Outer pressure	0.25	0.29
[2] Family 1	0.00	1.00
[3] Family 2	0.00	1.00
[4] Family 3	0.00	0.61
[5] Family 4	0.00	0.50
[6] Old Law	0.00	0.00
[7] New Law	0.00	0.50
[8] Animististic Religion	0.00	0.00
[9] Mythic Religion	0.00	0.50
[10] Chief	0.00	0.00
[11] Judge / Ruler	0.00	1.00

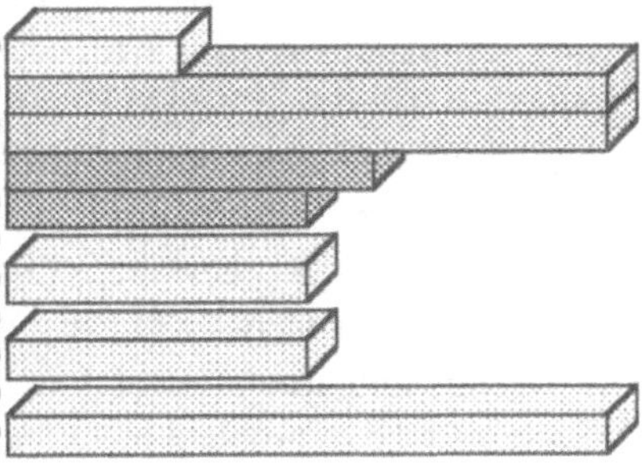

Figure 2: The result of the simulation after 50 cycles

Interpretation:

The unit outside pressure is externally activated with the value 0.25.[7] The external activation determines the dynamic process within the IN. In other words, the activation spreads. After 50 cycles the network reaches a steady state. Because of the activation values it is obvious that the social inequality, represented by the individual families, is enlarged and distinct. Units F1 and F2 reach the maximal possible value of 1.0. Units F3 and F4 achieve lower values with respect to the hierarchical structure. Additionally, only the new principles of the society, represented by the units new law, mythic religions and judge/ruler, have been successful, and have pushed away the old structures represented by the units old law, animistic religions and chief. Technically, this result could be explained

[7] The final state of the model contains the new activation value of the unit outside pressure of 0.29. This is due to the fact that all units are directly connected and therefore the first unit is changed also by the processing of the network. The new value has no direct social meaning.

because of the feedback process of the network, activating strongly the units which have strong positive connections to each other. Because of the inhibitoric values and the lower weight values from the units F3 and F4 to the old principles the activation of these units is not possible. The final activation of the units old law, animistic religion and chief is in each case 0.0 (the minimal possible value). In conclusion, this simulation demonstrates that Eder's assumptions are well transformed into the weight matrix, bearing out the initial theses concerning hierarchical structures and the influence of these structures on other social units.

This example supports the theses formulated by Eder. The transition from tribal societies into State-organised civilisations could have happened this way. For our model it was necessary to make the author's imprecise assumptions more explicit. In the process we have demonstrated that the IN is quite useful for developing simulations dealing with social problems. By means of such models and simulations theories of well known social processes can be reconstructed.

The most important shortcomings of INs and similar formal systems is the fact that although they are self-organising, they are not adaptive. But central to social systems is the fact that they are self-organising, adaptive and evolutionary. Accordingly, the Eder model was expanded to take these general characteristics of social systems into account, which we then used to develop simulations to deal with the second type of problem mentioned above.

4.3 The Genesis of State Organised Societies II

The following model contains two aspects of systems theories (Buckley, 1968; Miller, 1978; Willke, 1982):

(1) The adaptive capability of systems

The term "adaptive capability" refers to the ability to respond to changing environmental demands. The simulation with the IN in the previous section showed only a closed self-organised system which is able to reach a steady state, as a result of its inner recursive procedure. Adaptive capability, however, cannot be modelled with an IN alone. The adaptive competence of a system involves the system's capacity to change not only its states by its own rules (self-organisation) but also its rules of self-organisation. In logical terms this means that the system contains not only rules of self-organisation, but also meta-rules which change the rules of the system according to particular demands of the environment. To measure these demands the system also must possess an evaluation function by which it compares its states with environmental demands. The technical consequences of these considerations are twofold. On the one hand one has to make a simple IN "hybrid" (see section one) which means that in addition to the IN a meta-system must be constructed which changes the weight matrix of the IN (the weight matrix is the formal equivalent of the social structure of our system).

On the other hand, the mathematical measure must be defined in order to enable the hybrid system to evaluate the success of its adaptive processes.

As the meta-system a genetic algorithm (GA) is introduced, whose task is to change the connections between the units of the IN. The GA is an optimisation algorithm, developed by Holland (1975), which simulates biological evolution. The GA consists of two genetic operators, namely crossover ("heterosexual reproduction") and mutation. These two operators act upon the weight matrix of the IN. The results of these operations are new weight matrices and that means new networks.

At this point we have to fix the goal of the optimisation process of the system. This is done by providing an "environmental vector" (EV) that represents the different demands of the environment. For the sake of mathematical simplicity the EV has the same dimension as the "state vector" of the IN. The state vector is nothing other than the vector of activation values of the IN units. As our IN contains "principally" ten *social* units (see below), the EV also has a dimension of ten. In this case the EV is obtained by the activation values of the simulation of the manually constructed weight matrix (see picture 2):

EV=(1.0, 1.0, 0.61, 0.50, 0.0, 0.50, 0.0, 0.50, 0.0, 1.0)

The reason behind this definition is as follows. The aim is to model adaptive processes, which entails enabling the formal system to do by itself those tasks which were done manually in the first case. Thus, the formal system takes the result of the "manual" simulation as its goal which it must reach by itself.

The program then generates not one but ten different weight matrices at random. Strictly speaking, by this method we develop ten different INs. Of course these networks are externally activated with the same activation values and run as many time-steps as they need to reach an attractor state. The obtained activation values are interpreted as the specific activation vector (AV) of each simulation. The selective criterion for choosing the best results of the simulations are measured with the Euclidean distance (d) between the different AVs and the EV:

$$d=|EV-AV|$$

According to this specific criterion the attractor states are classified whether the networks have reached a "good" result or not. The five best matrices are taken and their weight values are recombined by crossover with the result of ten new matrices. These matrices are changed by mutation, the resulting INs are run to their attractor states, which get evaluated, the best five matrices are taken again to produce ten new matrices, and so on until the environmental criteria are fulfilled (if, in fact, that is possible).

This hybrid system allows us to observe the behaviour of the system with respect to its capacity to adapt. It is possible to measure this capability in mathematical terms, the so called "meta-parameters" (see Klüver this volume).

For example, we can analyse in different experiments how many degrees of freedom regarding changing rules are favourable versus disadvantageous (e.g. Stoica, 1999).

(2) The capability of a system to enlarge itself

The necessity of enlarging is present if, e.g., a system's existing rules (structures) are insufficient for adaptation, i.e. for reaching its goals. In this case a system is forced to find additional rules, and thus new ways to get a solution according to the demands of the environment. The technical realization of this problem goes with our formal systems as follows: the IN starts only with the units representing the social categories before social transformation takes place. In our example the units are: outside pressure (as the obligation to change something); the families 1 to 4; the latent existing hierarchical structures; the old law order; animistic religion and the role of the chief. The IN starts the simulation and compares the achieved activation values with the "environmental vector" (EV) described above. The program "notes" that the EV contains more units than the actual IN. Consequentially , the program expands the numbers of the units of the IN until that number corresponds to the number of the units given by the EV. In the following example the *new* units are not specified (that should be done by the user). The system then has the problem of integrating the new rules (in our case the new connections - weight values) into the ensemble of the old ones.

The result of the simulation based on this model is represented in figure 3.

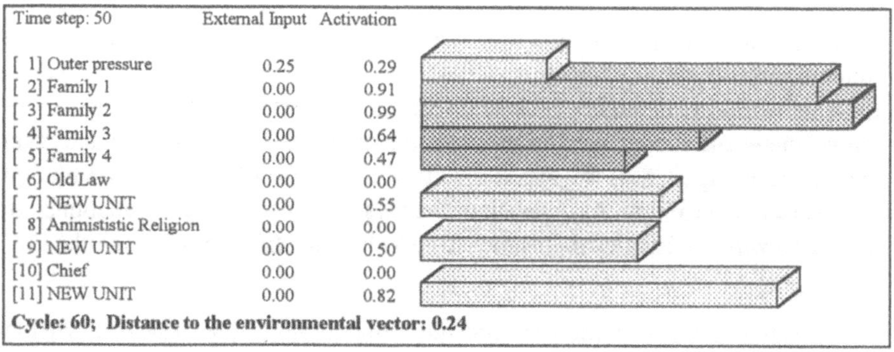

Figure 3: The hybrid system IN/GA

Interpretation:

The result shows a satisfactory state (measured in the distance to the environment vector) after 60 cycles. It is clear that this result could be interpreted like the result obtained with the matrix constructed manually (see picture 2).

More important is the fact that this hybrid system simulates the possibilities of a system to enhance the inner complexity by means of changing and enlarging structures, if the environment forces new orientations. The hybrid system IN/GA makes it possible to analyse particular aspects of systems theory, especially the conditions which make a system able (or unable) to adapt itself to environmental demands. Some parameters that are important for describing the dynamical process relevant to social systems' capacity to adapt are described in this volume by Jürgen Klüver.

Conclusions

The possibilities of using interactive networks or hybrid interactive networks for simulations of processes of social self-organisation and adaptation have been demonstrated in this paper by a single example. This example shows the technique of top-down modelling. I have constructed other simulations of very different processes (Stoica, 1999) which I can only mention here. Although it is not always easy to give precise sociological meaning to the structure of the network, INs seem nevertheless to be nearly universally applicable. Therefore, the technique of modelling social dynamics with an IN or hybrid IN, using differential equations or the newer techniques of e.g. cellular automata, should be taken as seriously and investigated as a standard modelling procedure. Because the art of modelling and simulating social dynamics is a rather new one each promising technique should be investigated thoroughly.

References

Benfer, R. A., Brent, E. E. & Furbee, L. (1991): Expert Systems, Newbury Park, London.

Bossel, H. (1992): Modellbildung und Simulation. Vieweg Braunschweig, Wiesbaden.

Buckley, W. (1968): Society as a Complex Adaptive System. In: Ders. (Hrsg.), Modern System Research for the Behavioral Scientist. Chicago, 490-513.

Eder, K. (1976): Die Entstehung staatlich organisierter Gesellschaften. Suhrkamp, Frankfurt.

Gallant, S. I. (1993): Neural Network Learning and Expert Systems. MIT Press, Cambridge, Massachusetts, London.

Giddens, A. (1988): The Constitution of Society. Outline of the Theory of Structuration. Cambridge, Polity Press.

Gilbert, N. & Conte, R. (eds.) (1995): Artificial Societies. UCL Press, London.

Goonatilake, S. & Khebbal, S. (1995): Intelligent Hybrid Systems. John Wiley & Sons, New York, Toronto.

Halpin, B. (1998): Simulation in Sociology: A Review of the Literature. To be published in: American Behavioural Scientist.

Holland, J. R. (1975): Adaptation in Natural and Artificial Systems. Ann Arbor: University of Michigan Press.

Kintsch, W. (1988): The Role of Knowledge in Discourse Comprehension: A Construction-Integration Model. Psychological Review 95, 2, 163-182.

Klüver, J. (ed.) (1995): Soziologie als Computerexperiment. Vieweg, Braunschweig.

Klüver, J. (1996): Sociological Discourses in Virtual Reality. In: Social Science Computer Review 14, 3, 280-292.

Klüver, J. & Stoica, C. (1995): Konnektionistische Simulationen - Neuronale Netzwerke. In: Klüver, J. (ed.): Soziologie als Computerexperiment. Vieweg, Braunschweig, 28-44.

Klüver, J., Stoica, C. & Schmidt, J. (1997): Simulations of the self modeling of social systems by a hybrid system. In: Bandilla, W. & Faulbaum, F. (eds.): SoftStat '97. Lucius & Lucius, Stuttgart, 411-417.

Klüver J. (1999): The Dynamics and Evolution of Social Systems. To be published at Kluwer Academic Publishers, Dordrecht (NL).

Kruse, H., Mangold, R., Mechler, B. & Penger, O. (1991): Programmierung Neuronaler Netze. Eine Turbo Pascal Toolbox. Addison-Wesley, Bonn, München.

Manhart, K. (1989): Können AI-Programme als Theorien betrachtet werden? In: Retti, J. & Leidlmeier, K. (eds.): Proceedings 5. Österreichische AI-Tagung. Springer, Berlin, 346-358.

Miller, J. (1978): Living Systems. McGraw, New York.

Ohly, H. P. (1994): Wissensautomatisierung in den Sozialwissenschaften. In: Best, H., Endres-Niggemeyer, B., Herfurths M. & Ohly, H.P. (Hrsg): Informations- und Wissensverarbeitung in den Sozialwissenschaften. Beiträge zur Umsetzung neuer Informationstechnologien. Westdeutscher Verlag, Opladen, 207-224.

Pfaffenberger, B. (1988): Microcomputer Applications in Qualitative Research. Newbury Park, London.

Stoica, C. (1999): Die Vernetzung sozialer Einheiten. Hybride Interaktive Neuronale Netzwerke in den Kommunikations- und Sozialwissenschaften. (Networks of Social Units. Hybrid Interactive Neural Nets in the Social Sciences) To be published in: Berlin: G+B Verlag Fakultas O.P.A. (Overseas Publishers Association)

Thagard, P. (1989): Explanatory coherence. Behavioral and Brain Science 12, 435-502.

Thagard, P. (1992): Adversarial Problem Solving: Modelling an Opponent using Explanatory Coherence. Cognitive Science Bd. 1, 16, 123-150.

Troitzsch, K. G. (1990): Modellbildung und Simulation in den Sozialwissenschaften. Westdeutscher Verlag, Opladen.

Waltz, D. L. & Pollack, J. B. (1985): Massievely Parallel Parsing: A strongly interactive model of natural language interpretation. Cognitive Science 9, 51-74.

Willke, H. (1982): Systemtheorie. Fischer Verlag, Stuttgart.

Part II

Multi-Agent Based Simulations

5. SMASS: A Sequential Multi-Agent System for Social Simulation

Wolfgang Balzer[1]

5.1 The Goal of SMASS

SMASS is a simple simulation program which can flexibly deal with many different forms of individual behavior. The combination of these features: simplicity, multiplicity of rules of behavior for one individual actor, and flexibility for the user to switch between different applications with different rule sets are rarely found in existing programs.

Up to now, social simulations are done along the following lines. On the one hand, there are programs for cellular automata, like (Hegselmann, 1996, Liebrand & Messick, 1996, Mark, 1998, Nowak & Lewenstein, 1996), investigating the macro effects which occur when all individuals perform one single rule of behavior. Second, there are simulation studies of populations of individuals in the spirit of evolutionary game theory in which all members of a population perform one single rule of behavior, and in which the relative success of such a population in comparison with other populations is at stake, (Axelrod, 1984, Schuessler, 1990). A third, recent approach uses object-oriented software like (Epstein & Axtell, 1996) or SWARM. Here, different rules ("methods") can be added to each agent ("object"). These systems are implemented in variants of object-oriented C, so it is difficult for social scientists to manipulate the code. Last not least, there are systems including models of a full cognitive apparatus for each agent. In these systems, an actor can apply many different rules, but the systems, exemplified best by Carley & Prietula (1994), are far from simple.

In any real social system a person may perform actions of several different kinds, and the person usually has some choice of which kind of action to engage in. In order to become more realistic, social simulations therefore must include the possibility of behaving according to different rules. As this may drive the system towards great complexity and make it difficult to use, special attention has to be given to securing flexibility and simplicity of application. One of the main goals in the design of SMASS is to strike a balance between simplicity and flexibility of changing the rules on the one hand and of coverage of a space large enough to comprise the most central types of human social interaction on the other hand.

---

[1] University of Munich.

In SMASS each individual commands several different rules of behavior among which it can choose in a given situation. But the user of SMASS is not restricted to those rules which presently are implemented. SMASS is built such that, first, among a range of implemented rules, the user may choose any arbitrary subset and run a simulation in which the actors are restricted to rules from that subset, and such that, second, it is relatively easy to invent and plug in new rules of behavior without changing SMASS's global "architecture". A comparison with object-oriented systems providing similar facilities SWARM, (Uhrmacher, 1997) still has to be made.

Up to now I have not concentrated on finding new, "interesting" simulation results, but focused on the development of the frame program and on reproducing different existing programs in that frame. At the present stage, SMASS is able to simulate systems of actors which can perform quite a number of different kinds of actions but I did not carry out systematic simulations with mixed sets of actions. What I have done up to now in applying SMASS is to reproduce several existing "one-rule" studies by simply inactivating all other rules of behavior present in the system. I think that this indicates the potential of SMASS and justifies the description of some of its major components even in the absence of new, "interesting" simulation results. My long term goal in developing the system is to perform simulation studies of more realistic kinds of social systems, like social institutions. A comprehensive formal theory of social institutions which can serve as a basis for implementation is described in (Balzer, 1990, 1993).

5.2 What Triggers Individual Action: BDI Versus Rule Governed Behavior

A widely used approach to the construction of multi-agent systems is the so-called BDI-approach (from belief, desire, intention). In a BDI model each actor is equipped with components representing his or her beliefs, desires and intentions (Cohen & Levesque, 1990, Rao & Georgeff, 1991). I use the term "BDI-approach" here in a broad sense so as to comprise approaches which do not use all three items. The BDI-approach has its sources in "classical" Bayesian decision theory and in the possible worlds approach of formal logics. Main applications of BDI systems are in the technical construction of robots which can interact with other robots and humans. The BDI approach is also attractive to social scientists for it centrally uses notions which humans themselves use in natural language to reason about their actions.

However, it can be argued that the BDI-model is deficient for the simulation of social systems, and that it has to be enriched by components for purely rule or norm-based behavior. There are at least four arguments for this claim, a theoretical, a methodological, a practical, and an "internal" argument.

The theoretical argument focuses on the central "actor-loop" in which input from the environment is used to produce output-behavior of an agent. In the BDI-model this loop always, i.e. for each action, includes some use of the BDI apparatus. In other words, there is no action that was not preceded by some deliberation. This assumption is central to the rational choice approach. However, in sociology there is also another paradigm, that of homo sociologicus, which emerged from the work of, for instance, Durkheim and Parsons. Homo sociologicus does not always use his BDI apparatus, a great deal of his behavior consists in following rules and norms. The picture roughly is this: An actor has stored many rules and norms which can be applied in specific situations, and which are given together with specific conditions of applicability. Whenever a person finds herself in a situation fitting to some application condition of a rule she will apply that rule and behave accordingly.

These approaches lead to different actor loops. In the "pure" BDI case the actor in each loop takes a decision according to the principles of Bayesian decision theory (or some weakened version of it), whereas homo sociologicus in some loops can get along without decision. In some loops he may simply check which of his rules applies in the respective situation. This check is not a decision, it is a process of fitting or of recognition or an application of a production system.

These two different kinds of loops cannot be realized simultaneously, a person cannot at the same time choose an action on the basis of decision theory and "choose" the same action by matching the situation with the application conditions of her stored rules. In the second case it is not appropriate to speak of a choice at all. It seems to me that both kinds of behavior (BDI guided and rule guided) are incommensurable in a precise sense (Balzer, 1985, 1989). Without going into details, the point is this: In a decision there have to be at least two alternatives among which one can choose whereas in a rule guided situation it may well be that only one application condition of one rule matches with the situation. Defenders of the BDI approach will say that the case in which there is just one alternative can still be regarded as a "limit" case of choice. In defending the sociological paradigm I would say that if there is no alternative, there is no choice, and there is no decision. What seems to be a quarrel about the use of words for a trivial borderline case in fact indicates a central schism between two ways of looking at the social world (Balzer, 1992a). Because of the far reaching implications of looking at things from these different points of view there is no easy way to settle this dispute -as we know from the work of Kuhn and Feyerabend.

In spite of the difficulty of this controversy it is not difficult to enrich the actor-loop so that both types of behavior (BDI guided and rule guided) can be realized by the same agent in different situations. The person simply may decide in some loops, and act on the basis of rules in other loops.

A variant of this discussion is obtained when the notion of action is specialized to plan-based action so that action is performed in order to execute some step in a plan. In such a setting rule guided behavior trivially occurs once a plan has been adopted for execution. The above arguments in this context get their bite when "decision" is restricted to "choice of a plan". They show that the pure BDI model is too strong even in this restricted field. Even the choice of plans may be a matter of routine rather than an application of the BDI apparatus.

The methodological argument against the pure BDI model points to the well known fact that belief, desire and intention are mental predicates and are not directly observable. Moreover, the theory-guided access to them is severely restricted and can be achieved only in laboratory situations from which there is no reliable inference to behavior in other, real-life situations. Therefore, the use of BDI components yields a restriction when simulation results are to be compared with real data, for the latter do not cover the central parts of the BDI model in a reliable way.

The practical argument consists in pointing to simulation programs, like SMASS, in which rule- and norm-based behavior is primary, and "rational", BDI features are secondary. However, "full rationality" is not excluded. The system at present can reproduce many of the existing simulations of cluster- and group formation on game theoretical or other principles (Hegselmann, 1996, Mark, 1998, Nowak & Lewenstein, 1996).

Finally, the "internal" argument points to the fact that rule- and/or norm-based behavior has proponents both in sociology (as documented, say, by the classics of Durkheim and Parsons) and increasingly also in DAI, for instance (Altermann & Zito-Wolf, 1993, Müller et al., 1996, Wooldridge & Jennings, 1991, Wooldridge et al., 1995).

In SMASS, a person has no explicit BDI apparatus. Her beliefs, desires and intentions are present only implicitly, namely in her rules of behavior which may -but need not- refer to principles of (bounded) rationality. Primarily, an actor's behavior in SMASS is social-habitual, or "situated" as some authors say. Each actor has a character which is given by a list of weights, one for each mode of behavior. If the system is run, say, with 5 modes the character of actor A has the form [C_1,..., C_5]. In each given situation, if the person is in the active state, she chooses one of the five modes (action-types) M_1,..., M_5, say M_i, with probability C_i, and then tries to perform an action of the chosen type according to the rule of behavior which is present for that type. If, for instance, M_1 is "rest" and C_1 is 0.5 the person is very lazy.

The characters are automatically created by means of discrete distributions whose weights can either be automatically created or set by hand. In a future version the choice of modes will be achieved by a mixture of character and a set of social norms (see "check-environment" in the "kernel" predicate, line (10) in the appendix).

5.3 Actions: Types and Tokens

In dealing with a multiplicity of actions of different types a problem is given by the token-type distinction. An actor can usually perform different action tokens of one type of action. The type "exerting power" (Balzer, 1992b) can be realized by many different action tokens, ranging from Hans' beating Fritz here and now to Helmut Kohl's signing the "re-unification" treaty for West-Germany in 1989. The type "playing a prisoner's dilemma game" can be realized by Marco and Pietro being questioned by the public prosecutor here and now or by my quarrelling with my neighbour about dealing with the garbage ten years ago. The type "exchange" may be realized by my buying some cherries from grocer Kuhnt on 12.12.1984 or by my signing a document at the notary's office of Thaler on 3.3.1986. At first approximation, tokens may be seen as "elements" making up a "class" which is the "type". However, as is well known from the philosophy of action this view ultimately is untenable the reason being that the notion of action is soaked with that of propositional attitudes and in the domain of propositional attitudes the principle of extensionality breaks down (Goldman, 1970, Quine, 1961, Schiffer, 1987)[2]. Put differently, the problem is that one action token under two different descriptions may "belong to" two different types. A standard example are the two action types "I wanted to shoot the thief" versus "I wanted to shoot my drunken friend" both describing what we would like to call "the same" token (something like my intentionally pulling the trigger after having heard noise in my home late at night and having seen some person moving in the dark). One reaction to this opaqueness of action tokens has been to deny them the status of scientifically reputable entities which in turn, and ultimately, would mean to give up any serious science dealing with human actions.

An alternative reaction is to acknowledge the contested status of action tokens but nevertheless use them with sufficient precaution in social theorizing; this is the philosophy for SMASS. Yet this leads to another problem. The number of action tokens that may be relevant in a social system is usually so large that it is practically impossible to store all possible tokens in a computer's memory. Also in humans, the tokens are not explicitly represented. Rather, they are created in the course of action and interaction. In the real course of events in which action tokens get realized there is an irreducible element of chance. No action token in the real world can be completely described before it has taken place. In SMASS, action tokens are introduced accordingly and created in the course of program execution. In their creation random elements are used with varying degrees depending on the action type and the level of abstraction.

SMASS uses action types and action tokens. The type of an action is given by a certain syntactic format which may vary from action-type to action-type. By contrast, action tokens are created in each situation when the actor has decided to perform an action of a distinct type. The creation of a token of that type is part of the rule which governs that type of behavior. In most cases a token is created by a mixture of random elements and of some deliberation in the way of bounded rationality.

The action type "exertion of power", for instance, is linked to a syntactic scheme of the form [A,B,IA,OA,IB,OB], with variables A,B,IA,OA,IB,OB. A denotes the person exerting power, B the person over whom power is exerted, IA, OA denote the input and output (integer) values of the interaction for A, and IB, OB the in- and output values for B. When A and B denote particular persons, a token is created from such a type by assigning concrete numbers to IA,OA,IB,OB.

[2] This insight did not have much impact on AI or so it seems.

Intuitively, A may choose some token in which "her" input-output relation IO-IA is large irrespective of the values IB,OB, as long as it is feasible for B to perform such a token. If power is exerted, for instance, by A's ordering B to carry a load for A then the value of A's input is just the value, or cost in this case, of her uttering the order while A's output value is the value -which we may imagine in monetary terms- of the load being carried. On the other side, B's input value is the value (cost) of his effort or labour of carrying the load, and B's output value is his cost of getting exhausted, tired and loosing part of his lifetime. Of course, representing these "values" numerically is a strong idealization, which however is not worse than other such representations central to exchange- and game theory. Integers are used for reasons of simplicity.

5.4 Active and Reactive Behavior

An actor in SMASS is in one of two different states. In the active state he is free to perform any action he may choose. In the reactive state, she must react according to some protocol that has been released in a previous interaction. For instance, in a previous interaction of exchange her partner may have agreed to exchange definite quantities of definite commodities, and may have himself adjusted his state so that, for him, the exchange is finished. This partner then has released a protocol telling her that he has done his part and that she now has to do her part. When she finds this protocol in a given situation she must act accordingly.

SMASS gives priority to the reactive states (see the predicate for "kernel", lines 10-13 in the appendix). Whenever an actor is called up in a simulation run at a given time he first checks whether any protocols have been activated for him, that is, he is in the reactive state. When he finds such activated protocols he will execute them and this is all he will do at that time. He gets into the active state only when at the time he is called up no protocols are activated for him. In an extreme case an actor at all times may be the target of many protocols and thus never become really active. Such cases do not seem unrealistic, though.

5.5 Rules of Behavior

The different types of action which the actors can perform in SMASS are called modes. For each mode SMASS contains a rule of behavior. When an actor is in the active state he will choose a certain mode. The rule for that mode then is called up and the actor performs an action according to that rule. Typically, a rule of behavior for a given mode comprises different steps which need not be independent of each other. Also, the order in which these steps are carried out may be different for different rules, and in some rules steps may be missing.

Step 1: The person chooses another actor with whom she will interact; in some cases several other actors are chosen.

Step 2: A token of the mode (action-type) is chosen or created. This is the token which the person ultimately will perform (if she succeeds).

Step 3: She checks whether interaction with the actor(s) chosen in 1) is feasible. In exchange, for instance, if a token is chosen representing the purchase of a certain quantity of a certain good, a partner must own a sufficient quantity of that good.

Step 4: If the preceding steps succeeded the person "performs" the action token. She adjusts her own state accordingly, and she releases a protocol that triggers the partner's corresponding reactions. This release consists of noting a fact which activates a particular protocol that is already programmed. When a protocol is activated it will be carried out in future actions by her partners and/or herself.

Each protocol is part of a corresponding mode and its code belongs to that for the mode. Various protocols that have been proposed in the literature, like the contract net protocol (Davis & Smith, 1983), SANP (Chang & Woo, 1992) or the proposal in Kreifelts & v.Martial (1991) can potentially be used in SMASS in connection with different modes. A great advantage of SMASS over many other simulation systems is that, being written in PROLOG, the message contents which are handed over in the protocols practically don't need any specific format. Any PROLOG term -which may be a whole PROLOG program itself- may serve as message content.

Not all rules involve protocols. One rule of behavior presently implemented is "bodily exercise". Performing an action of that type simply changes the actor's bodily strength. When all four steps - up to the activation of the protocol- succeed, the actor has acted in the given mode, he has performed an action-token of the kind represented by the given mode.

5.6 Updating

SMASS provides means for synchronous and asynchronous updating. Each mode has "its" own updating procedure which consists of two parts. One part belongs to the core program and is called up at the end of each period of time (see the "adjust" examples in the appendix). The second part is implicit in the rules of behavior linked with each mode. The second part may be absent for modes which require strict synchronous updating. For instance, in the mode "exchange" asynchronous updating is quite natural. After each exchange both partners adjust their endowments. However, even in this case at the end of a period a synchronous updating is called up which replaces each actor's final (in the given period t) endowment by the same (initial) endowment for the following period t+1[3].

[3] It may be noted that in the distributed version of the system, DMASS, all the problems of updating simply vanish. A kind of complementary problem arising in the distributed system, namely to report the agents' states to some external device at the end of certain real-time periods, is less likely to cause artificial effects, and much easier to handle technically.

5.7 Flexible Choice of Systems of Rules

A salient feature of SMASS is its flexibility of picking out a set of modes (action-types), and running a simulation with just these modes. This flexibility has two aspects. First, the user may choose any subset of those modes which are already implemented, that is, for which corresponding rules of behavior and adjustment have been programmed. As a special case of this aspect, the user may pick out just one single mode, and in this way repeat or vary many of the "one rule" simulation studies found in the literature. Second, the user may himself formulate new rules of behavior and add them to the system by programming and adding corresponding rules of behavior and updating rules. If new variables are introduced which require new initial data, it is also necessary to add these data or some program creating them automatically. Adding a new mode may require from two hours' work in simple cases up to several hours or even a couple of days for complicated modes involving complicated protocols, provided the programmer is fit in PROLOG and has got acquainted with SMASS.

SMASS contains a module for administrating the links between the modes on the one hand and the rules of behavior, the rules of updating, and the variables on the other hand. In this module a list of links between the modes chosen by the user and their corresponding rules of behavior, updating, and variables is created when SMASS is started. The main program is formulated invariantly relative to these link lists. All procedures in the main program can work with different such link lists, and the execution of the main program takes the same form for different lists of links.

Put differently, when the user has chosen a particular set of modes to be simulated, SMASS will create a corresponding link-list, and then run the main program. If in the next session the user changes the set of modes he wants to use, SMASS creates new link-lists, and with these runs the main program which is the same as before (and in all other sessions).

5.8 A Brief Synthesis

SMASS is written in SWI-PROLOG (Wielemaker, 1993) (which is free under UNIX, for instance in the LINUX package and also can be freely used with permission for academic applications under DOS) and laid down in several files. The "main" file, *sim*, which actually is among the smallest, contains the module for the creation of link-lists and the main program. A second file, *para*, contains parameters which must be set by the user before the simulation is started. The parameters include things like the number of runs (for statistics), the number of periods for one simulation run, the number of actors, the type of neighbourhoods used, and/or the number of neighbours, the number of commodities, as well as game parameters (payoffs), gridwidth (for cellular modes), ranges of initial physical strengths of the actors and other ranges for other variables. Most importantly, *para* must contain a list of the modes which are chosen by the user. This is just a list of names from the total list of modes which are implemented (16 at present). Also, *para* contains lists of weights for the creation of characters.

For each mode or rule of behavior, the file *rules* contains the code for that rule including code for adjustments in updating as far as there are special adjustments for the rule.

When sim is started, para is consulted, and on the basis of the parameters present in para, initial data are automatically created. The code for data creation resides in another file create which is consulted when needed. Alternatively, if a file of initial data is provided by the user, SMASS will take the data from data and will not create his own data. The data in any case are stored in an external file also called data. Next the link module creates link-lists for the modes it finds in the para file. With these link-lists the main program is started. In the main program, in each statistical run the files data and rules are consulted, and a simulation run is started. A simulation run simulates a fixed number of periods of time (as laid down in para). In each period, each actor is picked once (in random order) and activated. He checks whether there are protocols for him to perform, and if so, he executes them. If no protocols are activated for him he is set in the active state. He chooses one of the modes (those which the user chose for the present run) according to his character, and then acts in that mode according to the rule of behavior for that mode. This may involve the release of a protocol to be executed in the next period. In either case, when having reacted to a protocol, or acted, the actor is done and the next actor is called up. At the end of the period, individual updating is executed for each of the modes as laid down in rules and global updating is made as described in sim. When all periods are finished, all facts created in these periods are deleted, and the next statistical run is started. At the end of each simulation run (and if wanted, at the end of each period), the relevant data and new facts are stored in an external file results.

With results one can do the usual statistics and take the output, or results itself, as input for a graphics program for the visualization of the results. I use XPCE (Wielemaker, 1996) for this purpose which also is programmed in PROLOG. Permission to use XPCE under UNIX is free for academic applications, the DOS version costs about 200 dollars.

References

Agre, P. (1997): Computation and Human Experience. Cambridge University Press, Cambridge/Mass.

Alterman, R. & Zito-Wolf, R. (1993): Agents, Habitats, and Routine Behavior. IJCAI-93 Vol.1, Chambery, France, 305-10.

Axelrod, R. (1984): The Evolution of Cooperation. Basic Books, New York. Balzer, W. (1985): Incommensurability, Reduction, and Translation. Erkenntnis 23, 255-67.

Balzer, W. (1989): On Incommensurability. In: Gavroglu, K. et al. (eds.): Imre Lakatos and Theories of Scientific Change. Kluwer, Dordrecht, 287-304.

Balzer, W. (1990): A Basic Model of Social Institutions. Journal of Mathematical Sociology 16, 1-29.

Balzer, W. (1992a): Game Theory and Power Theory: A Critical Comparison. In: Wartenberg, T. (ed.): Rethinking Power. SUNY Press, Albany, 56-78.
Balzer, W. (1992b): A Theory of Power in Small Groups. In Westmeyer, H. (ed.): The Structuralist Program in Psychology. Hogrefe and Huber, Bern, 191-210.
Balzer, W. (1993): Soziale Institutionen. de Gruyter, Berlin.
Balzer, W. & Brendel, K. (1996): DMASS: A Distributed Multi-Agent System for Social Simulation. manuscript.
Carley, K. M. & Prietula, M. J. (1994): ACTS Theory: Extending the Model of Bounded Rationality. In: Carley, K. M. & Prietula, M. J. (eds.): Computational Organization Theory. Lawrence Erlbaum, Hillsdale NJ, 55-87.
Cohen, P. R. & Levesque, H. J. (1990): Intention is Choice with Commitment. Artificial Intelligence 42, 213-61.
Davis, R. & Smith, R. G. (1983): Negotiation as a Metaphor for Distributed Problem Solving. Artificial Intelligence 20, 63-109.
Epstein, J. M. & Axtell, R. (1996): Growing Artificial Societies. MIT Press, Cambridge MA.
Gilbert, G. N. & Doran, J. (eds.), (1994): Simulating Societies. UCL Press, London.
Goldman, I. A. (1970): A Theory of Human Action. Prentice Hall, Englewood Cliffs.
Hegselmann, R., Müller, U., Troitzsch, K. G. (eds.), (1996): Modelling and Simulation in the Social Sciences from the Philosophy of Science Point of View. Kluwer, Dordrecht.
Hegselmann, R. (1996): Cellular Automata in the Social Sciences, Perspectives, Restrictions, and Artefacts. In 17., 209-33.
Kreifelts, T. & Von Martial, F. (1991): A Negotiation Framework for Autonomous Agents. In: Demazeau, Y. & Müller, J. P. (eds.): Decentralized A.I.-2, 71-87.
Liebrand, W. B. G. & Messick, D. M. (1996): Computer Simulations of Sustainable Cooperation in Social Dilemmas. In 17., 235-47.
Man Kit Chang & Carson C. Woo (1992): SANP: A Communication Level Protocol for Negatioations. In: Werner, E. & Demazeau, Y. (eds.), Decentralized A.I.-3, Elsevier, Amsterdam, 31-54.
Mark, N. (1998): Beyond Individual Differences: Social Differentiation from First Principles. American Sociological Review 63, 309-330.
Müller, J. P., Wooldridge, M. J. and Jennings, N. R. (eds.), (1996): Intelligent Agents III. Springer, Berlin.
Nowak, A. & Lewenstein, M. (1996): Modelling Social Change with Cellular Automata. In: 17., 249-85.
Quine, W. v. O. (1961): Two Dogmas of Empiricism. In: Quine, W. v. O., From a Logical Point of View. Harvard University Press, Cambridge, Mass., 20-46.
Rao, A. S. & Georgeff, M. P. (1991): Modelling Rational Agents within a BDI-Architecture. In: Fikes, R. & Sandewall, E. (eds.): Proceedings of Knowledge Representation and Reasoning (KR+R-91). Morgan Kaufmann, 473-84.
Schelling, T. C. (1971): Dynamic Models of Segregation. Journal of Mathematical Sociology 1, 143-86.

Schiffer, R. (1987): Remnants of Meaning. MIT Press, Cambridge/Mass.
Schüssler, R. (1990): Kooperation unter Egoisten: Vier Dilemmata. Oldenbourg, München.
SWARM: http://www.santafe.edu/projects/swarm.
Uhrmacher, A. M. (1997): Concepts of Object- and Agent-oriented Simulation. Transactions of the Society for Computer Simulation International 14, 59-68.
Wielemaker, J. (1993): SWI-Prolog 1.8, Reference Manual. University of Amsterdam, Dept. of Social Science Informatics.
Wielemaker, J. (1996): Programming in XPCE/Prolog. University of Amsterdam, Dept. of Social Science Informatics.
Wooldridge, M. J. & Jennings, N. R. (eds.), (1991): Intelligent Agents. Springer, Berlin.
Wooldridge, M. J., Müller, J. P. and Taube, M. (eds.), (1995): Intelligent Agents II. Springer, Berlin.

Appendix: A Miniature Example

Here is the code of SMASS restricted to the application of three very simple rules. All features of the handling of data output and technical details pertaining to test and debugging are omitted as well as the automated creating of link-lists mentioned in Sec.8. Compare [32] for explanations of the built-in PROLOG predicates which can not be given here. Loading the different files in a directory accessible to PROLOG, compiling *sim*, and entering "start" plus return should run SMASS. The results can be seen in a file *results* that will be newly created by SMASS, and the data which were created and used should be present in a new file *data*. Changing the list of arguments of the "modes" predicate in line /* 1 */ will produce simulations in which only those rules present by names in that list are applied.

File *para* (parameters for SMASS)

```
/* 1 */ modes([takeweak,donothin,schellin]). runs(2). periods(10). actors(47).
use_old_data(no). gridwidth(8). variables_in_rule(donothin,[wealth]).
variables_in_rule(schellin,[location_schelling,schelling_colour]).
variables_in_rule(takeweak,[location,neighlist,strength,wealth_weak]).
weights(donothin,1,[101]). weights(schellin,1,[101]).
weights(takeweak,1,[101]).
```

modes([takeweak,donothin,schellin]) in line (1) specifies the list L=[takeweak,donothin,schellin] of modes (action-types) which are used in the present simulation.

File sim (core program)

```
/* 2 */ start :- consult(para), consult(pred), consult(rules),
  ( delete_file(results) ; true), use_old_data(X),
  ( X = no, create_data;  X = yes, consult(data) ), begin.
create_data :- ( delete_file(data) ; true ), consult(create),
  modes(L), make_global_data(L), make_variable_list(L,L1),
  length(L1,E),
  ( between(1,E,N), nth1(N,L1,VAR), make(VAR), fail ; true ), !.
make_global_data(L) :- actors(AS), make_characters(AS,L).
```

```
make_variable_list(L,L1) :- asserta(variable_list([ ])), length(L,E),
  ( between(1,E,X), nth1(X,L,M), variables_in_rule(M,L2),
    build_variable_list(L2), fail ; true ),
variable_list(L1),!.
build_variable_list(L2) :- variable_list(L), append(L,L2,L3),
  retract(variable_list(L)), asserta(variable_list(L3)),!.
```

R denotes a statistical run, T a period of time. RR is the number of statistical runs, TT that of periods of time for each simulation run, AS the fixed number of actors. results is the external file in which the raw data are written. All dynamical variables (those whose values may change during a simulation run) are written in the format fact(R,T,var(X1,...,Xn)) where R,T are as above, and "var" varies in the names of variables attached to the given modes (like "wealth", "strength" etc.). The loop in (4) produces TT executions of "kernel", that is, one simulation run covering TT periods. After each such run in (4) all dynamical facts are deleted, and the original data are reconsulted in (3) for the next run. (7) loops over all actors. In each step one actor A is randomly drawn from list L (6) and activated. After "activate(R,T,A)" (9) is executed actor A had his opportunity in period T as described in the following, A is deleted from the actor list (9), and another actor is called up in (7).

```
begin :- runs(RR), periods(TT),
  ( between(1,RR,R), mainloop(R,TT), fail; true ), !.
mainloop(R,TT) :-
/* 3 */ consult(data), findall(X,fact(0,0,X),L), length(L,E),
  ( between(1,E,Z), nth1(Z,L,FACT), append(results),
    write(fact(R,1,FACT)), write('.'), nl, told,
    retract(fact(0,0,FACT)),
    asserta(fact(R,1,FACT)), fail; true ), append(results), nl,
    told,
/* 4 */ ( between(1,TT,T), kernel(R,T), fail; true ),
/* 5 */ retract_facts,!.
retract_facts :- ( fact(X,Y,Z), retract(fact(X,Y,Z)), fail; true ).
kernel(R,T) :- actors(AS), findall(I,between(1,AS,I),L),
/* 6 */ asserta(actor_list(L)),
/* 7 */ ( between(1,AS,N), choose_and_activate_actor(R,T,N), fail;
  true ), retract(actor_list(L1)),
/* 8 */ adjust(R,T),!.
choose_and_activate_actor(R,T,N) :- actor_list(L), length(L,E),
/* 9 */ Y is random(E)+1, nth1(Y,L,A), activate(R,T,A), delete(L,A,L1),
  retract(actor_list(L)), asserta(actor_list(L1)),!.
```

When actor A gets activated he first checks his environment (10). This yields a possibility for immediate reactions to external (non-social) changes which are not implemented in the present version. Next, A executes protocols (11), if there are such for her. The protocols "protocol(M,A,R,T)" for all modes M are found in the rules file, and succeed only if a previous entry has been made in the fact base signalling that the protocol should be executed by A in the next period. If no protocols are activated in this way A switches to the active state. She chooses a mode M and acts in that mode (12), (13). The predicates "act_in_mode(M,A,R,T)" are found in the file *rules*. When in a given period all actors have "acted" once, in (8) the adjustment is called up. First, in (14) for each mode an individual adjustment is made, if necessary. The corresponding predicates are found in the *rules* file. Second, in (15) a global adjustment is made. This includes writing all facts present in the *results* file.

```
/* 10 */ activate(R,T,A) :- check_environment(R,T,A),
/* 11 */ ( execute_protocols(R,T,A)
          ;
/* 12 */ choosemode(R,T,A,M),
/* 13 */ ( act_in_mode(M,A,R,T) ; true)
          ),!.
check_environment(R,T,A) :- true.
/* 11 */ execute_protocols(R,T,A) :- protocol(M,A,R,T).
/* 8 */ adjust(R,T) :- modes(L), length(L,E), actors(AS),
/* 14 */ ( between(1,E,X), individual_adjust(X,R,T,AS,L), fail; true ),
/* 15 */ global_adjust(R,T), append(results), nl, told,!.
individual_adjust(X,R,T,AS,L) :- nth1(X,L,Z),
/* 14 */ ( between(1,AS,A), adjust(Z,A,R,T), fail ; true),!.
/* 15 */ global_adjust(R,T) :- T1 is T+1, repeat,
  ( fact(R,T,FACT), retract(fact(R,T,FACT)),
    asserta(fact(R,T1,FACT)),
    append(results), write(fact(R,T1,FACT)), write('.'), nl, told,
    fail; true ),!.
```

The character C of A is found in the file *data* which was downloaded in (2).

```
choosemode(R,T,A,M) :- fact(R,T,character(A,C,SUM)), length(C,K),
  modes(L), Z is random(SUM * 1000)+1, asserta(aux_sum(0)),
  between(1,K,X), do1(X,Z,C,Y), Z =< Y , nth1(X,L,M),
  retract(aux_sum(SS)),!.
do1(X,Z,C,Y) :- aux_sum(S), nth1(X,C,C_X), Y is S + (C_X * 1000),
  retract(aux_sum(S)), asserta(aux_sum(Y)),!.
```

File *pred* (auxiliary predicates)

Creates random numbers normally or discretely distributed, as well as von Neumann- and Moore neighbourhoods.

```
normal_distribution(N,AS,L,U,SI) :- MU is L + (0.5 * (U-L)),
  ( between(1,AS,A), determine_nd_value(N,MU,SI,L,U,A), fail; true),!.
  determine_nd_value(N,MU,SI,L,U,A) :- repeat, X is
  random(10001)+1,
  X4 is (1/10000) * (((X-1) * U)+(10001-X) * L), W is integer(X4),
  PI is pi, X1 is 2 * (PI * (SI * SI)), X2 is (1 / sqrt(X1)),
  X3 is (-((W-MU) * (W-MU))) / (SI * SI), Y is X2 * exp(X3),
  W1 is random(10001)+1, Z is (W1-1)/10000, Z =< Y, between(L,U,W),
  asserta(nd_expr(N,A,W)), !.
make_discrete_distribution(N,AS,EX,LIST) :-
  ( between(1,AS,A), determine_dd_value(N,A,EX,LIST), fail; true ),!.
determine_dd_value(N,A,EX,LIST) :- X is random(100)+1,
```

```
 between(1,EX,Z), nth1(Z,LIST,W_Z), X < W_Z, assert(dd_expr(N,A,Z)),!.
calculate_sum(L,S) :- asserta(counter(0)), length(L,E),
 ( between(1,E,X), auxpred(L,X) , fail ; true), counter(S),
 retract(counter(S)).
auxpred(L,X) :- nth1(X,L,N), counter(C), C1 is C+N,
  retract(counter(C)), asserta(counter(C1)), !.
make_nbh(moore,N,I,J,L) :- gridwidth(G),
  asserta(auxlist([[I,J]])),
  ( between(1,N,X), moore_nbh(X,G), fail; true),
  auxlist(L1), sort(L1,L2), delete(L2,[I,J],L),
  retract(auxlist(L1)),!.
moore_nbh(X,G) :- auxlist(L), length(L,E),
 ( between(1,E,Y), nth1(Y,L,[I1,J1]), moore_nbh1(Y,G,[I1,J1]), fail
 ; true),!.
moore_nbh1(Y,G,[I1,J1]) :- moore_nbh_1(G,I1,J1,Ln),
  auxlist(L), append(L,Ln,Lnew), retract(auxlist(L)),
  asserta(auxlist(Lnew)),!.
moore_nbh_1(G,I,J,L) :- recalculate_neg(G,I,1,Im),
 recalculate_neg(G,J,1,Jm), recalculate_pos(G,I,1,Ip),
 recalculate_pos(G,J,1,Jp),
 L = [[I,Jm],[Im,Jm],[Im,J],[Im,Jp],[I,Jp],[Ip,Jp],[Ip,J],[Ip,Jm]].
recalculate_neg(G,I,H,I1) :- X is I-H, ( ( 0 < X, I1 is X
 ; 0 =:= X, I1 is G ) ; X < 0, I1 is (G+I)- H ),!.
recalculate_pos(G,I,H,I1) :- X is I+H, ( ( I < G, X =< G, I1 is X
 ; I =:= G, (H > 0, I1 is H; H =:= 0, I1 is G) ) ; I < G, X > G,
  I1 is (H+I)-G ),!.
make_nbh(von_Neumann,N,I,J,L) :- gridwidth(G),
  asserta(auxlist([[I,J]])),
  ( between(1,N,X), vNnbh(X,G), fail; true),
  auxlist(L1), sort(L1,L2), delete(L2,[I,J],L),
  retract(auxlist(L1)),!.
vNnbh(X,G) :- auxlist(L), length(L,E),
 ( between(1,E,Y), nth1(Y,L,[I1,J1]), vNnbh1(Y,G,[I1,J1]), fail
 ; true),!.
vNnbh1(Y,G,[I1,J1]) :- von_Neumann_nbh_1(G,I1,J1,Ln),
  auxlist(L), append(L,Ln,Lnew), retract(auxlist(L)),
  asserta(auxlist(Lnew)),!.
von_Neumann_nbh_1(G,I,J,L) :- recalculate_neg(G,I,1,Im),
 recalculate_neg(G,J,1,Jm), recalculate_pos(G,I,1,Ip),
 recalculate_pos(G,J,1,Jp), L = [[I,Jm],[Im,J],[I,Jp],[Ip,J]].
decompose(Y,I,J,G) :- between(1,G,Z), Y =< Z * G, Z1 is Z-1, I is Z,
 J is Y-(Z1 * G),!.
```

File *create* (generates characters and data)

The characters for the actors are created and written to the *data* file. For each variable initial data are created and written to the *data* file.

```
make_characters(AS,L) :- build_up_characters(AS,L),
```

```
export_results(AS).
build_up_characters(AS,L) :- length(L,E),
  ( between(1,E,X), make_distribution(X,L,E,AS), fail ; true ),
  ( between(1,AS,A), collect_characters(L,E,A), fail ; true),
  retractall(dd_expr(M1,M2,M3)), !.
make_distribution(X,L,E,AS) :- nth1(X,L,M), weights(M,EX,LIST),
  make_discrete_distribution(M,AS,EX,LIST),
  retract(weights(M,EX,LIST)),!.
collect_characters(L,E,A) :- asserta(character(A,[ ])),
  ( between(1,E,X), nth1(X,L,M), add_character(M,A), fail; true ),!.
export_results(AS) :- ( between(1,AS,A), export(A), fail; true),!.
export(A) :- character(A,L2), calculate_sum(L2,SUM), append(data),
  write(fact(0,0,character(A,L2,SUM))), write('.'), nl, told,
  retract(character(A,L2)),!.
add_character(M,A) :- dd_expr(M,A,C), character(A,L1), append(L1,[C],L2),
  retract(character(A,L1)), asserta(character(A,L2)),!.
make(wealth) :- actors(AS), domain_of_wealths(L,U),
   sigma_wealths(SI), normal_distribution(wealth,AS,L,U,SI),
   ( between(1,AS,A), nd_expr(wealth,A,W), append(data),
     write(fact(0,0,wealth(A,W))), write('.'), nl, told,
     retract(nd_expr(wealth,A,W)), fail ; true ),!.
make(wealth_weak) :- actors(AS), domain_of_wealth_weak(L,U),
  sigma_wealth_weak(SI),
  normal_distribution(wealth_weak,AS,L,U,SI),
   ( between(1,AS,A), nd_expr(wealth_weak,A,W), append(data),
     write(fact(0,0,wealth_weak(A,W))), write('.'), nl, told,
     retract(nd_expr(wealth_weak,A,W)), fail ; true ),!.
make(strength) :- actors(AS), weights(strength,LIST),
  expressions(strength,EX),
  make_discrete_distribution(strength,AS,EX,LIST),
  ( between(1,AS,A), dd_expr(strength,A,W), append(data),
    write(fact(0,0,strength(A,W))), write('.'), nl, told,
    retract(dd_expr(strength,A,W)), fail; true ),!.
make(location) :- actors(AS), gridwidth(G), G1 is G * G,
   findall(X,between(1,G1,X), L), asserta(cell_list(L)),
   ( between(1,AS,A), locate(A), fail ; true),
retractall(cell_list(L2)),!.
locate(A) :- cell_list(L), length(L,E), X is random(E)+1, nth1(X,L,Y),
  gridwidth(G), decompose(Y,I,J,G), append(data),
  write(fact(0,0,location(A,I,J))), write('.'), nl, told,
  asserta(fact(0,0,location(A,I,J))), delete(L,Y,L1),
  retract(cell_list(L)), asserta(cell_list(L1)),!.
make(neighlist) :- type_of_neighbourhood(TYPE,DEGREE), actors(AS),
  ( between(1,AS,A), make_neighbourhood(A,TYPE,DEGREE), fail ; true).
make_neighbourhood(A,T,D) :- fact(0,0,location(A,I,J)),
   make_nbh(T,D,I,J,L), length(L,E), asserta(aux_list(A,[ ])),
```

```
( between(1,E,X), collect_neighbours(A,X,L), fail ; true),
  aux_list(A,L2), sort(L2,L3), append(data),
  write(fact(0,0,neighlist(A,L3))), write('.'), nl, told,!.
collect_neighbours(A,X,L) :- nth1(X,L,Z), Z=[I,J],
  fact(0,0,location(N,I,J)), aux_list(A,L1), append(L1,[N],L2),
  retract(aux_list(A,L1)), asserta(aux_list(A,L2)),!.
make(location_schelling) :- gridwidth(G), actors(AS),
  L=[[1,1],[1,G],[G,1],[G,G]], asserta(auxlist([ ])),
  ( between(1,AS,A), schelling_locate(A,G,L), fail; true),
   retractall(auxlist(LL)),!.
schelling_locate(A,G,L) :- auxlist(L1), repeat, I is random(G)+1,
  J is random(G)+1, not member([I,J],L1), not member([I,J],L),
  asserta(fact(0,0,schelling_loc(A,I,J))), append(data),
  write(fact(0,0,schelling_loc(A,I,J))), write('.'), nl, told,
  append(L1,[$[I,J]$],L2), retract(auxlist(L1)),
asserta(auxlist(L2)),!.
make(schelling_colour) :- actors(AS),
  ( between(1,AS,A), set_colour(A), fail ; true),!.
set_colour(A) :- fact(0,0,schelling_loc(A,I,J)), N is I+J,
  N1 is N mod 2, (  N1 =:= 0, append(data),
  write(fact(0,0,colour(A,white))), write('.'), nl, told;
  append(data), write(fact(0,0,colour(A,black))), write('.'), nl,
  told ), !.
```

File *rules*

RULE 1: “donothin”. The person intentionally does not do anything. A fixed sum (3 * E) is deducted from her wealth in each period.

```
domain_of_wealths(50,500). sigma_wealths(20). exist_min(20).
act_in_mode(donothin,A,R,T) :- feasible(donothin,A,R,T),
  chooseaction(donothin,A,R,T), perform(donothin,A,R,T),!.
feasible(donothin,A,R,T) :- fact(R,T,wealth(A,W)), exist_min(E),
  E1 is 3 * E, W1 is W-E1, W1 > 0,!.
chooseaction(donothin,A,R,T) :- true,!.
perform(donothin,A,R,T) :- fact(R,T,wealth(A,W)), W1 is W-5,
  retract(fact(R,T,wealth(A,W))),
  asserta(fact(R,T,wealth(A,W1))),!.
protocol(donothin,A,R,T) :- fail.
adjust(donothin,A,R,T):- true.
```

RULE 2: “schellin”.

One of the first programs leading to emergent patterns, here: the clustering of persons of equal colour. For explanations consult (Schelling,1971).

```
act_in_mode(schellex,A,R,T) :- feasible(schellex,A,R,T),
  chooseaction(schellex,A,R,T), perform(schellex,A,R,T),!.
feasible(schellex,A,R,T) :- true.
chooseaction(schellex,A,R,T) :- gridwidth(G),
  fact(R,T,schelling_loc(A,I,J)),
scan_neighbourhood(A,G,I,J,R,T,ANSWER),
```

```
( ANSWER=yes ; calculate_move(A,R,T,G) ),!.
scan_neighbourhood(A,G,I,J,R,T,ANSWER) :-
make_nbh(moore,1,I,J,L),
  findall(N,neighb(N,L,R,T),L1), length(L1,E1),
  findall(N1, equal_colour(N1,A,L,R,T), L2), length(L2,E2),
  ( ( ( E1 =< 2, 1 =< E2; 3 =< E1, E1 =< 5, 2 =< E2 )
  ; 6 =< E1, E1 =< 8, 5 =< E2 ), ANSWER=yes ; ANSWER=no),!.
neighb(N,L,R,T) :- member([I,J],L),
fact(R,T,schelling_loc(N,I,J)).
equal_colour(N,A,L,R,T) :- member([I,J],L),
fact(R,T,schelling_loc(N,I,J)),
  fact(R,T,colour(N,CN)), fact(R,T,colour(A,CA)), CA=CN.
calculate_move(A,R,T,G) :- G1 is G * G, between(1,G1,X),
  decompose(X,I,J,G), not fact(R,T,occupied(B,I,J)),
  not fact(R,T,schelling_loc(B1,I,J)),
  scan_neighbourhood(A,G,I,J,R,T,ANSWER), ANSWER=yes,
  asserta(fact(R,T,occupied(A,I,J))),!.
perform(schellex,A,R,T) :- true.
protocol(schellex,A,R,T) :- fail.
adjust(schellex,A,R,T) :-
    ( fact(R,T,occupied(A,I,J)),
      fact(R,T,schelling_loc(A,IA,JA)),
        retract(fact(R,T,schelling_loc(A,IA,JA))),
        asserta(fact(R,T,schelling_loc(A,I,J))),
        retract(fact(R,T,occupied(A,I,J))) ; true ),!.
```

RULE 3: "takeweak" (take from the weaker).

Each actor tries to find a neighbour which is physically weaker, and to take away some part of that persons wealth. The amount taken away is randomly chosen from a pre-specified range (3 * SS). In (17) a protocol is formulated which will be performed by the addressee in the next period. The message content for that protocol is "handed over" by asserting it in (16) and by being read in (17) by the addressee. More complicated protocols basically work in the same way, all necessary regulations being described and handed over (perhaps several times) as "messages" in the way of the example.

```
exist_min_weak(20). domain_of_values(20).
domain_of_wealth_weak(100,500).
sigma_wealth_weak(40). expressions(strength,4).
weights(strength,[10,50,90,100]).
type_of_neighbourhood(moore,1).
act_in_mode(takeweak,A,R,T) :- feasible(takeweak,A,R,T),
  chooseaction(takeweak,A,R,T), perform(takeweak,A,R,T).
feasible(takeweak,A,R,T) :- exist_min_weak(MIN),
  domain_of_values(SS), fact(R,T,neighlist(A,L)), length(L,E),
  fact(R,T,strength(A,SA)), between(1,E,X),
  investigate(R,T,X,L,MIN,SA,SS).
investigate(R,T,X,L,MIN.SA,SS) :- nth1(X,L,N),
```

```
   fact(R,T,strength(N,SN)),
   SN < SA, fact(R,T,wealth_weak(N,WN)), W1 is WN-(3 * SS), !,
   MIN =< W1,
   T1 is T+1, not fact(R,T1,give_the_stronger(N,Y)),
   asserta(neighb(N,WN)).
chooseaction(takeweak,A,R,T) :- true.
perform(takeweak,A,R,T) :- neighb(N,WN), retract(neighb(N,WN)),
   fact(R,T,wealth_weak(A,WA)), domain_of_values(SS), S1 is 3 * SS,
   X is random(S1), WA1 is WA+X,
   retract(fact(R,T,wealth_weak(A,WA))),
   asserta(fact(R,T,wealth_weak(A,WA1))), T1 is T+1,
/* 16 */ asserta(fact(R,T1,give_the_stronger(N,X))).
/* 17 */ protocol(takeweak,A,R,T) :-
  fact(R,T,give_the_stronger(A,X)),
  fact(R,T,wealth_weak(A,WA)), W1 is WA-X,
  retract(fact(R,T,wealth_weak(A,WA))),
  asserta(fact(R,T,wealth_weak(A,W1))),
  retract(fact(R,T,give_the_stronger(A,X))).
adjust(takeweak,A,R,T) :- true.
```

6. PECS: A Reference Model for the Simulation of Multi-Agent Systems

Christoph Urban[1]

Abstract

There is currently a multitude of design approaches all concerned with building simulation models for multi-agent systems. The majority of the designs, however, focus on the requirements of specific application domains. We suggest that it would be desirable to develop a modelling methodology which can act as a generic platform from which to construct simulation models for multi-agent systems. In order to promote standardisation of modelling methodology, this paper presents a first-cut reference model whose feasibility will subsequently be demonstrated by a number of applications.

6.1 Intelligent Agents and Multi-Agent Systems

Modelling real application domains as multi-agent systems has come to the fore not only in computer science but also in other areas of science and engineering (Sundermeyer, 1993). This particular modelling technique has its origins in distributed artificial intelligence (DAI) which concerns itself with societies of intelligent and interacting artificial individuals (agents).

Distributed AI has its main roots in traditional AI – the main difference is that traditional AI is concerned with a single agent's views and behaviour, whereas DAI is particularly interested in the behavioural aspects arising from the interactions of groups of individuals in such agent systems. Communication and cooperation play a particularly important role in multi-agent systems.

6.1.1 Main Characteristics of Intelligent Agents

Scientists working with agent technologies have produced many competing definitions of agents in order to explain the notion of agent in their specific application context. Maes (1995) describes an agent as a computational system which exists in some complex and dynamic environment and which perceives its environment through sensors. Furthermore, it can act autonomously in its environment and it can achieve goals and carry out various tasks defined in its design.

[1]Chair for Operations Research and Systems Theory, University of Passau, Innstraße 33, D - 94032 Passau. email: urban@fmi.uni-passau.de

There is at present no universally accepted definition for intelligent agents. We therefore list some of the main features frequently regarded as essential characteristics of agents (Sundermeyer, 1993; Moffat & Frijda, 1994):

- *Autonomy*
  Each agent has its own behavioural set-up which is composed of its individual behaviour and its induced behaviour. Individual behaviour describes the behaviour displayed by the agent as a result of its predefined objectives and motivations. Induced behaviour describes how the agent will react to environmental stimuli.
- *Individual point of view*
  Agents will usually play their part embedded in a larger system. This environment provides a number of influences on the agents' behaviour. It is therefore necessary for agents to gain knowledge about their environment in order to adapt and react adequately when conditions change. This view of their environment may have its idiosyncrasies, which means it is affected by the agent's overall properties and capabilities, it is incomplete and it may even have defects.
- *Intelligent and rational behaviour*
  Many application domains will require agents who display a certain amount of intelligence. Intelligent behaviour includes, for example, the ability to learn, draw logical conclusions, or the capability to find one's bearings when faced with a new previously not encountered environment or situation. An agent displays rational behaviour or decision-making capabilities when it is able to evaluate possible actions in a situation, and then chooses an appropriate option. This will take into account the objectives of the agent as well as its internal states and environmental factors.
- *Social Behaviour*
  Frequently, an individual agent belongs to a larger system of agents. This necessitates the capacity of an agent to display some form of social competence in order to integrate within a society of agents. This concerns communicative competence on the one hand (i.e. exchanging information with other agents), and cooperative capacities on the other. Cooperation mechanisms allow us to coordinate the actions of different agents and to put together individual, partial solutions which will contribute to the solution of a larger problem posed at a global level.
- *Mobility*
  A number of application domains (e.g. in sociology or biology) require the agent to be mobile within its environment. This means we have to model each agent as occupying a specific position within its surroundings, and that position can obviously be changed over time.

6.1.2 Aims and Objectives of Multi-Agent Systems

There are typically two principal reasons for opting for a multi-agent approach when modelling a real system:

- *Modelling processes of distributed problem solving*
  Agents existing and acting within a common environment make their contributions to the solution of a globally posed and complex problem by using their individual capabilities and cooperation. Basic issues arising in this context concern methods and mechanisms for interaction and cooperation among agents.
- *Investigating emergent behaviour*
  Microscopic agent interaction can sometimes result in a stable system in which novel, macroscopic behavioural properties emerge. The novel behavioural patterns are then investigated in order to find relationships between the system's high-level properties and the individual and social patterns exhibited by the agents.

6.1.3 Advantages of Multi-Agent Models

One attraction of the multi-agent approach is its natural affinity with large and complex systems, encountered frequently for example in the social sciences (Huhns, 1994). Recent examples for the use of multi-agent systems in social science and economics have been the formation of social groups in prehistoric humans (Doran et al. 1994) and models for the emergence of price levels in oligopolistic markets (Caldas & Coelho, 1994). Both studies use agent-based simulation and micro-simulation to model and explain the emergence of higher-level phenomena.

In contrast to modelling dynamic behaviour of populations through deterministic or stochastic equations (e.g. world-models, macro-simulations), multi-agent models allow for individual actors to be built and observed. This enables analysis of global behaviour as well as the causes and consequences of each individual's actions. Compared with traditional simulation techniques, multi-agent systems offer the most sophisticated representation at the micro-level of modelling.

6.1.4 Outline of the Following Sections

In the following sections we try to give a survey of the reference model PECS which in our opinion can act as a generic platform for the construction of multi-agent simulation models.

First of all section 6.2 addresses the question of how to construct multi-agent models. We therefore deal with some architectural considerations and list the main tasks of creating multi-agent models. Starting with a summary of the current state of the art we give a brief description of the main goals of our approach.

Section 6.3 describes the main features of the reference model. At the beginning we take a look at the design of the PECS agent world. After that we focus the internal architecture of the PECS agents that are the active parts of the model world. As communication is essential for cooperation of agents, section 6.3.3 explains the basic mechanisms provided for interchanging information. Furthermore the agents' environment is discussed as a collection of all aspects that are external to the agents.

In the last section we show the feasibility of the PECS reference model in different application domains. The case studies investigated there are taken from game theory, ecology and economy.

6.2 Building Multi-Agent Models

This section addresses the question of how multi-agent models can be built. We will start with some architectural considerations for agents. Then we discuss the main tasks of creating a multi-agent model. We close with a summary of the current state of the art and a brief description of the objectives of our own work in this area.

6.2.1 Agent Architectures

Depending on context, the term "architecture" can be associated with many semantic interpretations. Therefore, we need to clarify up-front our understanding of the term with reference to the work reported here.

According to Maes (1991), an architecture describes a methodology for the construction of agents. It specifies the set of components which constitute the agent, and it describes how the components interact. The combination of the modules and their interactions determine the emergence of the actions an agent will be capable of doing. These actions will depend on the agent's sensory inputs and the agent's current internal state. In addition, the architecture specifies techniques and algorithms supporting the particular design methodology.

Many and very widely differing agent architectures have been proposed in the literature. One way of classifying these architectures is by looking at the mechanisms used for their internal control. Sundermeyer (1993) suggests a distinction between *reactive*, *deliberative* and *hybrid* architectures with the latter group exhibiting both reactive and deliberative elements of control (see section 6.2.3 for a more detailed discussion of the terms *reactive* and *deliberative*). There are of course other classification schemes for agent architectures as well, using different criteria.

6.2.2 Building Multi-Agent Simulation Models

Building a multi-agent model poses three principal design challenges: firstly, one has to decide on the properties of the agents. This is done within the framework of the agent architecture which pre-determines the component structure of agents as well as defining the control mechanism which the agents will follow while pursuing their own activities (see section 6.2.1 above).

Secondly, interaction and planning in multi-agent systems requires methods which define the exchange of information between agents. This includes the provision of appropriate communication mechanisms and protocols, as well as the definition of how the society of agents will be organised – for example with respect to hierarchical structures or concerning the social roles played by individual agents.

Finally, many application domains require explicit modelling of the environment in which the society of agents exists. This environment will provide part of the external conditions for the agent's behaviour. In ecological systems, the environmental component could be a model of the biotop whose inhabitants are being investigated. One could imagine to model various external parameters such as the climate or topological properties of the biotop.

6.2.3 State of the Art and Objectives of this Work

The challenges mentioned in section 6.2.2 have been addressed by a number of theoretical and practical studies. Many of the theoretical works focus on the issues of agent architecture and construction of agent societies. A problem associated with many of the results is that the concepts and terms used are closely related to the application domain. This makes the transfer of the techniques to other application domains difficult or even impossible. As a result, systems built on those premises can only be used for very specialised tasks and as a rule, turn out to be rather inflexible. Users of such systems – which have a fixed structure and predefined dynamic behaviour – can only influence their simulation runs by changing parameter values. This limitation subsequently means that existing models can no longer be used once new questions are being addressed and the old model cannot cater for the new concepts. In those cases, the previous modelling approach has to be reconsidered and it is often necessary to consider a new model implementation from scratch. See Agah & Beckey (1996), Agre & Chapman (1987), and Drogoul & Ferber (1994) for examples of such systems.

A further weakness of current multi-agent simulation models is the use of general-purpose, high level programming languages such as C, Prolog or LISP for their implementation. On the one hand, the programmer's only restriction when implementing the model is the expressivity of the chosen programming language, which gives the implementor maximum flexibility. The big disadvantage of this approach, however, is that important features available in state-of-the-art simulation packages have to be built from scratch and lead to unnecessarily large amounts of effort spent on non-core functions. These include model maintenance, environments for experiment set-up, analytical processing of results, graphical representation of simulation data, and integrated animation tools.

Using high-level programming languages for model implementation also requires a longer training period and in addition, forces the model builder into dealing with language-specific computational issues and concepts which are far removed from the core problem s/he should be focusing on. A typical example of this type of problem is "Swarm", a library of classes developed at Santa Fe Institute (Hiebeler, 1994). While it does provide a great deal of functionality as well as a programming framework for the development of multi-agent simulations, users can only make effective use of its concepts once they have mastered the underlying programming language Objective C.

To address the above problems of current systems and to work further towards standardisation of the methodology for modelling multi-agent systems, this paper proposes a modelling concept which can form the basis for the specification of multi-agent simulation models which are applicable in diverse application domains in science as well as engineering.

The proposal is for a reference model (e.g. Schmidt, 1996) which is aimed at providing a standard methodology of building multi-agent models. Using such a model - which draws on interdisciplinary experience – will reduce the amount of effort otherwise spent developing new modelling concepts for each novel application domain. A more uniform modelling scheme also improves the understanding of applications built in this way and facilitates discussion and critique of implemented models because there is less argument about discrepancies arising from unnecessary methodological variations. Furthermore, models standardised in this way stand a better chance of being reused, and this opens up the possibility of exchanging models between end users.

The reference model's software implementation is based on the integrated simulation system SIMPLEX III (Schmidt, 1995). This package supports all phases of a simulation study which means users have access to state-of-the-art support tools for model development, carrying out simulation experiments, analysis of results, as well as for presentation of those results. This avoids many of the problems mentioned earlier, which are typically associated with the use of general purpose programming languages for simulation model development.

6.3 The PECS Reference Model

As pointed out by Schmidt (1996), reference models are a way of standardising the construction of simulation models which belong to certain classes of applications. We shall now give an overview of the structure and functionality of the reference model PECS (Physis, Emotion, Cognition, Status) for the modelling of multi-agent systems.

6.3.1 Structure of the Agent World

There are three principal components making up the PECS agent world: the component "Environment", the component "Connector", and the agents.

The "Environment" component is used to model all the external settings and factors which are of interest in terms of agent actions and behavioural patterns.

Communication plays a vital role in multi-agent systems, building the basis for potentially many forms of interaction. It must therefore be possible for any agent to get in contact with any other agent. The component "Connector" is introduced as a central switchboard which organises information exchange between the agents.

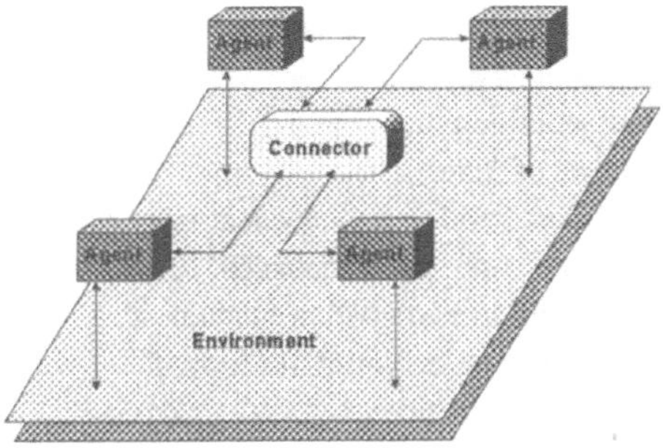

Figure 1: Structure of the agent world

The main components are the agents themselves as the active parts of our model world. We need to ensure that the reference model is able to cater for the modelling and integration of different types of agents in the multi-agent system.

6.3.2 PECS-Agent Architecture

Using the classification mentioned in section 2.1 (reactive, deliberative, and hybrid agent architectures) the PECS system can be classified as a hybrid architecture. PECS-agents can follow reactive as well as deliberative behaviour specifications when selecting an action plan.

System-theoretic Fundamentals of Agent Dynamics

The description of the dynamic behaviour of PECS-agents is based on system theory and draws its ideas from hierarchically structured dynamic systems. The basic concept taken from system theory is that the observable state transitions and outputs of a system depend on its internal state as well as on external input coming from the environment of the system.

Figure 2: System-theoretic structure of a dynamic system

According to Wunsch, (1986) a dynamic system S can be represented as a 9-tuple

$$(T,\ X,\ Y,\ Z,\ \mathbf{X},\ \mathbf{Y},\ \mathbf{Z},\ F,\ g\ )$$

where

- $T \subseteq \mathbb{R}$ is a partially ordered ($\leq$) set of time values
- X is any set of inputs
- Y is any set of outputs
- Z is any set of states
- $\boldsymbol{X}$ is the set of valid time-dependent input functions, $\boldsymbol{X} \subset X^T$
- $\boldsymbol{Y}$ is the set of valid time-dependent output functions, $\boldsymbol{Y} \subset Y^T$
- $\boldsymbol{Z}$ is the global state transition function, $\boldsymbol{Z}: T \to Z$
- F is the local state transition function, $F: (T \times Z \times X) \to Z$
- g is the output function, $g: (T \times Z \times X) \to Y$

Given $X(t) \in \boldsymbol{X}$ and $t \in T$, any time-dependent output function $Y(t) \in \boldsymbol{Y}$ can be described by the relation $Y(t) := g(t, \boldsymbol{Z}(t), X(t))$.

The local state transition function F describes how the state of a system at a given point in time changes to a state at the next observed point in time. Starting at an initial state of the system we get the progression of the global state transition function $\boldsymbol{Z}$ by evaluating the local state transition function F for all time values in consideration of a given time-dependent input function $X(t)$. Let us annotate here that the state transition function can be discrete as well as continuous and that even combined discrete and continuous descriptions for the dynamic behaviour of systems can be applied.

The state transitions of discrete systems can be written as

$$\boldsymbol{Z}(t_{n+1}) := F(t_n, \boldsymbol{Z}(t_n), X(t_n))$$

whereas the state transitions for continuous systems are of the form

$$\boldsymbol{Z}'(t) := F(t, \boldsymbol{Z}(t), X(t)).$$

The dynamic behaviour of a system is influenced by the initial state of the system on the one hand and on the other hand by the time-dependent input function $X(t)$. For the definition of the input function of a system also the output function of another system can also be used. An output value $y_i^{(S')}$ of a system S' may then act as an input value $x_j^{(S)}$ of a system S as long as their ranges correspond with each other.

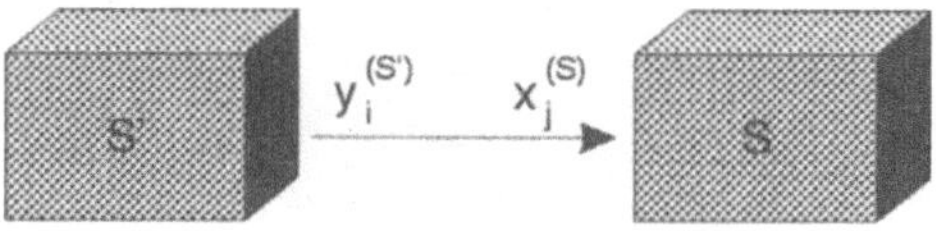

Figure 3: Composition and decomposition of systems

By this approach we gain the possibility of hierarchical modular modelling. Thereby we are able to compose complex systems from subsystems with reduced complexity that are easier to handle.

Fundamentals of the PECS Architecture

When designing the dynamic behaviour attention must be given to the internal states of agents. This basic understanding suggests the use of the system-theoretic concept of dynamic systems (see section 6.2.1 above) for the modelling of agents. The distinguishing features of dynamic systems are therefore carried over to the architecture of PECS-agents. The components "Sensor" and "Perception" of the agents correspond with the input of the dynamic system. The internal state of the system is distributed over the agent's components "Cognition", "Emotion", "Physis" and "Status". Note that state transitions are specified by discrete transition rules or algebraic respectively differential equations within the agents. Finally, the output function of a dynamic system is taken over by the agent's component "Behaviour" which contains the agent's competence for action and which provides the component "Actor" with specific task descriptions. The agent's "Actor" component therefore corresponds with the actual output of the system.

PECS takes the view that in general, an agent is a psychosomatic unit which has certain cognitive capabilities. This is borne out in the representation of an agent's internal state. A PECS-agent is described by a set of attributes representing its internal physical, emotional, and cognitive state. For software engineering reasons, these are modularised into distinct components ("Physis", "Emotion", "Cognition") but there is tight interaction between them. This can be seen in the transition rules which describe state transitions within one component and whose condition-part refers in many cases not only to values of the component's own attributes but also to properties of other internal components.

This idea of viewing an agent as a unit consisting of physical, emotional, and cognitive capabilities is also reflected in the component "Behaviour". When selecting action responses, the agent does not only take into account environmental stimuli as perceived through its sensors: it also uses information provided by all of its own internal components. The necessity for this type of modelling can be demonstrated with the simple scenario below.

We create an agent representing a person who wants to buy things in a shop. The purchasing decision may yield different results depending on the agent's knowledge, emotions, and physical state. For example, the physical component of the agent has a state variable "Hunger", the emotional component has a variable representing the "Attitude towards salesperson", and the "Cognition" component contains "Information about prices". The decision rules shown in Figure 4 may apply for this agent.

The condition part of the behavioural rules uses the values provided by attributes "Hunger", "Attitude" and "Price" which are controlled by the components "Physis", "Emotion" and "Cognition", respectively. The example shows that in general, the overall internal state of an agent can influence its behaviour, and that we need to look at the agent's state as a whole.

```
Physis:      Hunger    ∈    } enormous, medium, little {
Emotion:     Attitude  ∈    } kind, nemtral, hostile {
Cognition: Price       ∈    REAL

WHENEVER Stock < Minimal_Stock
DO
  IF )Hunger=enormous AND Attitude=hostile AND Price<Up_Limit(
  DO Buy_Pieces (5(;
  END

  IF )Hunger=enormous AND Attitude=kind AND Price<Up_Limit(
  DO Buy_Pieces (6(;
  END

  IF )Hunger=little AND Attitude=hostile AND Price<Up_Limit(
  DO Go_To_Other_Shop )(;
  END

  IF )Hunger=little AND Attitude=kind AND Price<Up_Limit(
  DO Buy_Pieces (2(;
  END
END
```

Figure 4: Internal state influencing agent behaviour

The modelling technique shown in figure 4 follows the reference model SSA which we will discuss in more detail in the next section. For example, the term "WHENEVER Stock < Minimal_Stock" constitutes the Strategy Call Condition, the individual behavioural patterns are conditioned by the IF-statements and the call "Buy_Pieces ()" is the task specification which is finally communicated to the "Actor" component.

Internal Structure of PECS-Agents

On the basis of a block diagram we will now discuss each of the components of a PECS-agent, the role of the component, its characteristic features, and the way it interacts with other components.

Component "Sensor"

The "Sensor" is the agent's input interface. It acts like a basin collecting environmental inputs and making them accessible to the internal components for further processing. Any kind of stimuli, triggers, or sensory impressions which are computed in the "Environment" component and are of relevance to the agent's tasks can be imagined as input. The "Sensor" is also responsible for receiving messages from other agents or from other sources of information.

The "Sensor" also has an internal state which allows us to model for example, quality and quantity-dependent differences of processing of sensory input. The following scenario shows that the ability to model the internal state of a sensor is an important requirement: A person taking strong drugs to fight some disease may suffer from side-effects such as a partial loss of sight or a sight impairment. In other words, there can be many reasons why environmental stimuli may not always be perceived identically by agents. In these cases it is necessary to represent the context-dependent performance differences of the "Sensor" with the aid of internal states defined in the model.

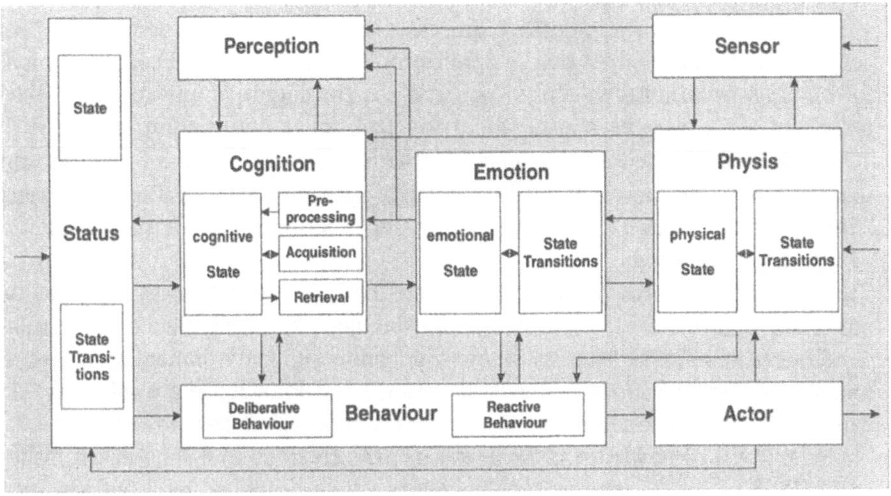

Figure 5: Internal structure of a PECS-agent

<u>*Component "Perception"*</u>

The "Perception" component processes all environmental inputs collected in the "Sensor" according to the individual ("subjective") cognitive set-up of the agent. Zimbardo (1988) describes human perception as follows: firstly, the individual's sensory organs collect the dynamically changing and often chaotic input from external energy sources which the individual is exposed to. Perception helps us to organise this multitude of stimuli into more stable perceptive chunks, aided by such processes as decision making, estimation, remembering, and association. This whole process happens in several steps.

The process starts with sensory perception which transforms the physical impact of forms of energy such as light or sound waves into neural activity in our brain cells. At the same time, information is collected about the kind of stimulus that was received by the receptor organs and some low-level transformations of the incoming sensory information are carried out. The PECS reference model represents these tasks in the "Sensor" component.

The next stage is a more specific cognitive process. The external stimulus is represented internally through the use of pattern recognition and estimation mechanisms. Also, the newly gained information gets integrated with existing knowledge.

Finally, the information gets classified. That is, the properties of the perceived objects get associated with known categories. For example, round objects get associated with footballs, oranges, clocks or other previously known objects exhibiting some similarity with the new ones. In this phase, the cognitive apparatus works out the function of the perceived object, leading to the resulting classification – a cognitive unit ("percept").

The modelling of these cognitive processes can be done in the component "Perception" if this is required by the application. Analogous to the cognitive model for humans, building the agent's "percept" i.e. the cognitive unit for some external event there may be a need for using additional information stored in the "Cognition" component of the agent. Furthermore, there is scope for influencing the perceptive processes of agents through specific states of the components "Physis" and "Emotion" which may affect the selective attention of the agent.

Component "Cognition"

This component stores all the information that is available to an agent at this particular point in time. The information includes knowledge about the agent's own internal state as well as knowledge about its environment. An agent's knowledge about its surroundings and even it's self-knowledge are subjective and can - in general – be incomplete or even faulty representations of the facts.

Modelling the knowledge of an agent is currently done in a state-based fashion in PECS. The agent has a current knowledge state described by a defined set of relevant attributes.

Updating an agent's knowledge base and making it consistent with actions carried out and inputs received can be done through some knowledge pre-processing. This pre-processing can either be activated by the "Perception" component by moving relevant environmental information into the knowledge base of the agent, or by performing behavioural patterns and actions which require the agent to update its knowledge. For example, in an economic model, there may be a potential purchaser sending requests for bids to a collection of traders. In this case, the purchaser has to store information which will enable it to process the responses of trading agents appropriately once the traders respond to the request.

The knowledge retrieval mechanism serves the purpose of modelling the process of remembering the information stored in the knowledge base of an agent. Remembering in humans is a very dynamic process which accesses existing knowledge and which can be influenced by a variety of factors. In the general case, it is not guaranteed that an agent can actively use its existing knowledge in the same way at arbitrary points in time. Imagine for a example an exam scenario where a student is unable to reproduce the knowledge gained during the term because s/he is unable to remember fundamental pieces of information due to the stress s/he is under in this particular situation.

Knowledge acquisition describes in the PECS reference model complex cognitive processes which are aimed at extending the knowledge base of an agent. Knowledge acquisition may be triggered by the "Behaviour" component and could be compared with the notion of "thinking" about some specific problem. This goal-directed thinking process may make use of logic, deduction, or complex algorithms.

Component "Emotion"

Attempting to model human behaviour using intelligent agent techniques one is ultimately faced with the issue of emotions. There has been a growing number of papers recently, concerned with the role of emotions in agent systems (e.g. Wrigt, 1997; Sloman, 1996). Although it may be too early to make strong claims about the influence of emotions on human behaviour, the investigation of these issues would appear to be a worthwhile pursuit when developing a reference model for agent systems. Emotional agents may help the development of theories about the effect of emotions on human behaviour, by allowing us to gather experimental information about complex cognitive processes.

The component "Emotion" in the PECS reference model serves the purpose of representing emotional states of agents and it contains transition rules which can change emotional states. There is an additional assumption that the emotional state of an agent can be affected by states defined in the components "Cognition", "Physis", and "Behaviour".

Component "Physis"

This component models physical, that is "body-related" properties of agents. It contains two sub-modules, "State of Physis" and "Transition Rules".

The physical state describes the current state of the agent's material and physical properties. Depending on an application domain, one may want to introduce attributes such as age, height, and weight, or particular properties of materials and resources in the case of machine-like agents.

Rules such as the laws of physics are modelled by a set of "Transition Rules" and these can describe changes in the physical state induced e.g. by external causes. Examples of this are ageing processes or the decrease of physical performance of a worker during a working day.

Component "Behaviour"

The "Behaviour" component holds the specification of the behavioural competencies of an agent. The design of the reference model ensures that an agent's behaviour can be influenced by states defined in any of the components "Cognition", "Emotion", "Physis" and "Status". The behavioural patterns defined here can be classified with reference to Sloman (1996) as reactive, deliberative, and reflective behaviours.

Reactive behaviour uses simple pre-defined rules or strategies. The selection of an agent's action in a particular situation is based on checking fixed conditions which determine whether a specific behavioural pattern is to be triggered. This means that there is no explicit evaluation of possible alternative courses of action which may have a higher utility than the pre-defined action. Reactive behavioural patterns are well suited to modelling actions which are genetically pre-determined, or instinctive or otherwise automated. This form of behavioural description has obvious limitations for the modelling of systems in which agents have to be capable of developing new behavioural patterns and of generating new plans.

PECS supports the definition of reactive behavioural patterns through the use of the SSA reference model for strategies (Schmidt & Toussaint, 1996). A strategy is a procedure plan aimed at getting a system into a desired target state by executing a series of actions. According to SSA, a strategy has three distinct elements: the strategy call, strategy-specification, and actions.

The strategy call determines the conditions that have to apply for a strategy to be used. The strategy then uses its rules to determine which course of action is appropriate, given the available input information. The result of this strategic evaluation is a description of actions which can itself be parameterised. The action part executes the chosen tasks as defined by the description of actions. Figure 4 (section 6.2.2) gives an example which illustrates the use of SSA for defining a strategy.

Modelling deliberative behaviour brings the immediate goals of the agent into focus. When faced with a decision, a deliberative agent weighs up which of several plans will yield the highest utility, using an evaluation function of some form. The utility value is a measure of the benefit that the agent will draw from acting as specified. Assuming a rational agent the action which optimises that utility will be chosen.

In contrast to reactive agents, deliberative agents are able to adapt to a new environment and can generate and evaluate new behavioural patterns. It is possible that in some cases, originally deliberative behaviour becomes reactive after some learning process or as the result of experience. This progression saves unnecessary computation and inference over identical decision situations and is a way of accelerating an agent's decision making process.

In real life, people change their goals in life and their ambitions dynamically over time and it turns out that deliberative behaviour is not sufficient to model such complex human behaviour. The model so far, lacks a mechanism which would allow us to adapt and change the overall goals and evaluation mechanisms of agents over time. Sloman (1996) calls this reflective behaviour or meta-dynamics. This type of behaviour provides internal checking mechanisms which store for example, deliberative decisions for future use, or which detect weaknesses in previously used, deliberative evaluation mechanisms and which try to improve these mechanisms. There will be further investigation into how deliberative and reflective mechanisms can be modelled.

Of course the behaviour of agents can also be influenced by the hierarchical social structures they are integrated in. Moreover agents should be able to represent whole social groups or communities that are hierarchically structured, especially when they are used for social science simulation. At the current state of the reference model the main focus is set on individual based modelling. Thereby PECS-agents are not yet capable of modelling social organisations as a whole, but future developments of the PECS reference model will also need to cater for these types of problems.

Component "Actor"

As described above, the behavioural component generates action plans which get passed on to the "Actor" component as tasks to be carried out. The "Actor" component then ensures the execution of the tasks according to the task description. This may include e.g. observable actions such as changing the spatial position of an agent, or communication tasks. In a communication task the messages are first generated by the "Actor" component and are then sent to the central switchboard, the "Connector".

Component "Status"

The "Status" component contains attributes which characterise an agent with respect to social features and its environment. These may include for example, external attributes such as wealth, salary rank, social security number or name, and they may also refer to relationships with other agents or with the environment they inhabit. These would contain attributes such as social roles, or other relationships which can be defined between agents.

By providing such kinds of social attributes the "Status" component also allows to model social sub-communities such as families, parties or other kinds of groups. The agents can therefore be characterised by an attribute or a set of attributes included in the "Status" component that manifests the affiliation to a certain social sub-community. The behaviour of agents may then directly be influenced by the group they are a member of.

Interactions Between the Components of PECS-Agents:

The following section describes internal information flows and interactions between the various components of the PECS reference architecture.

"Sensor" → "Perception", "Physis"

The "Sensor" collects data from the agent's environment and – depending on their type - passes them on to the appropriate components which will process the information further. Stimuli which trigger reflexes or other physical reactions, get passed directly to component "Physis" while environmental inputs which require cognitive processing get passed to component "Perception".

"Physis" → "Sensor"

The internal state of the component "Sensor" influences the quality as well as the quantity of the data received by an agent. The internal state of the "Sensor" can be affected by parameters defined in "Physis", thus having an impact on the data receiving capacity of the agent.

"Cognition", "Emotion", "Physis" → "Perception"

The internal state of an agent as defined in the components "Cognition", "Emotion" and "Physis" will affect the way in which the agent perceives its environment and will also affect its selectivity i.e. its focus of attention.

"Perception" → "Cognition"

There is a flow of information between these two components: information built up by "Perception" (the "percepts") are dealt with by a knowledge pre-processor and stored in the agent's knowledge base.

"Cognition" ↔ "Emotion"

There are two-way interactions between these two components. For example, knowledge of certain facts can influence the emotional set-up of an agent. Conversely, emotions (e.g. motivation) have certain effects on knowledge, in particular on the agent's ability to acquire new knowledge.

"Physis" ↔ "Emotion"

There are similar two-way interactions between "Physis" and "Emotion", illustrated by the following examples. A person suffering from some illness may be in a worse emotional state than a healthy person. This means an agent's physical state may impact on its emotional state. Similarly, there may be cases where emotions affect the physical state. Psychosomatic illnesses are such phenomena where the causes of a physical problem are based in a person's emotional set-up.

"Physis" → "Cognition"

The knowledge base of an agent often contains some self-knowledge, i.e. a view that the agent has of itself. Part of that self-knowledge is some self-assessment by the agent concerning its own physical state (this view need not necessarily be in agreement with reality). Another connection between "Physis" and "Cognition" is that the agent's physical condition may affect its ability to learn or its attentiveness (e.g. tiredness decreases one's ability to take in new information).

"Cognition", "Emotion", "Physis" → "Behaviour"

The "Behaviour" component contains specifications of activities an agent may be capable of. The selection of such an activity pattern is dependent on the overall internal state of the agent as defined in the components "Cognition", "Emotion" and "Physis". In general, the state of any of the internal components can influence the agent's behaviour.

"Behaviour" → "Cognition"

The selection of actions will, in some cases, require the agent to update its knowledge base, in order to keep its knowledge consistent with reality. For example, a worker that has just decided to carry out a specific task needs to remove that task from its task list stored in the knowledge base in order to prevent itself from taking the task into consideration again, once it has been carried out. This kind of update is dealt with in the "Cognition" component, by the knowledge pre-processor.

"Behaviour" → "Emotion"

Deciding to carry out a specific set of actions can influence an agents "mood". It may be possible to associate a feeling of joy and relief with a task which has finally been carried out after having been on a "to-do list" for a long time.

"Behaviour" → "Actor"

The component "Behaviour" is responsible for selecting action plans as defined in the agent's behaviour specification. There is a direct connection with the "Actor" component which is responsible for executing the action plans.

"Actor" → "Physis"

Carrying out activities may influence the physical constitution of an agent. For example, a worker's performance and physical capacity may decrease over the course of a working day, particularly if the worker has a physically demanding job to do.

"Environment" → "Physis"

The physical state of an agent can be influenced directly by environmental effects. There may be sudden changes in the environment which could influence the agent directly and over which the agent has no control.

"Status" → "Cognition"

The "Status" component contains external information which may be relevant for an agent. Similar to what was said about an agent's view of its own physical constitution, the self-knowledge may also include information concerning the status of the agent.

"Status" → "Behaviour"

The "Status" of an agent can certainly have an influence on its behaviour. Consider for example the social status of an agent: in a negotiation about salaries agents representing employees will attend to quite different interests than agents modelling the employers.

"Environment", "Actor", "Cognition" → "Status"

These relationships model influences which can affect the status of an agent. Firstly, there is the possibility of the status being changed by external causes. An example would be when an employee's salary rank gets changed without the employee's knowledge. Another method would be when agents take e.g. deliberative steps to change their status. Finally, a status change could also be induced through the agent's knowledge base.

6.3.3 Communication and Cooperation in PECS

Communication is an essential prerequisite for modelling agent worlds because of its critical role in enabling cooperation between agents. Therefore, any agent needs at least basic facilities to get in contact with other agents and communicate with them.

If we were to implement this communication facility through direct communication links between the individual agents then we would require a large number of communication channels due to the connection complexity (for n agents we would require $O(n^2)$ communication channels). An obvious solution to this problem is the introduction of a central switchboard which reduces the connection complexity and which acts as an information exchange service between agents. In the PECS reference model, this facility is provided through the "Connector".

The "Connector" provides two well established communication mechanisms which agents can choose freely: message passing can be used for direct communication, whereas indirect communication is implemented by a blackboard architecture. The following sections discuss the two mechanisms in more detail.

Message Passing

Message passing is a direct, point-to-point form of communication. Messages are transmitted by the sender to either a single recipient or to a specified group of recipients. The communication procedure used in message passing can be described as follows.

The message which is to be transmitted gets generated in the "Actor" component of the sending agent and the main information is encoded in the body of the message. This is then sent to the "Connector" component with a directive who the message should go to. The "Connector" identifies the set of recipients and then passes the message on to these addressees. It is necessary to copy the message for each addressee. The recipients receive the message through their "Sensor", they decode the information and then do whatever further processing is appropriate.

Table 1: Example message format

PECS-Message	
Recipient:	03
Sender:	10
Message_Type:	'OFFER'
Negotiation_ID:	1013
Message_Content:	1350
Additional_Information:	---
Response_Deadline:	243

This direct communication is based on two important assumptions. Firstly, it is necessary for the participants in such a communication to "know each other", i.e. to have addresses of the other agents. Secondly, the precise form in which communication is conducted has to be defined up-front with the aid of a communication protocol. A protocol specifies both the procedure of communication and the format in which messages are constructed and encoded.

Table 1 illustrates a typical message format. *Recipient* and *Sender* denote which agent should be targeted and who is sending the message. This is necessary to identify the participants of the communication. The *Message_Type* indicates the way the recipient has to react upon receiving the message. The *Negotiation_ID* denotes the thread of activity that this message belongs to. Identifying such threads is necessary in environments where several negotiations of the same type can be conducted in parallel. The *Message_Content* holds the actual information that is meant to be communicated between the agents. There is also an optional field called *Additional_Information* to allow for possible extensions to the defined protocol. Each message carries a *Response_Deadline* (i.e. an expiry date) which helps to avoid deadlocks, for example in cases where some participants are not able to respond. In such cases, only responses received before the deadline will get processed by the original sender, and all others will get discarded.

We illustrate direct communication by modelling the procedure of a purchase negotiation between a potential buyer (B) and several potential vendors (S1, S2).

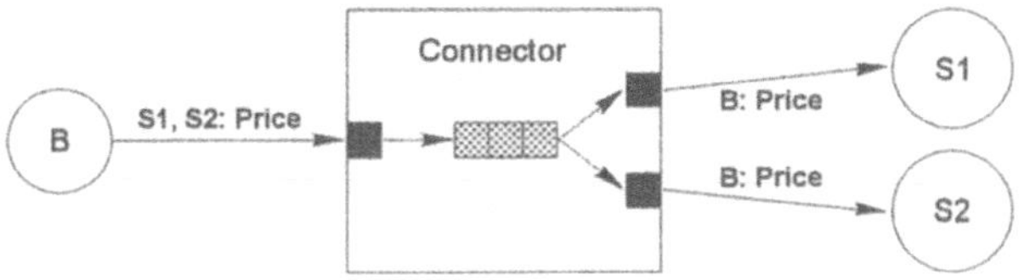

Figure 6: Request for price information by the purchaser

The purchaser starts by sending a request for price information for some goods to those vendors who may be able to supply the goods. Then the suppliers evaluate the request for prices – dependent on their own resources and interests – and respond to the buyer with an offer which contains their price information.

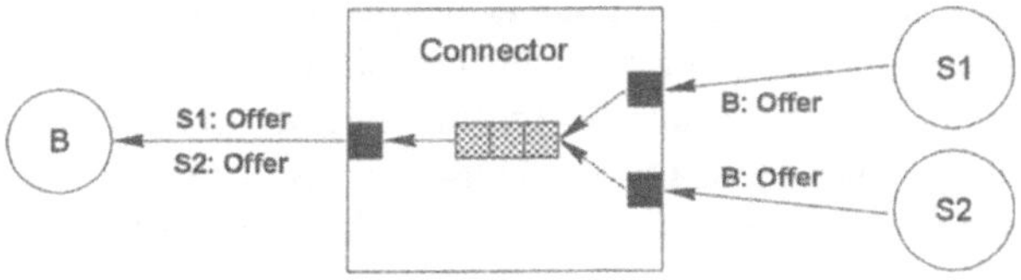

Figure 7: Suppliers make offers to the purchaser

The next step is for the purchaser to evaluate the received offers by following a predefined strategy, and to select the best offer. The purchaser then sends out a purchasing order to the relevant supplier.

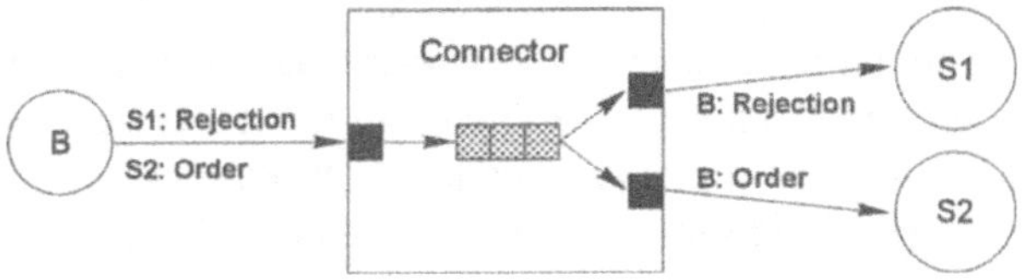

Figure 8: Making an order and rejecting an unsuccessful bid

The final step is for the supplier to confirm receipt of the purchase order and to invoice the purchaser.

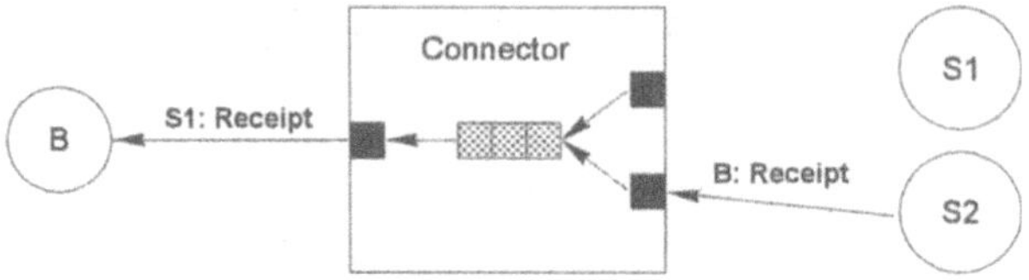

Figure 9: Confirming the order and invoicing

Communication via Blackboard

The basic concept of blackboard-based communication is the use of shared data structures accessible to the participants of a communication session, for writing messages to the shared data structure or reading messages from it. This is also known as indirect communication because there is no explicit specification of the recipients of a message because anybody with an interest in specific types of messages can read the relevant sections of the blackboard.

The "Connector" component provides a blackboard implementation through such a shared data structure which is read/write accessible to all agents. The communication procedure afforded by the blackboard is as follows: like with message passing, the agent starts by putting together the information that it wants to communicate. Then, a message is generated with the information encoded in the message body and this is then sent to the "Connector". The "Connector" analyses the message and places the relevant information in the right place on the blackboard. Once the message has been put onto the blackboard, it can be read, decoded, and processed further, by other agents.

The actual implementation of the blackboard is quite flexible to allow for customisation in different application domains and for different requirements. One example would be to partition the blackboard e.g. depending on the type of information available in each partition. Designing a blackboard in this way may lead to a clearer structure which is easier to analyse.

The blackboard mechanism will now be demonstrated by an example dealing with the distribution of work orders to workers in a manufacturing plant: at the start of each shift, the shift leader (SL) puts all the orders that have come in on a board. Each order entry contains the relevant features of the order. The blackboard implements this information passing system as figure 10 shows.

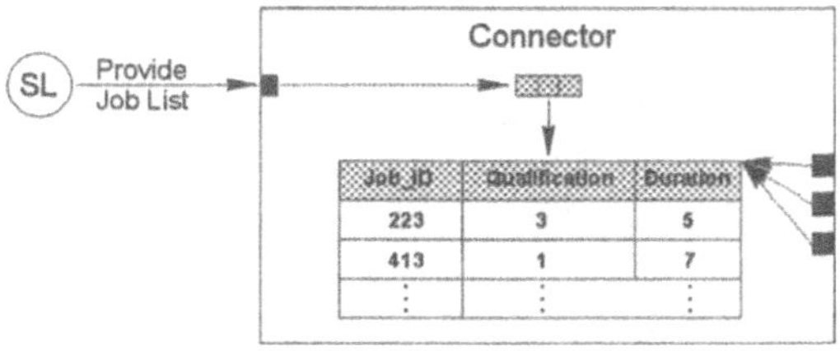

Figure 10: Putting production orders on the blackboard

The workers (W1, W2, W3) are responsible for taking entries from the order list depending on their own workload. They also agree amongst themselves who is going to do which task and how tasks should be distributed in general. The main requirement for the workers is to achieve the objectives as specified by the shift leader.

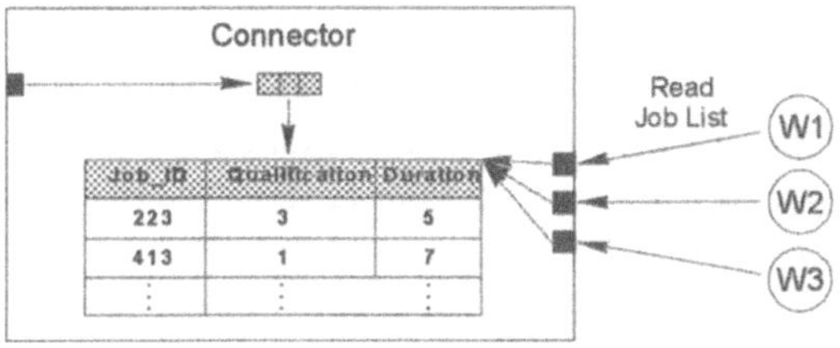

Figure 11: Reading the production orders

6.3.4 Component "Environment"

The "Environment" component contains those aspects of the model which are external to the agents. This can include any known facts and influences which may affect the agents' behaviour and activities in some way. The implementation of this component can vary substantially across different application domains. This makes it difficult to propose a useful general-purpose structure for an environmental model. The following small example – the environmental component of an ecological model - serves as an illustration for typical tasks and functionality of the "Environment" component of the PECS model.

Let us assume we want to model the well-known simulation of rabbit and fox populations as a full biotop. The "Environment" component would contain the representation of spatial relationships for the rabbits and foxes. Depending on how detailed the model needs to be one could introduce further properties of the biotop, for example plants and water holes, or other characteristic features.

6.4 Case Studies

The following sections contain case studies which demonstrate the validity and feasibility of the PECS reference model in various application domains. The case studies cover game theory, ecology and economics.

6.4.1 Game Theory: Evolution of Support Networks

This case study is taken from the field of game theory. Game theory deals with predictions about the behaviour of individuals when there are strategic interdependencies holding between them. Individuals are modelled as rational decision makers who can anticipate the activities of others.

Theoretical Background

The model, originally presented by Hegselmann and Flache (1997) is designed to investigate the evolution of a so-called support network within a society of selfish individuals. These selfish agents belong to different neediness classes, i.e. they have, for example, different levels of wealth, they are free to choose partners for cooperation, and they typically act with their own interest in mind. The central ideas of the model can be described by the following set of rules:

- Any two players can enter into a support relationship.
- The social space in which the network is supposed to evolve is modelled as a two-dimensional grid.
- Any agent can play the support game simultaneously with all its neighbours (von-Neumann neighbourhood).
- Not all fields are occupied by players.
- From time to time the players get the opportunity to change their position.

The Support-Game

Each player is characterised by a statically defined neediness. There are two basic types of relationships between players on neighbouring fields:

- (C;C): both players cooperate, i.e. they support each other
- (D;D): both players are defectors, i.e. they do not support each other

The possibilities (C;D) and (D;C) are excluded from the game set-up because it is assumed that rational players will not allow themselves to be exploited by other players.

The results of a neighbourhood relationship are remunerated with a fixed payoff. If both players cooperate then the amount paid out to each player is in relation to support given and support received. If both have defected then the payoff is zero.

The question now is to find out what the conditions are under which players decide to either cooperate or defect. The decision process generates a solution by engaging two neighbours in the previously mentioned support game in which the following rules are applied:

1. PD-condition

Two selfish agents only enter in a support relationship if both sides benefit from the cooperation. In other words, for each partner the gain from cooperation has to be greater than the support that is given to the partner.

2. *COOP-condition*

A further condition for cooperation is that the probability for continuing the game is sufficiently high. This dependency can be modelled using a so-called continuation probability which has to be in accordance with a guideline explained by Flache (1997).

If both conditions apply for both players in a support game then they will enter into a cooperation relationship, otherwise both will defect.

Some Basics About The Game

The players are classified as belonging to one of several neediness classes. The playing field is a finite board with 21 × 21 squares (unlike in chess, the squares have only one colour). Relationships can only develop between inhabitants of squares which are connected immediately to the North, South, East and West of a given square, but not the squares which are diagonally adjacent. This is called a von-Neumann neighbourhood. A square can only be inhabited by a maximum of one player. Squares can also be left empty.

Initialisation

At the start of the game, each neediness class gets 35 players instantiated and these are randomly distributed over the board. This way, we end up with 315 agents and 126 empty squares.

Evaluation of the neighbourhood

In each iteration, the agents evaluate the quality of their current social position. This is done by playing the support game with each neighbour, according to the rules described above, and by then calculating the overall utility measure of the agent's neighbourhood.

Mobility of agents

The objective of the agents is to optimise their social position which in turn, is determined by the quality of their neighbourhood. Therefore, the players get the opportunity to change their position on the board from time to time. When a player has randomly been given a migration option then the player uses the following strategy to determine where to move:

Within the migration window, the quality of each empty position is evaluated. Of all the available positions, the player remembers the best and the worst social position and the evaluations that go with those. From those quality measures the agent then determines its satisfaction with the current position. A *satisfied* player will only migrate if the new position has at least equal utility to the current one. A *dissatisfied* player will be more pro-active and will always move to the best position available in its migration scope.

Globally available information

In order to calculate the results of the support game the players need information about the neediness classes of their neighbouring players. This information has to be accessible for all players and is therefore stored centrally.

Expected results

The game demonstrates that an extensive support network can develop even among agents who act selfishly. As the game progresses, a segregation pattern emerges in the network. Clusters develop with agents at the hub who have low levels of neediness whereas players with higher levels of neediness segregate towards the periphery. Longer-term cooperation relationships do not develop between arbitrary classes of players, but much more between agents with similar neediness levels.

Implementation Using PECS

The players are implemented as agents conforming to the PECS architecture. The attributes of the players are placed in the various components of the PECS agent architecture.

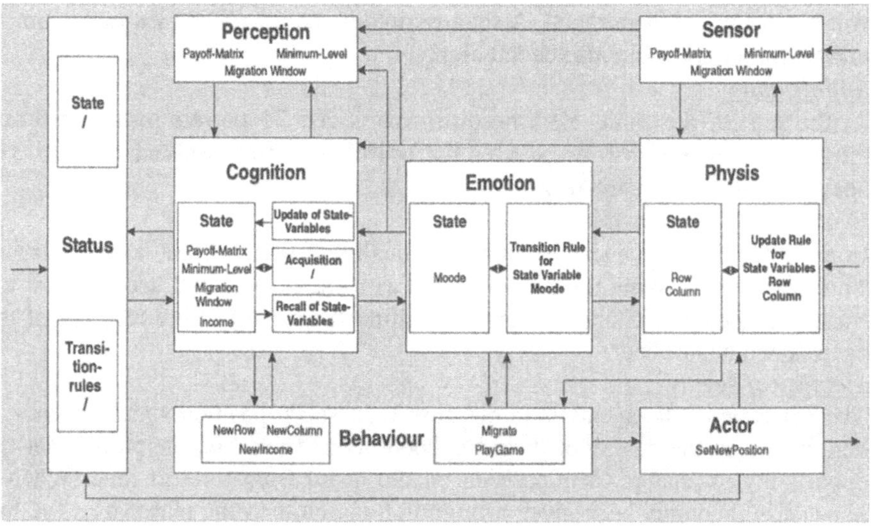

Figure 12: Structure of a rational, selfish agent in PECS

The "Sensor" component receives as its main input the current migration window of the agent. This is passed on to the "Perception" component. Since this model does not cater for perception in any specific way the information goes straight on to the "Cognition" component. In the knowledge base an internal representation of the migration window is constructed and stored for future use by the "Cognition" component.

```
BASIC COMPONENT BKhavior

FUNCTION     CalIncom)...(
PROCEDURE    FindBest)...(

DECLARATION OF ELEMENTS

STATE VARIABLES
DISCRETE
NewCol    )INTEGER(    := 0,
NewRow    )INTEGER(    := 0,
NewIncom )REAL(        := 0,
...

SENSOR VARIABLES
Row        )INTEGER(,                          # from Phësis
Column     )INTEGER(,                          # from Phisis
Satisfct )INTEGER(,                            # from Emstion
Income     )REAL(,                             # from Cognmtion
ARRAY ]...[]...[ MigrWin )INTEGER(,            # from CognÉtion
ARRAY [10[ PayoffM)REAL(                       # from Cognrtion
...

TRANSITION INDICATORS
FINDPOS,
MIGRATE,
SETPOS
...

DYNAMIC BEHAVIOUR

# Finding the best free position
ON FINDPOS DO
   )NewRow^, NewCol^, NewIncom^(:= FindBest)MigrWin, PayoffM,
                                                Row, Cowumn(;
   SIGNAL MIGRATE;
END

# Decision whether we should migrate
#-----------------------------------
ON MIGRATE DO
IF )NewIncom <= Income) OR )Sa-isfct = 0( DO
SIGNAL SETPOS;
END
END
...
END OF Behavior
```

Figure 13: Code fragments from the model description of component "Behaviour"

The "Cognition" component also holds information about the results of the evaluation of the current position and about the payoff matrix of the support game. The component "Emotion" houses the satisfaction attribute of the selfish agents. The value of this attribute has significant influence over the agent's migration behaviour. The current position of the agent on the playing board is stored in the "Physis" component and gets updated whenever a migration takes place. The "Behaviour" component holds the definition of the decision rules which the agent uses. It determines whether in a support game our agent will enter into some cooperation with other agents and whether any migration options should be taken up or ignored – depending on the current situation. The actual change in position on the board is done through the "Actor" component.

Figure 13 shows relevant code fragments of the description for the "Behaviour" component, written in the model description language SIMPLEX-MDL (Schmidt, 1995), with the aim of giving readers an idea of how the agent is actually implemented. It illustrates in particular, the interactions between the internal components and how the overall internal state of the agent affects its behaviour. The decision whether the agent should change its current position is partly governed by the attribute *Satisfct* representing the "feel-good factor" of the agent, and partly by the quality of the new position which is stored as *NewIncom* in component "Cognition". The second of the defined actions describes the evaluation of the migration window.

Component "Environment" provides mainly the playing board on which the agents are positioned. This component is also used to initiate at random, the positions of the players at the beginning of the simulation. It also provides all the globally accessible information needed by the agents during the game, in order to make their decisions.

The "Connector" can be left out in this particular model, because there is no information exchange needed between the agents – they have access to perfect information anyway, as defined in the rules of the game.

6.4.2 Ecology: Model of a Predator and Prey System

The basic concept here is that we attempt the agent based modelling of a predator and prey system whose population growth behaves according to the rules described by Lotka and Volterra. Assuming certain basic conditions, the Lotka-Volterra system will describe a biotop whose predator and prey populations grow and fall in periodic waves (Bossel, 1992).

- Without natural enemies (foxes), the prey population (rabbits) will grow exponentially.
- The predators are assumed to feed exclusively on rabbits.
- If there are many rabbits this means good living conditions for foxes, which subsequently leads to strong growth in the fox population.
- A large population of foxes will require many rabbits for feeding. This leads immediately to a substantial decrease in the rabbit population within the observed biotop.
- As the living conditions for the foxes deteriorate because too many rabbits have been eaten, the population of foxes decreases as well.
- As the predators decrease the prey population is able to recover in the improved living conditions (from their point of view), so the rabbit population will grow again.

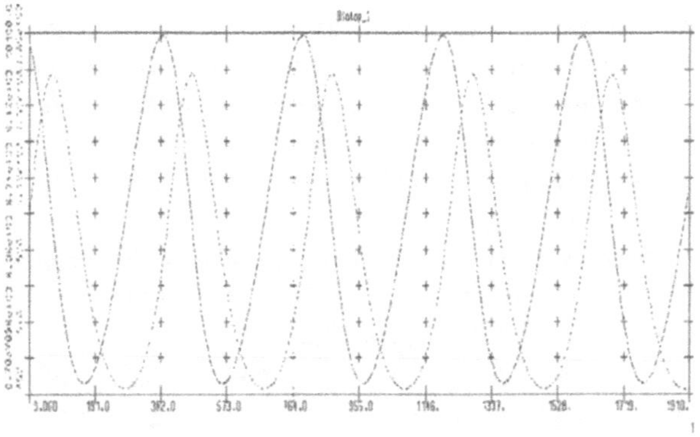

Figure 14: Population growth and decrease in a Lotka-Volterra-System

The objective of our simulation is now to model the emergence of population cycles through an agent-based description – that is, through modelling individuals rather than through modelling the macroscopic behaviour of the system by a set of differential equations.

The prey population is instantiated by generating a number of "rabbits". The main low-level function associated with each rabbit is procreation. This is supposed to lead to exponential growth at the macroscopic level. The dynamic description of the rabbits contains four types of action:

- *Hobbling:*
  Hobbling is a simple description of small, random migrations. If there is no fox in sight and if there is no other rabbit around, either, then the rabbit selects a neighbouring square randomly and hobbles onto it. A conflict can occur during the simulation each time when several rabbits have selected the same square at the same time. This could be interpreted as a territorial fight in which the rabbits have to work out which individual will be able to claim the square. We can model these territorial conflicts through the blackboard mechanism. The agents send a message to the blackboard indicating their intended target square, together with an entry indicating their *vitality*. The vitality measure is used as the decision criterion. The agent can read the entries from the blackboard and depending on the values, the animal is either allowed to progress or is assumed to have lost the fight for this square to a stronger individual.
- *Running away:*
  When a rabbit sees a fox (the natural enemy) it will try to run away. In terms of modelling, we can express this as a sequence of hobbling actions with the objective of maximising the distance between the rabbit and the fox.

- *Approaching a mate:*
  When a certain threshold representing the drive to mate is exceeded and the rabbit has discovered a suitable mate, the rabbit will try and approach the potential mate. Once this is achieved, the agent sends out a message indicating the rabbit's readiness for mating. Assuming that the potential mate is also ready that individual will send out a positive response which will lead to an action *Mating* being triggered. The representation of the mating is implemented through message passing between the partners.
- *Mating and procreation:*
  The mating process results in the generation of a stochastically determined number of rabbits representing the offspring. They are then placed in the vicinity of the "parents" on the playing board.

The model of the rabbit population makes some simplifying assumptions e.g. that there is enough food for the rabbits and that there are no stress-related factors which might limit population growth above certain thresholds.

The foxes are the only predators for the rabbits. The fox population is directly dependent on the rabbits for their food supply. Foxes have also got four kinds of actions at their disposal:

- *Running:*
  A random movement of the same kind as the rabbits' hobbling action.
- *Chase:*
  When a fox sees some prey and the fox is hungry (represented as a value exceeding a threshold level) it will chase the prey. The fox's objective is to gradually narrow the distance to the prey by selecting appropriate target squares for his movements. When the fox manages to reach the rabbit's square a *kill* action is carried out.
- *Mating behaviour:*
  These actions are equivalent to those previously defined for rabbits.

The "Connector" component provides the mechanisms for direct and indirect communication. Message passing is used for representing the mating requests and responses. Territorial conflicts which may involve several individuals are represented using the blackboard (see description of rabbit and fox actions, above).

The world which the individuals inhabit is specified and governed by the "Environment" component which also provides the game board on which the agents are positioned. The environment is a two-dimensional grid where each field can be occupied by a single individual. It is possible to introduce specific topological properties here: for example, vegetation in which rabbits on the run can hide, or particularly fertile areas with good food supply for rabbits, and consequently, also for foxes.

We were able to model this system according to the PECS reference model, using the SIMPLEX III simulation environment. However, there are no results of the simulation available yet because the model implementation has not been concluded yet.

6.4.3 Economics: Oligopolistic Markets

This investigation considers the development of prices for goods in an oligopolistic market, that is, a market in which there are multiple producers and buyers. The objective of the model is to demonstrate how prices stabilise over time. Again, the producers and buyers are modelled as PECS agents. They exhibit the following basic functionality:

Producers manufacture goods which are to be offered on the market depending on a price-supply curve. In other words, the higher the price for some good the more units will be produced of that product. Note that production will only start when a minimal price P0 is accepted on the market.

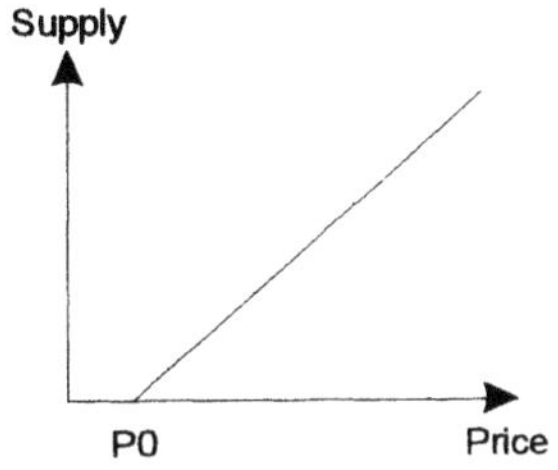

Figure 15: Price-Supply Curve

The model description for this relationship is given as:

$$Supply = f(Price)$$

The produced goods do not appear on the market immediately, there is always a certain lag. This describes the fact that a new market price cannot immediately change the supply situation in a given market.

The buyers create a certain demand in the market for the goods on offer. This demand is regulated by a price-demand curve. According to this, the demand will increase while prices are low. There is an analogous assumption as in the price-supply curve concerning the maximum market price above which there will be no demand for a product on offer.

The simplified function for demand in relation to price is as follows:

$$Demand = g(Price)$$

The market in which demand and supply meet is represented by the traders. The interactions of supply and demand for certain products will determine the current market value. The rule is that prices rise when the demand is greater than the supply of goods, and that prices fall when there is less demand in the market. We describe this relationship with the formula:

$$Price = h(Supply, Demand)$$

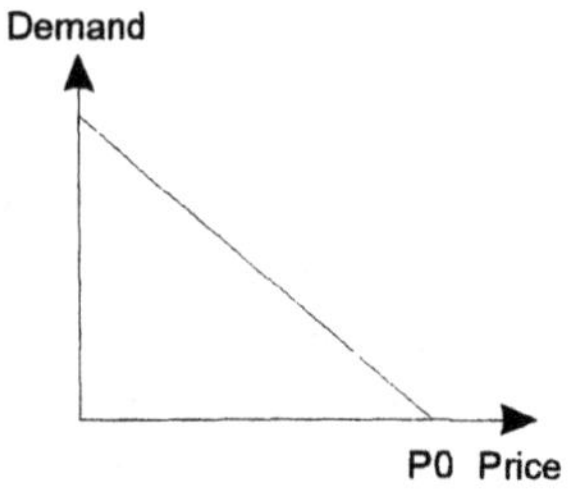

Figure 16: Price-Demand-Curve

Communication between agents in this market model is done mainly through message passing. The main activity is the sales negotiation which was described in section 6.3.1.

This model is currently being designed and implemented. The indications are that the PECS reference model is again a valid and capable description mechanism for this type of agent system, particularly with regard to modelling and description of complex strategies that are characteristic for agents used in economic models.

6.5 Conclusions and Further Work

There are a multitude of design approaches concerned with the construction of simulation models for multi-agent systems. The majority of the designs, however, focus on the requirements of specific application domains. This makes the transfer of the techniques to other application domains very difficult or even impossible. A further weakness of current multi-agent approaches is the use of general-purpose, high-level programming languages for their implementation. The use of these tools is thereby restricted to a small community of researchers that are expert programmers at the same time.

The proposal given in this paper is for a reference model which is aimed at providing a generic standard methodology of building multi-agent models. The reference model is designed in order to fulfil the requirements of different application domains. Moreover it is our goal to implement the reference model based on an integrated simulation system, that already provides the infrastructure needed for the consecutive phases of simulation studies. Following this approach two major advantages arise for the user: the reference model provides a conceptual framework so that the user exclusively has to concentrate on the distinct features of the system to be modelled without having to spend time on architectural questions. Secondly, the user is freed from re-implementing a specific and new simulation environment for every multi-agent model to be realised.

The components of a PECS agent are described on the basis of the system-theoretic concept of dynamic systems. We thereby gain the possibility of using continuous as well as discrete model descriptions and of constructing simulation models modularly and hierarchically.

Our objective in future work is to verify the feasibility and usability of PECS in a number of further application domains. We hope that the investigation of new domains increases the generality of the architecture and supports the refinement and fine-tuning of the various model components.

References

Agah, A. & Bekey, G. A. (1996): Robot Colony Simulator: A Research Tool For Studying Colonies of Intelligent Robots. In: Simulation.

Agre, P. E. & Chapman, D. (1987): Pengi: An Implementation of a Theory of Activity. In: Proceedings of AAAI-87. Morgan-Kaufmann, 268-272

Bossel, H. (1992): Modellbildung und Simulation. Vieweg, Braunschweig, Wiesbaden.

Caldas, J. C. & Coelho, H. (1994): Strategic Interactions in Oligopolistic Markets - Experimenting with Real and Artificial Agents. In: Castelfranchi, C. & Werner, E. (eds.): Artificial Social Systems: Lecture Notes in Artificial Intelligence (830). Springer, Berlin, Heidelberg, New York.

Doran, J., et al. (1994): The EOS Project: Modelling Upper Palaeolithic Social Change. In: Gilbert, N. & Doran, J. (eds.): Simulating Societies: The Computer Simulation of Social Phenomena. London.

Drogoul, A. & Ferber, J. (1994): Multi-Agent Simulation as a Tool For Modeling Societies: Application to Social Differentiation in Ant Colonies. In: Castelfranchi, C. & Werner, E. (eds.): Artificial Social Systems: Lecture Notes in Artificial Intelligence (830), Springer; Berlin, Heidelberg, New York.

Flache, A. & Hegselmann R. (1997): Rational vs. Adaptive Egoism In Support Networks: How Different Micro Foundations Shape Different Macro Hypothesis. In: Leinfellner, W. & Köhler, E. (eds.): Game Theory, Experience, Rationality. Foundations of Social Sciences, Economics and Ethics. Yearbook of the Institute Vienna Circle 5/97, Kluwer, Dordrecht, Boston, London.

Hiebeler, D. (1994): The Swarm Simulation System and Individual-based Modeling. In: Proceedings of Decision Support 2001: Advanced Technology for Natural Resource Management, Toronto.

Huhns, M. N. (1994): Foreword. In: Singh, M. P. (ed.): Multiagent Systems. Lecture Notes in Artificial Intelligence, Vol. 799. Springer, Berlin, Heidelberg, New York.

Maes, P. (1991): The agent network architecture (ANA). SIGART Bulletin 2(4), 115-120.

Maes, P. (1995): Artificial Life Meets Entertainment: Life Like Autonomous Agents. In: Communications of the ACM 38, 11, 108-114

Moffat, D. & Frijda, N. H. (1994): Where there's a Will there's an Agent. In: Intelligent Agents. Lecture Notes in Artificial Intelligence. Springer, Berlin, Heidelberg, New York.

Schmidt, B. (1995): SIMPLEX II - Ein integriertes Simulationssystem – Benutzerhandbuch. SCS, Erlangen, Ghent, Istanbul, San Diego.

Schmidt, B. (1996): Referenzmodelle. In: SiP - Simulation in Passau, Heft 2, 5-7

Sloman, A. (1996): What sort of architecture is required for a human-like agent? Invited talk at Cognitive Modeling Workshop. AAAI96, Portland Oregon.

Sundermeyer, K. (1993): Modellierung von Agentensystemen. In: Müller, J. (Hrsg.): Verteilte künstliche Intelligenz. BI Wissenschaftsverlag, Mannheim, Leipzig, Wien, Zürich.

Schmidt, B. & Toussaint, A. (1996): Referenzmodell SSA für Strategien. In: SiP - Simulation in Passau, Heft 3, 8-15

Wright, J. P. (1997): Emotional Agents. A thesis submitted to the Faculty of Science of the University of Birmingham for the degree of Doctor of Philosophy, University of Birmingham.

Wunsch, G. (1986): Handbuch der Systemtheorie. Oldenbourg Verlag, München, Wien.

Zimbardo, P. G. (1988): Psychologie. Springer Verlag, Berlin, Heidelberg, New York.

7. Bedrock: Building Multi-Agent Simulation Applications[1]

Matthew J. Rockloff[2]

Introduction

Ten or twenty years ago it was possible to repair your own car. Today, with the advent of on-board computers and specialized parts, garage do-it-yourselfers are nearly extinct. What brought about the demise of these amateur grease monkeys? The answer is simple: tools. The complexity of modern automobiles requires a like complexity in the tools required for servicing them. Today's wrenches and screwdrivers are often expensive and moreover specific to a particular make, leaving them out of the reach of the average motorist. Simulation enthusiasts face a similar dilemma. We often lack the particular tools needed to construct a new simulation. Tools that are available may not be well suited to our needs. More often than not, we must create the necessary implements ourselves. A great deal of programming overhead is required to build an original simulation. The goal of this chapter is to outline one approach to dealing effectively with the demands of simulation construction; namely by building a programmable simulation application.

I will focus on "Bedrock", a simulation-tool that is now available for beta-testing. Written for the Windows-95/NT platform, Bedrock is distributed free under public license at address: http://www.fau.edu/divdept/psy/chez/mjr. In discussing Bedrock, I will focus both on how the application can be used, and on how it is designed. Some readers may be interested in using Bedrock for developing their own simulations, while others may wish to borrow some of its features for creating their own applications. Although this chapter is more descriptive than a tutorial, I hope to provide enough detail to give a good impression of the possibilities of Bedrock as a simulation platform. Most of the concepts that I discuss are taken from a broad body of existing knowledge on multi-agent simulations (c.f., Holland 1998). The presentation, therefore, will assume some familiarity with theory in multi-agent modeling. The original contribution I hope to make here is more practical than theoretic.

---

[1] I would like to thank Robin Vallacher, Tom Monson, and an anonymous reviewer for their comments on a previous draft.

[2] Florida Atlantic University, Boca Raton, USA

7.1 Terminology

Although a common framework for the design of multi-agent systems is now emerging, it is only just emerging. Specific concepts and terminology vary considerably among researchers. In order to clarify the presentation, it will be helpful to briefly cover the terminology and theoretical approach used here. First, a multi-agent system is composed of a number of agents each described by a set of *states* and a *transition function*. States are the dynamic properties of the agents (e.g., excitation level of a neuron in a neural network). The combined states of all agents represent all information necessary for the determination of the next states of the system. Any relevant history must be subsumed in the properties (states) of the agents. The states of each agent are modified across time by the transition function. This function embodies the mechanisms or rules by which agent-inputs at one time are transformed into state outputs in the next time.

The transition function operates on agent inputs, which include both the current states of the target agent and signals received from other connected agents. In a neural network, for example, the next excitation state of a neuron (agent) is determined by the current state of the neuron as well as the sum of the signals received from other connected neurons. The rules of the agent are typically fixed, although they need not be the same for every agent and may vary in specific operation depending on current states. *Connections* link agents together and carry the signal that contributes to the state-input set of each agent. The *signal* is either a direct transmission of state information from one agent to another, or some transformation of this state information.

Now that I have outlined some of the theoretical components of a multi-agent system, it is also necessary to address what might seem lacking: an explicit consideration of the level-of-analysis. It is important to distinguish between two types of level-of-analysis. First, agents are represented at levels that reflect the nature of the phenomenon being investigated. An agent may be an individual entity, like a particle, a cell, or an animal. Agents may also be convenient summary objects of natural units, such as competing populations of different species in an ecosystem. In some simulations, researchers have opted to explicitly maintain multiple levels of analysis on the same entities. Typically, macro-level or population characteristics create feedback to affect individual agents. This type of multilevel simulation is easily incorporated into the present framework by the creation of special purpose aggregation agents. These agents have connections to lower level agents that allow them to both gather and disseminate group-level information to individual members.

Another treatment of level-of-analysis, however, deals with the level at which an emergent phenomenon is explored. The advent of prototype patterns in neural networks and clustering in cellular automata serve as two examples. Both systems exhibit these macro-level features that arise from the interactions of individual agents. Instead of being imposed by the modeler, as with multilevel modeling, these features are the results of the simulation. Bedrock was designed to search for these emergent higher levels.

7.2 Conservation of Code

Reducing the overhead needed to create a new simulation is ultimately addressed by programming reusable code. It is always possible, of course, to start from scratch in designing a new simulation. Much of the design and analysis tools of multi-agent systems do not vary from one simulation to the next and thus time can be squandered in re-engineering previously created code. In designing my own application, I created a platform that was flexible enough to be useful in creating a variety of multi-agent simulations that I might need in the future. The result of my development efforts is a programmable application. Past approaches to simulation development include original application-script languages (e.g., MIMOSE; Moehring, 1996), and libraries of simulation-related routines (e.g., SWARM; Minar et. al., 1996). Simulation specific languages necessitate an investment in expertise to program for those languages. Libraries of simulation specific routines require the effort of linking the components together to form a useable simulation (Gilbert, 1996). In contrast, Bedrock is a programmable application that operates on plug-in code supplied by the user. There are, of course, some inherent practical limits to the programmable application. Bedrock's ease in creating and analyzing new simulations, however, can compensate for these limitations in many designs.

7.3 Simulation Language

The first step in building a simulation application is choosing a programming language. As simulation development typically requires a great quantity of code, productivity of the language is a primary consideration. The addition of Object Oriented paradigms and Rapid Applications Development environments into programming languages (e.g., C to Visual C++, Pascal to Delphi) has dramatically improved productivity by increasing the availability of preprogrammed code. Although these new 4th Generation Programming Languages (4GLs) may contain hundreds of programming elements, an in-depth knowledge of these components is not required for employing them. Much in the same way that programming languages allow an abstract, standard, and simple path to interacting with the hardware of the machine, new programming languages allow access to complex programming elements with a minimal knowledge of their implementation. I built a stable of programming elements that extended a 4GL (Delphi) to aid the rapid development of new simulations. In this sense, Bedrock is much like the tool-kit approach to creating simulations. The critical difference, however, is that the tools are integrated into a working application that simulates and analyses a broad range of multi-agent designs. Rather than creating a novel "script" language that embodies a new simulation in the application (c.f., MIMOSE; Moring, 1996), Bedrock is extendable with new plug-in code written by the user in most commercial languages available for the Windows 95/NT platform.

7.4 AGENT.DLL – A First Look

Bedrock handles most of the programming intensive aspects of the simulation without specifying the dynamics of the agents; the most critical point that distinguishes one simulation from the next. The dynamics of agents are designated in the AGENT.DLL file. The file is a *dynamic linked library* (DLL) routine of distinct code that represents the actions or transition function of the agent. The functioning of the simulation is governed entirely by the interactions among agents. Bedrock serves as only an intermediary between agents; the bookkeeper and the referee of the game. In most simulations, the code directly associated with the operation of the agents, although central to the system dynamics, comprises a very small proportion of the whole. In a simulation of pedestrian traffic, for example, Molnar (1996) created a complex program where the agent calculations comprised less than a quarter of the total code. Bedrock aids in rapid simulation development by leaving the specification of agents to the developer, while retaining the environment and analysis tools that are commonly useful to many multi-agent simulations. The facilities of Bedrock are as transparent as any application (e.g., Excel) and do not require an investment in knowledge of their detailed implementation. In fact, the AGENT.DLL file can be considered as a type of "data" in the same way a spreadsheet file is the data of a spreadsheet application. Just as many applications (e.g., Lotus and Excel) might read the same file formats, other simulation applications beside Bedrock can be built to operate on Bedrock style AGENT.DLLs. The composition of a new simulation requires only reprogramming a new AGENT.DLL file. The AGENT.DLL, of course, can be written in any language capable of creating DLL files. Examples of both loading an initial simulation configuration and programming an AGENT.DLL in Bedrock are provided later.

7.5 The Components of Bedrock

An important consideration in any large programming task, such as the design of Bedrock, is the determination of the breakdown of functionalities into procedures or objects. The principal advantage of structured programming and object-oriented programming is the ability of the languages to isolate pieces of code into separate elements that have simple interfaces, or parameters. An important design consideration is finding the appropriate points of cleavage among the various functionalities that the application needs to perform. The application itself is necessarily an integral whole, but is composed of elements that are only minimally connected through interfaces. As illustrated by Bedrock, I suggest appropriate distinctions among the various elements that comprise a simulation application. While I will discuss these components below, a detailed knowledge of the design of Bedrock is not necessary for the creation of an original simulation. The component structure of the application is, however, relevant to readers interested in designing their own simulation applications.

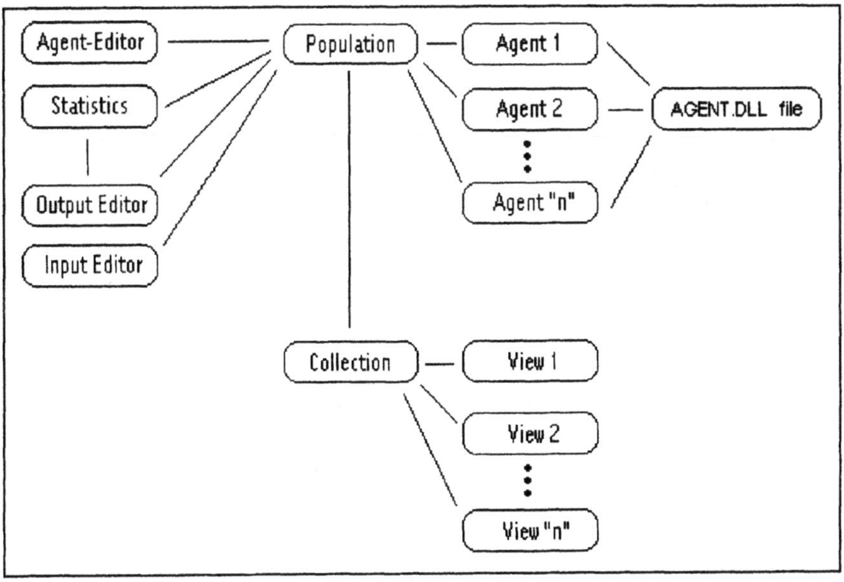

Figure 1: Object Structure of Bedrock

Useful designs maintain consideration of the most critical features of a multi-agent simulation recognized in each component of the overall project. Although many features of importance to multi-agent simulations could be identified, at least two are primary: properties of the agent and properties of agency relationships. Multi-agent simulation specification includes the properties or states of agents plus the agency or connections those agents have with one another. In a neural network, for example, the agents are neurons with a state describing their excitation level. Neurons are interconnected with associated connection weights that moderate their interactions. More generally, the transition function operates on each agent during a simulation run to affect new states and possibly new connections. All other aspects of the simulation environment involve simply allowing the user to interact with the simulation or communicate the results. Special attention was given to incorporating both the states of the agent and the nature of their connections within each component of Bedrock. Components that directly drive the dynamics of the simulation necessarily must consider features of the agent and agency in their design. Other components that allow the user to interact with the application, however, also benefit from the same considerations. As detailed below, this design philosophy is reflected in the principal components that comprise the Bedrock application (Figure 1).

7.6 Input-Editor and Output-Editor Components

The ability to save state-configurations and output-statitics of a simulation is an important convenience for the interactive exploration of a multi-agent model. In particular, the modeler often wants to try many variations of initial conditions and explore their impact on system dynamics. Bedrock has both the input-editor and the output-editor component descended from a simple text edit-box that can load and save physical files. The input-editor has methods for transferring a character-based representation of the simulation configuration to the application. Methods of the input-editor load a simulation configuration, including a specification of the agents, their initial properties and connections, into Bedrock's data structure. Although not strictly necessary, the input-editor allows an interactive edit of the information loaded into the simulator. The output-editor, meanwhile, is simply a chalkboard for the results of the simulation including statistics and state configurations. An example will best illustrate simulation specification in Bedrock, and will also be useful later in demonstrating other application features.

Conway's "life" is a cellular automata model that neatly demonstrates emergence with simple rules (c.f., Gardner, 1983). No entirely original results will be shown in this simulation, but Conway's "life" will be helpful in illustrating simulation construction in Bedrock. First, I will describe the simulation environment: a 5X5-cell square matrix of cells (agents), each with 8 neighbors defined as the cells immediately around the target. The square matrix space "wraps" around the borders in a torus. In other words, all top row cells have bottom row reciprocated neighbors. Likewise, far-left column cells have far-right column reciprocated neighbors. The transition function of life operates as follows: if a cell is empty (states –1) and three of eight neighbors are occupied (state 1), then that cell becomes occupied in the next round (state 1). If a cell is already occupied (state 1), however, and either two or eight of the cell's neighbors are occupied (state 1), the cell becomes empty on the next round (state –1). These two statements comprise a complete description of the transition function. Intuitively, cells live only if properly supported but not smothered by immediate neighbors. As illustrated later, these rules can give rise to the emergent phenomenon of "gliders", or self-perpetuating and repeating patterns of occupied cells that move diagonally across the matrix.

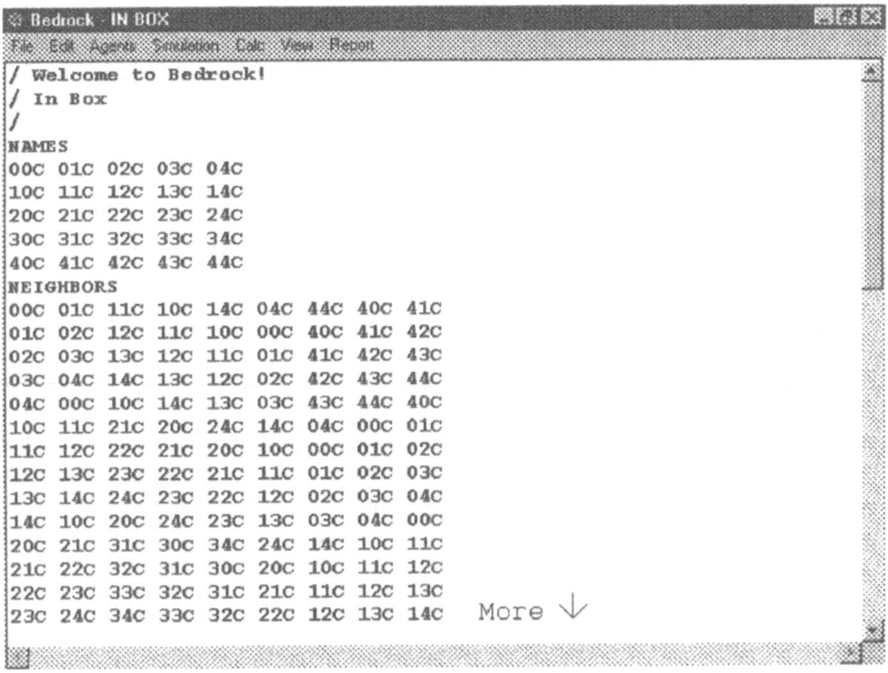

Figure 2: Input Editor of Bedrock

Figure 2 illustrates initial state specification for the Conway life simulation in text format within the input-editor. The first line indicates that the simulation configuration will contain agents with the target "names", or identifiers, on the lines that follow. Next, the input editor would interpret the command "neighbors" to mean that the trailing lines show the connections among agents. Specifically, each subsequent line first identifies the agent of focus, and next indicates the eight other connected agents. Connection weights, typically found in neural networks, are not needed for the life simulation. Lastly, although not shown in figure 2 due to space limitation, initial states (i.e., unoccupied –1; occupied 1) for each cell are loaded into the application's data structure. The input-editor provides a simple way to store a state configuration of a simulation. With a mouse-click, these states can be loaded into the application.

7.7 Agent-Editor Component

The initial specifications of agent states and connections are loaded into the application from the input-editor. The agent-editor component of Bedrock provides a complementary facility for the on-line viewing of the states of the agents already loaded in memory, and provides methods to interactively edit these agents.

Figure 3 illustrates Bedrock's agent editor. The user can select agents from the list and view states over the life-course of the simulation. Figure 3, for example, presents the states of agent "1A" for six topics (rows) over five simulation rounds (columns). The editor can serve as a quick device to track down aberrant simulation results, or inappropriate agent values. The buttons illustrated in figure 3 allow the user to interactively create, destroy, and alter the relationships among agents. Similar to the input-editor, the facilities of the agent-editor make it easy for the user to change the initial conditions of the simulation.

7.8 Population Component

In most business organizations there are some employees that deal directly with the customer, taking the product-orders for example, and others that do the job of actually building the widgets. The component organization of Bedrock is similar. I have already discussed how the user can interact with the input-editor and agent-editor to initialize a simulation configuration. The actual "work" of constructing agents and assigning states, however, is performed behind the scenes by the population component.

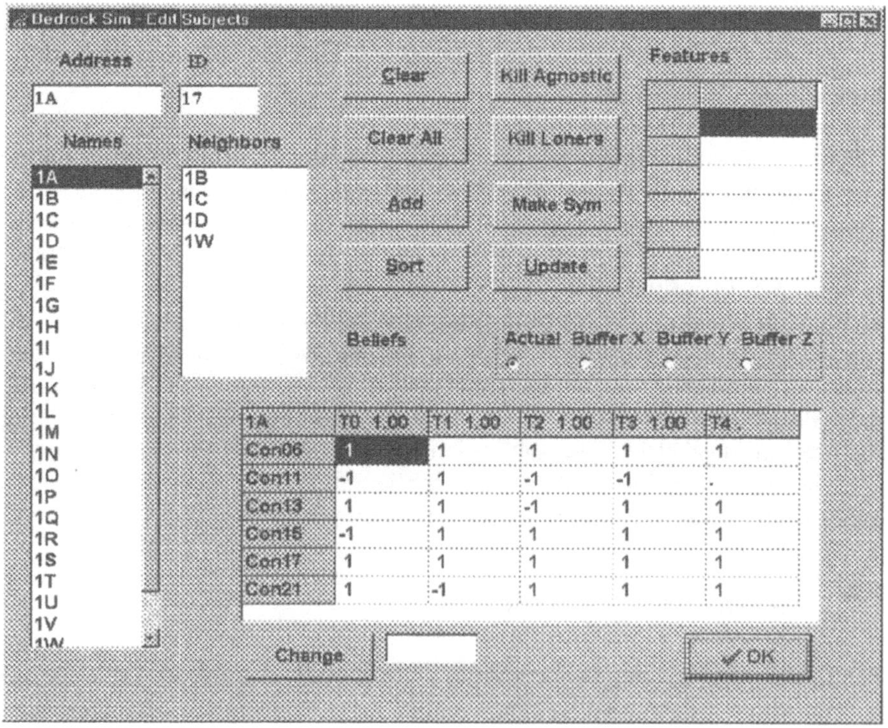

Figure 3. Agent Edit Screen of Bedrock

This component exclusively manages the agents of the simulation. Methods of the population create, destroy and sequence the updating of agents. Although the population component is not visible to the user, it is the focal point for information flow among the components of Bedrock. As the population is the only component with pointers to individual agents, it must pass agent properties indirectly to all other components. Thus the input-editor, output-editor, and agent-editor described above do not directly interact with agent objects (see figure 1), but simply request information and changes from the population. Additionally, the population mediates communication among agents during a simulation-run. Agents are not directly linked to one another, and thus pass relevant state information (the signal) to the population for transmission to their neighboring agents. Communication of information among agents is, of course, central to the dynamics of the multi-agent model. In a cellular automata model, for example, the mutual influence among cells typically creates a regional stratification, or clustering, of states. Communication in a neural network model commonly produces emergent prototype patterns that are useful for identifying structure in noisy data.

7.9 Agent Component and AGENT.DLL in Detail

The most central object to the dynamics of a multi-agent model is the specification of the agents themselves. Separate agent components store both simulated and actual data on individual beliefs, moods, attitudes, and motivations (states). In addition, an agent has select information about the states of other connected agents. In a multi-agent system, agents are repeatedly exposed to new information about the evolving states of other agents, and adjust their properties accordingly. In a game theory simulation, for example, an agent might decide to "cooperate" or "defect", based on the previously witnessed "cooperative" behavior of other agents (c.f., Axelrod 1984, Fischer & Suleiman, 1997). If agents must gather information from others, how do they have access to this information? How should agents communicate with one another through interfaces? An obvious solution is to let each agent have access to some subset of properties of every other agent. This can become cumbersome, however, as each agent must hold information on how to address or find every other agent. The population component, fulfilling its role as the administrator, acts to mediate relationships among agents. As illustrated by the example below, an agent sends to the population component its state-information, called the "signal", which may affect another agent. The population component in-turn can access other connected agents and, acting as a broker, transmit these results to connected agents.

Once the initial conditions of both states and connections in the simulation have been set, a simulation run can begin. A simulation run consists of a user-specified number of rounds. During each round, the population component cycles through its inventory of agents. For each agent, a string-based agent description is passed to the AGENT.DLL. To make the transmission of agent information as user-friendly as possible, Bedrock hands a 256-character string containing agent properties to the user-created DLL file. I call this string the "football." Holland (1998) provided the inspiration for this string-based description of the agent, although Bedrock's agent-strings differ in detail. Any of a large number of schemes might be appropriate for the encoding of agent properties and connections on a string. I selected a 256-character string because it should be easy to find support for this parameter type in any language. For simplicity in programming and debugging, I opted to have the football encode all numbers in text format (e.g., "-1") rather than as binary numbers. This arrangement, of course, is an inefficient use of the 256-characters and limits agents to 10 neighbors and 20 states at one time. Other more compact schemes may be implemented in the future. The format of the agent football is illustrated below (figure 4).

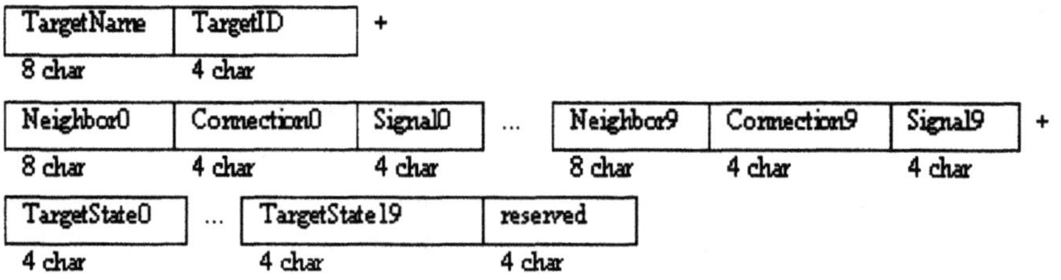

Figure 4. Agent-Description String (a.k.a. Football) of Bedrock.

The football compactly describes the current state of an agent to the AGENT.DLL. The transition function embodied in the AGENT.DLL file specifies how the agent of focus changes over a simulation round. Simply changing the agent description in the football modifies the state of the agent for the next round. Although seldom necessary, agents can both be dynamically renamed (TargetName Property) and given a new future execution order (TargetID Property). More commonly, however, connections to neighbors are added or deleted (Neighbor Property) and connection weights and states are modified. The signal property of the football is the input received from each neighbor (0 through 9) on the previous round (e.g., neural input). The signal, modified as appropriate by the DLL, will be sent back to neighbors for their prospective use.

For Conway's life-simulation, a sample AGENT.DLL is provided below. In brief, the program works as follows: The football is received containing the current states of the target cell (agent) from the Bedrock application. The input signal is read off the football from the 8 neighbors. Next, the program reads the current state (-1 empty, 1 occupied) of the present cell. As described previously, the new state of the cell is calculated from the two rules of life. Intuitively, cells live if supported, but not smothered, by immediate neighbors. The football is subsequently updated with the new state of the cell. Lastly, the signal property of each neighbor in the football is also modified to the new state. In other words, the same string-space that was received as input is now used to send the output signal (-1 empty, 1 occupied) to connected neighbors for their use on the next round. This example illustrates how a simulation can be programmed in Bedrock with a minimal amount of code.

7.9.1 AGENT.DLL: Conway's Life-Simulation Programmed in Delphi

```
library agent;
uses SysUtils, Classes; {standard includes}

function agentcall(football:shortstring):shortstring;export;
var Input:array [0..7] of integer;
   i,State,InputSum:integer;
begin
InputSum:=0;
for i:=0 to 7 do {get input signal from 8 neighbors}
 begin
 if copy(football,25+i*16,4)<>'    ' then
  begin
  Input[i]:=StrToInt(copy(football,25+i*16,4));
  InputSum:=InputSum+Input[i]; {occupation in neighborhood}
  end; {if}
 end; {for i}

 State:=StrToInt(copy(football,173,4)); {get current state of cell}

{transition function follows}

if (State=-1) and (InputSum=-2) then
  begin
  delete(football,173,4);
  insert('   1',football,173);
  state:=1;
  end;

 if (State=1) and ((InputSum<>-4) and (InputSum<>-2))
  and (InputSum<9999) then
  begin
  delete(football,173,4);
  insert('  -1',football,173);
  state:=-1;
```

```
 end;

{output signal to neighbors}
for i:=0 to 7 do
 begin
 delete(football,25+i*16,4);
 insert(format('%4d',[state]),football,25+i*16);
 end; {for i}


agentcall:=football; {assign return string to function}
end;

exports
 agentcall;

begin
end.
```

7.10 View Component

So far I have described how a Bedrock simulation is specified and run. After completing a simulation-run, however, the application-user will be interested in examining the history of agent properties as well as relationships (i.e., connections and weights). This examination allows us to extract, for example, the emergent phenomenon of gliders in the Conway-life simulation.

Figure 5, below, shows a graphic representation of the states of cells (agents) for Conway's life simulation. The icons, or "views," each represent a single cell. The bars are color coded to show the configuration of states (empty/occupied) for each cell. Finally, the views are arranged in a matrix where cells are generally closer to one another if they communicated during the simulation. The icons represent the properties of the agent. The icons relative positions on the screen reveal their agency relationships or interconnections. Inspection of figure 5 reveals the emergence of a so-called glider. Round-by-round, a grouping of 5 occupied cells migrates down the matrix diagonal. As revealed by comparing the two rows constructed in figure 5, the specific cell pattern repeats itself each fourth round. Typical of many dynamic systems, simple rules operating on the agents (cells) gives rise to a surprising macro-level phenomenon. The views of Bedrock provide a window on emergence that is not easily obtained through numbers.

7.11 Collection Component

Views are arranged on the application canvas, as described above, to reveal both dynamics and emergent state relationships among agents. Specific view configurations are created interactively according to the demands of the application user. Many different compositions of views can be constructed to reveal system properties and the collection component is needed to manage these arrangements. The collection component has the primary responsibility to manage the life cycle of views in a manner highly analogous to the population component's stewardship of agents. The population component and collection component both descend from a common ancestor that more generally manages lists of objects. Collections, apart from populations, have the added functionality of arranging their stable of objects (views) on the screen relative to agency relationships or the demands of the user. Because the collection component must be able to find the properties of agents, the collection communicates with the population to retrieve agent information. The collection can arrange views, as already demonstrated in figure 5, to show the progression of states in a simulation over time. In Bedrock, other types of graphics (not shown) can be mixed on the canvas to present alternate forms of state information and connections.

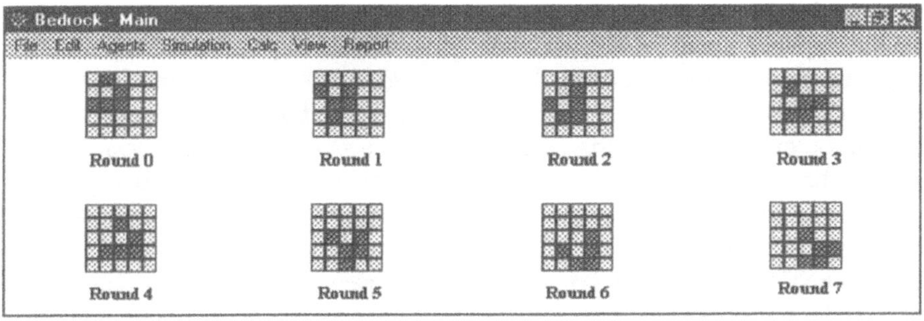

Figure 5. "Color Bar" icon view of agent data in Bedrock (labels added)

7.12 Statistics Component

A critical task for the user of a simulation application is finding ways to interpret the results. Graphic representations as provided jointly by the collections and view components may provide some information, but often the user would like to summarize and describe the system with numbers. Other statistics can provide a goodness-of-fit measure between simulation results and actual data. As statistics may involve complex computational routines, the functionality is best encapsulated in a new component.

The user may be interested in testing whether the relationships in the network have any effect on the properties of agents and the sizes of these effects. For example, the user might want to know if actual friendships (connections) have an effect on the development of shared beliefs (states). In Bedrock, a relatively simple realization of this statistic is obtained through a bootstrapping procedure (c.f., Mooney & Duval, 1993). First, a sum total count of all differences between agent properties is obtained for all relationships in the network. In other words, a procedure cycles through each agent and sums the state discrepancy between that agent and the others with which the agent has a connection. This difference count is a measure of the total discrepancy in the network for that agent property. Next, agents are temporarily and randomly assigned to new positions in the network. The properties of each agent remain the same, but each agent may occupy a new social role within the network. As before, a sum total count of the discrepancy between agent states and those of their neighbors is calculated for the entire network. This count is added to a compiled list of other random arrangements, repeatedly calculated to form a distribution. The distribution represents the expected property discrepancies among agents in the population if the original network configuration of relationships had no net effect on agent properties. By sorting the distribution (e.g., 1000 runs) from lowest to highest and finding the position of the original discrepancy count within the distribution, a probability value can be obtained for the given values in the network. If, for example, only 10 of the 1000 randomized runs obtained a discrepancy count lower than the actual discrepancy count in the network, then relationships affected net shared properties among agents, termed clustering, with a probability value of .01 (i.e., $p=10/1000=.01$). Special care, however, is taken to adjust for the fact that the distribution of random discrepancy counts often does not well approximate continuity. If the actual discrepancy count is encountered more than once throughout a range in the compiled distribution, the value is approximated at the midpoint of that range. Results using this bootstrap procedure are highly congruent with traditional ANOVA procedures when agents are arranged in discrete groups. The logic of the procedure parallels statistics devised in the social networks tradition (c.f., Wasserman & Faust, 1994).

The last critical task of the statistics component is to provide the user with some indication of goodness-of-fit between simulation results and actual data. Simulations can be considered as performing a function analogous to traditional linear statistical models in providing predictions of actual results. The statistics component has the task of assessing the goodness-of-fit between the predicted data and actual. A useful statistic realized in Bedrock is a simple average correlation between simulation and actual. Although correlation eliminates differences in means and standardizes variances before judging a relationship, the results of a correlation can be diagnostic in determining fit when the simulation model may be off by some linear transformation. The statistics component must interact with the population in order to collect data for analysis. The resulting statistics are communicated to the user through the output-editor.

7.13 Future Development

I have already hinted at the practical limits to the programmable application. As Bedrock's multi-agent framework can reproduce the functionality of a general purpose computer, in principle, any design can be programmed into the application (Holland, 1998). However, the facilities of Bedrock most naturally lend themselves to particular pursuits such as neural networks and cellular automata. Although the current approach is domain general and capable of easily reproducing many of the models of emergence explored so far, it is conceivable that another framework may prove more able for some simulation designs. Some specifics of Bedrock's present design, although not an indictment of the general simulation-application approach, can also frustrate modeling. The practical bounds of a 256-character football, for example, constrain the number of states and connections an agent can maintain. New conventions for passing state information to the transition function might expand these limitations. Perhaps most importantly, the views of Bedrock are currently limited in scope and number. A user may not find an appropriate or necessarily revealing way to explore their simulation results. In the future, I expect to make a system for plug-in code that allows the users of Bedrock to construct original view types. Another practical limit is the ease with which data and graphics can be transferred to other applications for further analysis. The ability to save simulation configurations, output statistics, and graphics in common formats (e.g., Excel) will increase Bedrock's ease of use. Despite the current limitations of Bedrock, I have found the application to be highly convenient for simulation development (c.f., Rockloff & Latané, 1996). I expect the usefulness and generality of the platform will dramatically increase given the planned enhancements.

Conclusions

It will never be possible to satisfy all simulation enthusiasts with one tool. Just as a garage full of gadgets is needed to service different makes of autos, many separate tools are required for the myriad of possible simulations. The requirements of developers necessarily reflect the variety of their interests and designs. The programmable application concept, however, suggests a common template for a compact description of agents (via the AGENT.DLL) that can be used across future applications. As demonstrated by Bedrock, analysis-tools, graphics, and statistics can be programmed independently from the transition function of the agent. Programmable applications of different origin can operate on the same agent descriptions, just as different spreadsheets and word processors can utilize the same types of data. Bedrock is a working example of a programmable simulation application. It serves as a potent illustration of the utility of applications for simplifying simulation construction and description. With this new type of tool in hand, we can more confidently drive into the future of multi-agent modeling.

References

Axelrod, R. (1984): The evolution of cooperation. Basic Books, New York.

Fischer, I. & Suleiman, R. (1997): The emergence of mutual cooperation in a simulated inter-group conflict. In: Gilbert, G. N., Müller, U., Suleiman, R. & Troitzsch, K. G. (eds.): Social Science Microsimulation: Tools for Modeling, Parameter Optimization, and Sensitivity Analysis. Wadern, Germany, Dagstuhl-Seminar- Report, 177.

Gardner, M. (1983): Wheels, life, and other mathematical amusements. Freeman, New York.

Gilbert, G. N. (1996): Environments and languages to support social simulation. In: Troitzsch, K.G., Müller, U., Gilbert, G. N. & Doran, J. E. (eds.): Social Science Microsimulation. Springer-Velag, Berlin, 457-458.

Holland, J. H. (1998): Emergence: from chaos to order. Addison- Wesley, Reading.

Minar, N, Burkhart, R., Langton, C. & Askenazi, M. (1996): The Swarm simulation system: a toolkit for building multi-agent simulations. Santa Fe Institute Working Paper.

Molnar, P. (1996): A Microsimulation tool for Social Force models. In: Troitzsch, K.G., Müller, U., Gilbert, G. N. & Doran, J. E. (eds.): Social Science Microsimulation. Springer-Velag, Berlin, 155-170.

Moehring, M. (1996): Social Science Multilevel Simulation with MIMOSE In: Troitzsch, K.G., Müller, U., Gilbert, G. N. & Doran, J. E. (eds.): Social Science Microsimulation. Springer- Velag, Berlin, 123-137.

Mooney, C. & Duval, R. (1993): Bootstrapping. Sage university paper series on quantitative applications in the social sciences, series no. 07-095. Sage Publications, Beverly Hills and London.

Rockloff, M. J. & Latané, B. (1996): Simulating the social context of human choice In: Troitzsch, K.G., Müller, U., Gilbert, G. N. & Doran, J. E. (eds.): Social Science Microsimulation. Springer-Velag, Berlin, 359-377.

Wasserman, S. & Faust, K. (1994): Social Network Analysis: methods and applications. Cambridge University Press, Cambridge.

8. Diversity in Strategies of Partnership Formation*

Rosaria Conte[1]

8.1 Rationality, Agent Theory, and Social Action

In this paper, the paradigm of rationality is examined from the point of view of designing, and possibly implementing, agents endowed with the capacity to interact in a multi-agent environment. The work that will be presented originates from a cross-fertilization between the fields of social simulation and Multi-Agent Systems (MAS) research. In the MAS field, models of *intelligent* agents in a multi-agent context have been formalized and implemented. The question then arises as to whether the model of *rational* agents may contribute fruitfully to construct and implement multi-agents systems.

In Castelfranchi & Conte (1998) and Conte (1998), some general merits of Game-Theory (GT) as an elegant and sound formal approach to social action are re-stated. In those works, however, some general critiques with regard to this paradigm were also formulated, both at the social and at the non-social level. Some of those critiques are far from original (computational limitations, multiple equilibria, etc.; cf. Gmytrasiewicz,1995). Others are perhaps less obvious and refer to three specific sub-features of the current notion of rationality, namely:

1) *formal/instrumentalist*: (in the sense meant by Moser 1990): owed to this property, no guidelines are provided about how real agents should be designed, how their utilities and their concrete alternatives for choice are formed;
2) *unsituated*: rationality is seen as an absolute notion. What is rational, with regard to what, and for which reason (see also Brandt, 1983; Bratman, 1987)?

---

* This work has been realized with the contribution of the Vigoni program on the "Evolution of elementary morals" between the Department of Philosophy of Bayreuth and the IP, Cnr of Rome.

[1] Division of AI, Cognitive and Interaction Modelling -PSS (Project on Social Simulation), IP/Cnr, V.LE Marx 15 - 00137 Roma.
email: rosaria@pscs2.irmkant.rm.cnr.it
http://pscs2.irmkant.rm.cnr.it/users/rosaria/home.html

3) *uni-dimensional*: although rationality is essentially based upon subjective utility functions and bounded by cognitive limitations, it is indifferent as to the various motives of the agents.

In the next section of this paper, some further aspects of the rational approach will be focused upon, related to the third aforementioned property of rationality, namely its uni-dimensional character. In particular, GT's failure to account for the agents' *qualitative*, *heterogeneous* motivations for action, assessed *independently of their choices.*

In the third and fourth sections, some amendments will be proposed. In particular, diverse strategies for decision-making will be distinguished: a classical *utility-based strategy*, in which agents tend to maximize their utility, and a *goal-based strategy*, in which *utility is subordinated to goal-satisfaction*. These strategies will be compared in terms of a simulation model of partnership formation. The related findings will be presented, and some general conclusions discussed.

8.2 Preferences As Interchangeable Goals

Game theory works under the assumption that agents try to maximize their utility *whatever their preferences are*. The notion of preference is an unsatisfactory *motivational* notion, for a number of reasons which will be briefly examined below. For a discussion about preferences as compared with goals, and about the advantages of the latter over the former notion, the reader is pointed to Conte (1996).

First, preferences are defined only in *quantitative*, rather than symbolic, terms. They are not used by the scientist to describe or specify an internal process, a series of mental operations which lead an agent to accomplish a given choice, but only to feed the function of utility.

Second, and consequently, preferences are *interchangeable*, while qualitatively defined motivations are not. Once the function of utility is applied to ordered preferences, preferences can be substituted to one another in order to maximize utility. If reading a book while listening the music gives me a higher overall utility than taking a nap, I should forget about sleeping no matter how tired I am. However rational, such a choice is simply unfeasible, because the goal of taking a nap and that of cultivating one's mind are not interchangeable! People do not *maximize their utility while satisfying their preferences.* Quite on the contrary, *they try to satisfy their goals while possibly maximizing their utility.* Actions may be controlled at a meta-level by utility maximization, but are always triggered and ruled by goals, that is, by mental symbolic (qualitative) and not interchangeable representations of wanted world states.

Finally, preferences are often *assessed post hoc*, that is to say, they are inferred from behaviors. If you choose A rather than B, you must have preferred A over B, other things being equal. This feature renders preferences a strongly behavioral notion, and endangers the falsifiability of the theory. One can always impute violations of expectations to unpredicted preferences, rather than to wrong or inadequate models. What is more, the predictive value of the theory is weakened: if the assessment of one's preferences depends upon one's displayed behaviors, behaviors will not be predicted but only rationalized.

8.3 Amendments: Diversity in Rationality

It has been argued (see Castelfranchi &Conte, 1998) that rationality is usually meant as a uni-dimensional notion, which excludes or neglects diversity in several different ways. It neglects the variety of context in which some feature or event is evaluated as rational; it is indifferent to the variety of goals, motivations, etc. within and between the agents. Motivations are considered interchangeable and always comparable on a quantitative scale.

A notion of rationality which does not rule out diversity is therefore called for.

There are many senses in which diversity must and can be integrated with a theory of rationality accounting for:

(a) *substantial differentiation*: qualitative heterogeneity among individual agents' goals; to do this, a goal-based rather than preference-based, view of motivations should be provided. The qualitative difference among goals will allow a *variety of social actions* to be described and predicted: in GT, only the alternative between defection and cooperation is possible. But social life is interspersed with different types of pro-social action, from *influencing*, to *exchange*, to *cooperation*.

(b) Architectural, or *formal, differentiation*: agents do not only differ in the contents of their goals but also in their architectural properties. Therefore, they do not only have different goals, but also different decision-making rules. (Notice that in evolutionary game-theory, different strategies for interaction are allowed but the principle of utility maximization in decision-making is never questioned!) In our view, a further principle (*goal-based rationality*) exists, in which *utility is subordinated to goal-satisfaction*. rather than vice versa. This shows that goals are not always interchangeable. A qualitative, symbolic representation of goals renders interchangeability possible but not necessary.

(c) A *context-bounded* notion of rationality, such that different contexts call for different rational strategies.

In the rest of the paper, our preliminary contribution to these objectives -- especially to the first and second - will be illustrated. The reader is also addressed to (Sichman et al., 1994; Conte & Sichman, 1995; Conte &Castelfranchi, 1996, etc.) for further details about the models presented here.

8.4 An Agent-Based Approach

In our view, a (social) *action* (Castelfranchi, 1997) is defined in terms of (social) goals (cf. Bratman, 1987).

A *goal* is a symbolic representation of the world, which the agent wants to be true. A goal is not yet an action. Goals are ordered according to precedence or preference relations; but a goal is not primarily defined by its order position. A social action is defined as aimed to modify another agent's mind.

A (social) *agent*, in a quite elementary sense, is defined as characterized by goals, actions, (planning) knowledge and resources (see also Doran, this volume). The notion of agent here referred to is defined as: (a) an ideal-type construct - not to be confused with the subjective, idiosyncratic individual of differential psychology; and (b) an AI-based notion, where the attention is drawn on the whole process that leads a system to acting, and on its internal makeup, i.e., the internal regulatory mechanisms and representations allowing a system to act adaptively in its environment.

8.4.1 Formal Diversity: Goal-Vs Utility -Based Strategies

There is *more than one type of rational agent* (a claim which is now shared by several authors; cf. (Hargreaves-Heap, 1993 and Binmore's (1987) argument that some games are solved by players "departing" from a presumption of common rationality); and moreover, that there is *more than one type of rational decision-making*. Actually, there are several strategies of decision-making depending on the criteria built into the deciding system. Let us examine at least one general criterion, compared with the classical utility maximization, namely: the *goal-based strategy*.

While in the utility-based principle *goal-satisfaction is subordinated to maximizing utility;* and the real goal is just to maximize utility,

$Max\ U(X_1, ..., X_n; I_t)$

where Max stands for the maximization principle; $X_1, ..., X_n$ *the set of alternatives, with their relative probabilities; and* I_t *the available information;*

in the motivation-based rationality the system tries to achieve the most-valued goal; utility is subordinated to goal-satisfaction: motivations are not equivalent/interchangeable. This strategy consists of choosing in favor of the most-valued goal,

$Most\text{-}V(G_1,...,G_n; K_t; A_1,...,A_n)$

where Most-V stands for a choice function; $G_1,...,G_n$ *for a set of alternative goals;* K_t*; the knowledge available to the agent at any given time (including-knowledge about the world, as well as planning knowledge);* $A_1,...,A_n$*;the set of actions available to the deciding agent.*

Therefore, given a set of alternative *active* goals (Castelfranchi, 1996), the agent will decide to achieve the most-valued one, provided it is believed to be achievable, convenient, and unfulfilled. Suppose, x has a set of ordered (active) goals: $g_1 > g_2 > ... > g_n$. Suppose that x prefers to go out with Jim over dining with John, but finds the latter alternative more beneficial as for the rate between cost and incomes (gain). The two principles allow for two different predictions: the utility principle leads to choosing John, while the goal principle enables x to choose Jim.

A *basic* theoretical question arises here: what is the difference, if any, between maximizing one's satisfaction and maximizing one's utility? etc. In the following section, this question will be addressed to some further detail.

8.4.2 The Interplay Between Goal-Satisfaction and Utility-Maximization

The distinction introduced above is rather more complex than might appear. Indeed, the two strategies in question are actually intertwined in the real world. In the following, a variety of substrategies will be analytically derived from the interaction between these main strategies.

(a) *Blind hedonism (or, addiction).* The most-valued goals will always be achieved: given $g_1 > g_2 > ... > g_n$, g_1 will always be chosen even if it has negative consequences on other goals

(b) *Prudent (or, self-defensive) hedonism.* The most-valued goals will be achieved only if they are not self-destructive.

(c) *Hard (or obstinate) hedonism.* If the most-valued goal cannot be achieved, exit.

(d) *Next-best hedonism.* If the most-valued goals cannot be achieved, then choose the next-best.

(e) *Failed hedonism* (or, utilitarianism by indifference). Utility is maximized only when hedonism is inapplicable. If there are no (significant)discrepancies in the intensity or urgency of desires, goals, needs, etc., then resort to utility.

(f) *Mixed: hedonistic choice +utilitarian execution.* The most urgent goal is chosen, but it is applied in a utilitarian way. The most-valued goals will be achieved in the most economic way. (The other way around does not make much sense)

(g) *Opportunistic hedonism.* As many goals as possible will be pursued, provided the most-valued one(s) can be achieved.

Let a very important difference between the two heuristics be stressed. *The principle of maximizing utility renders the agent indifferent to both the content and the context of her activity!* Since what counts is the final utility obtained, it should be totally indifferent *how* that value is realized. *The application of a general principle of utility maximization which is indifferent to the content and the context of decision is essentially counter-intuitive*, allows for too complex and useless decision-making. On the contrary, a mechanism of decision-making which is based upon *goal satisfaction* rather than utility maximization seems to avoid some of these consequences: *goals are not interchangeable.*

8.5 A Computer-Simulation Study of Partnership Formation: The MICROdep System

The present study aimed at exploring alternative mechanisms of decision-making and observed their consequences in the context of partnership formation.

In a previous work, a computational system, MICROdep (cf. Conte et al., 1998), has been developed on the tracks of DEPNET (for a description of this system, see Sichman et al., 1994) and applied to:

(a) *Describe a "social market"*, that is, a set of social agents as endowed with different goals and actions (*substantial* diversity) and calculate the emerging dependence relations among them. In fact, in a heterogeneous social world, agents which do not have all the actions needed to achieve their goals may depend on others, namely on those agents which can do the actions required.

(b) *Calculate the agents' preference lists* of possible partners, as determined by the dependence relations mentioned above. Each list of preference is formed according to either a *goal*- or a *gain*-oriented criterion - that is, either by considering how many goals the agent will have satisfied by each possible partner (independent of how much the agent pays for it) or by considering how much each agent can receive by the other compared to how much it should pay. The agents one is not dependent upon have a null score on one's preference list. Completely autonomous agents have a null preference list. Each item in a list is endowed with:
   - a *partnership value (p.v.)*,that is, the sum of each agent value (a.v.) in the preference list of her partner, and
   - the absolute difference between agents' values, that is the difference, considered in absolute value, between the agents' values as shown in its partner preference list;

(c) *Form real partnerships*. MICROdep simulates the formation of real partnerships with a *next-best* procedure, that is, by repeatedly pruning a general matrix formed by all agents' preference lists. In each subsequent pruning, the relative best choices are formed (highest partnership values and lowest difference between agents' values) and removed from the market. The matrix is updated after each pruning, and this function is applied again and again until partnerships are possible (dependent agents exist and match one another).

In our simulations, the agents' values are determined neither from the others' capacities to help them (the number of agents they depend upon), nor from their own capacities to help others (the number of agents which depend upon them) only, but from the intersection between these two factors. The more the agents which need one's help among those one is in need of, the higher the chances that one will find people with which it is both possible and convenient to interact. Consequently, the more likely one is to achieve one's goals. This intuition has been incorporated into the notion of *negotiation power*. A single agent's negotiation power (negot-power for short) is a function of the intersection between the set of agents one *OR*-depends upon (the set of agents each sufficient but unnecessary for one agent to achieve its goals) and the set of agents *CO*-depending upon one (each depending upon one to achieve their goals). The larger the intersection and the higher the agent's negot-power (see Conte & Castelfranchi, 1996). In order to easily compare the agents' negot-powers within one social market, the intersection is weighted against the total agents' number -1. (However, if agents from different markets need to be compared, the formula should be modified).

8.5.1 The Predictive Role of the Negotiation Power

MICROPdep has been worked out in order to simulate the formation of partnerships in social markets. First, the dependence network for a given social market and each agent's negot-power are calculated; secondly, an algorithm for identifying each agent's list of preferred partners is applied; finally, an algorithm for extracting actual partnerships out of the global list of preferences is applied.

In a previous study (see Conte et al., 1998), the central hypothesis underlying MICROdep - namely that the dependence network allows the formation of partnerships to be predicted - was confirmed. More specifically, each agent's negot-power in a given market was checked against:

(a) The frequency of that agent's appearance in the others' preference lists, that is, the degree of its social utility: the higher the agent's negot-power, and the more frequently it should appear;

(b) The probability of its appearance in the partnerships actually formed, that is, the extent to which the agent is able to form real partnerships: the higher its negot-power and the more chances it has to find partners;

(c) The rewarding potential of actual partnerships: the higher one's negot-power and the higher the outcomes one should get from the partnerships entered.

Findings from the simulations confirmed that the negotiation power shows a significant correlation with all these values. Let us describe with some detail what the system does

8.5.2 The Social Market and the Dependence Network

The system includes three main functionalities: (a) "reading" a social market, namely the agency which the user feeds the system with, and calculating both the dependence network corresponding to that market, and the single agents' negot-powers; (b) identifying the agents' preference lists; (c) calculating the actual partnerships.

Let us see each of these functions at some length.

MICRODEP reads the description of a given market, consisting of a given number of agents with their individual characteristics; agents may differ in terms of *what they have*, that is, their actions and resources, and *what they want*, that is, their goals.

Then it computes whether any of the agents is autonomous for reaching its goals. In our terms, an agent is socially autonomous when it has got all actions required to achieve its goals; an agent is socially dependent when it is lacking at least one action required by its goals and someone else has it. Within our perspective, an autonomous agent has a null negot-power, because it doesn't need to create partnerships: she calls itself "out of the game".

Table 1: External description of a 3-agent market

Market	Actions/Res.	Goals	Dependence
Agent 1	a	c	Y(2)
Agent 2	c	a	Y(1)
rrAgent 3	b	a	Y(1)

Agent 1 has the action/resource a, achieving the goals of both Agent 2 and Agent 3. But Agent 1 needs only Agent 2, whose action/resource fulfils Agent 1's goal. Therefore, Agents 1 and 2 should have the same negot-power.

MICRODEP computes now the NegotP of the three agents.

$NegotP_{(1)} = 0.5$,
$NegotP_{(2)} = 0.5$,
$NegotP_{(3)} = 0$.

8.5.3 Heuristics for Social Choice

MICROdep has also been applied to investigate two "characters" or heuristics which agents may resort to for choosing their partners (architectural diversity). These heuristics are treated as two distinct criteria for producing each agent's preference list of possible partners, the *goal*- and the *gain*-oriented criteria. In the former, each agent orders its potential partners according to the relative goal-satisfaction it can get out of each of them and, but only as a second-order factor, to the costs it should sustain for those partnerships. For example, if $agent_i$ obtains from $agent_j$ the satisfaction of four goals, and from $agent_k$ the satisfaction of three goals, it will prefer $agent_j$ over $agent_k$ even though the payoff of the former partnership happens to be lower (suppose she must reward $agent_j$ with, say, three actions) than the payoff of the latter (which we may suppose to cost, say, only one action). In the gain-oriented strategy, instead, the goal-satisfaction has no priority over the principle of utility. Therefore, in the above example, the preference order would be inverted, and $agent_k$ would be preferred over $agent_j$. With the strategy in question, we endeavored to account for one important difference between rational agency and other architectures of intelligent systems, namely the difference between an abstract, uni-dimensional function of utility as usually attributed to rational agents, and a variety of specific goals as usually implemented in intelligent natural systems such as humans. Usually, agents do not act in order to maximize their utility, but rather to achieve the most important[2] of their goals, if more than one.

By definition, the gain-oriented strategy is more advantageous than the other from a strictest utilitarian perspective. Overtime, the agents following this strategy should sustain lower costs for entering partnerships than the agents following the alternative option. The present model offers the possibility to compare the two heuristics in terms of negotiation power and partnership formation. Some major questions concerning possible differences and/or similarities between the two strategies could be explored, namely:

(a) Are goal-oriented agents significantly different form the gain-oriented ones? And, if a difference between the two heuristics emerges,
(b) What is their relative behavior with regard to partnership formation? In other words, who is favored in partnership formation, goal-oriented or gain-oriented agents?

8.5.4 MICROdep Preference Lists

Each agent has a preference list formed by all other agents, ordinally arranged, within the same market. In creating the agent's subjective preference list, either the goal- or the gain-oriented heuristics must be chosen, so that each list is formed either by considering how much each agent can receive by the others, or else, how many goals can be satisfied (independent of how much the agent pays for it). The agents one is not dependent upon have a null score on one's preference list. Completely autonomous agents have a null preference list. The preference lists of the agents are given by the decreasing ordered series of those agents which they gradually depend upon. In the goal strategy, for example, if $agent_i$ depends on $agents_{k,\ l,\ m}$, that respectively fulfil two, four and three goals of $agent_i$, its preference list will be serially formed by $_{l,m}$ and $_k$. In the gain strategy, if $agent_i$ must give them back respectively one, four and five actions, its preference list will be serially formed by $agents_{k,\ l,\ m}$.

[2] Unfortunately, goal-importance is only partially expressed by the goal-oriented strategy as implemented in MICROdep at the current level of its development, since the importance of the goal is transformed into the number of goals. Future advances of the system are meant to implement goals with different values. However, even within the present restrictions, some comparison between the two strategies could be made.

8.5.5 MICROdep Algorithm for Partnership Formation

Let us now consider a matrix M formed by n rows, corresponding to the preference lists of the n agents in the market. Each preference list is a vector containing at most n-1 non-empty items, that is, as many as the (n-1) agents in the market that a single agent can choose as partners for exchange.

To the non-empty items included in the first column (that is, to the "first best" in each list of preference), a value (*agent* value, a.v.) equal to n-1 is associated; to the non-empty items of the second column, an a.v. equal to n-2 (second-best) is associated, and so on up to the non-empty items of the last column, whose a.v. is equal to 1. The empty items have an a.v. of 0 (zero). In Table 2, the matrix of preferences relative to a market including seven agents is shown.

Table 2: The matrix (7,6) of preferences relative to Market A, including seven agents

Agent A	C (6)	E(5)	G (4)	 (0)	(0)	(0)
Agent B	A “	C“	E“	G (3)	 “	“
Agent C	D “	B “	F“	A “	E(2)	G (1)
Agent D	A“	G “	E “	B“	 (0)	(0)
Agent E	F “	G“	A “	C “	 “	 “
Agent F	G“	C “	B “	D“	A (2)	E (1)
Agent G	A “	C“	E “	B “	D “	F “

MICROdep simulates the formation of real partnerships by producing a list of all possible exchange dyads, and by repeatedly pruning this list. In each subsequent pruning, the relative best choice is formed and removed from the market. The matrix is updated after each pruning, and this function is applied again and again until partnerships are possible (dependent agents exist and match one another). Let us see how this is done.

Each possible dyad is characterized by (a) a global value called p.v. (partnership value), equal to the sum of the agents' values (a.v.), and (b) the absolute value of the difference between the agents' values[3] (|D a.v.|). Possible dyads in

---

[3]This algorithm allows partnerships to be formed starting from the agents' preference lists. Undecidable situations, namely those produced by equivalent partnerships where

which one of the partners is not included in the other's preference list are not considered.

In our market, the following possible dyads (see Table 3) with non-null p.v.s obtain:

Table 3: Possible partnerships in Market A

Dyads	p.v.
AC	9
AE	9
AG	10
BC	10
BG	6
CE	5
CF	9
CG	6
DG	7
EF	7
EG	9
FG	7

This is the situation of Market A's potential partnerships at time t_0. All the partnerships that will be actually formed will maintain the p.v. they have at time t_0.

MICROdep updates the matrix M (see Table 3), and produces M^1 (see Table 4), where only agents whose a.v.s, if summed up (S.a.v.), give a non-null value are considered (that is, agents which appear each in the other's list of preference). New a.v.s are now assigned to the items in the matrix.

In M^1, possible dyads are characterized by the following p.v.s, sum and absolute difference between partners' values (S a.v.; |D a.v.|; see Table 5):

preferences form loops, are prevented by selecting the last partnership with the highest p.v. and the lowest difference of a.v.

Table 4: M^1 the matrix corresponding to Market A at time t_1

Agent A	C (6)	E(5)	G (4)	(0)	(0)	(0)
Agent B	C “	G“	(0)	“	“	“
Agent C	B“	F “	A (4)	E (3)	G(2)	“
Agent D	G“	(0)	(0)	(0)	(0)	(0)
Agent E	F “	G(5)	A (4)	C (3)	“	“
Agent F	G “	C“	E“	(0)	(0)	(0)
Agent G	A“	C “	E“	B (3)	D (2)	F (1)

Table 5: M^1's values

Dyads	p.v. (t_1)	S a.v.	|D a.v.|	p.v. (t_0)
AC	10	(6+4)	2	9
AE	9	(5+4)	1	9
AG	10	(4+6)	2	10
BC	12	(6+6)	0	10
BG	8	(5+3)	2	6
CE	6	(3+3)	0	5
CF	10	(5+5)	0	9
CG	7	(2+5)	3	6
DG	8	(6+2)	4	7
EF	10	(6+4)	2	7
EG	9	(5+4)	1	9
FG	7	(6+1)	5	7

Now, MICROdep :

(a) Selects from M^1 the dyad(s) with the highest p.v. and the lowest |D a.v.|;
(b) Removes from the matrix all dyads which have formed real partnerships.

If some dyads are left, a new matrix, M^2, is produced (see Table 6).
In M^2, values are updated (see Table 7).

Now, we have two dyads with the same p.v. and |D a.v.|, namely AG and EF. Once these have been picked, and the four agents which they include have been removed from the matrix, no other possible partnerships are left. The computation stops.

Table 6: M^2 in Market A

Agent A	E (6)	G(5)		
Agent D	G			
	"			
Agent E	F "	G (5)	A (4)	
Agent F	G "	England & Wales "		
Agent G	A "	England & Wales "	D (4)	F (3)

Table 7: M^2's values

Dyads	p.v.(t_1)	S a.v.	|D a.v.|
AE	10	(6+4)	2
AG	11	(5+6)	1
DG	10	(6+4)	2
EF	11	(6+5)	1
EG	10	(5+5)	0
FG	9	(6+3)	3

8.5.6 Findings

In this study, MICROdep was applied to explore the difference between different strategies of decision-making in the formation of partnerships. Two such strategies were examined, the goal-oriented and the gain-oriented. The former is aimed at achieving as many goals as possible independent of their costs, and the latter is aimed at maximizing utility.

Several simulations have been carried out with markets of 10 agents. In Table 8, the average values of the findings (correlations between negot-power and four other values) are shown:

(a) SU1, which refers to the agents' Social Utility of type one, that is, to the frequency with which each agent appears in the list of *preference* of the other agents in the same market formed by applying the *goal*-oriented strategy;

(b) P.V.1, which refers to the values of the *partnerships* formed by following the *goal*-oriented strategy;

(c) SU2, which refers to the agents' Social Utility of type two, that is, to the frequency with which each agent appears in the list of *preference* of the other agents in the same market formed by applying the *gain*-oriented strategy;

(d) P.V.2, which refers to the values of the *partnerships* formed by following the *gain*-oriented strategy.

Table 8: Correlations in small-size markets

	SU1	*PV1*	*SU2*	*PV2*	*Negot-power*
SU1	10000	2745	8190	2509	3345
	p= ---	p=551	**p=0.02**	p=587	p=463
PV1	2745	10000	5601	9922	7751
	p=551	p= ---	p=191	**p=0.00**	**p=0.04**
SU2	8190	5601	10000	5957	7739
	p=0.02	p=0.191	p= ---	p=0.158	**p=0.04**
PV2	2509	9922	5957	10000	8316
	p=587	**p=0.00**	p=158	p=---	**p=0.02**
Negot-power	3345	7751	7739	8316	10000
	p=463	**p=0.04**	**p=0.04**	**p=0.02**	p=---

Marked correlations are significant at $p < 0.05$
N=10 (Casewise deletion of missing data)

As is shown by the Table, the predictivity of all the dimensions is not high. In particular, neither one's SU1 (that is, one's appeal for goal-oriented agents) nor PV1 (that is, the satisfaction values of the partnership really formed by goal-oriented agents) are correlated with one's negotiation power (significant correlations are in bold). SU2, instead, performs better: it is correlated with one's negotiation power, if not with partnership values. Therefore, at least in small-size markets highly goal-oriented agents may not have a high negotiation power, while highly gain-oriented agents do. A conclusion can be tentatively drawn from this finding: goal-oriented agents are requested for specific *qualitative* reasons, in order to achieve a specific (set of) goal(s). Therefore, it is not necessary that they have a large capacity for finding partners, since the others' needs may not be overlapping with their own: a given individual agent which is useful to satisfy others' goals may be depending on them to a higher degree (it may need much more than what it can give). Hedonistic agents are more demanding, less flexible than utilitarian ones: for them, both agents and goals are not interchangeable. In our experiment, where hedonists are of a next best type, obviously they will end up with partners which are not those mostly satisfying. In the complementary typology, this is almost impossible since agents as goals are interchangeable. Therefore, useful agents must significantly intersect with those who have a large potential for interaction. In other words, a goal-oriented strategy is less flexible and more demanding than a gain-oriented one. One could argue that by enlarging the size of the market, chances increase that satisfying agents find good partners for interaction. With middle size markets (30 agents), the first type of social utility never shows significant correlations (see Table 9). But with large markets (50 agents), not only the correlation between the negot-power and each of the remaining four dimensions are found significant (see Table 10), but also the correlations between any dimension and each of the others turned out to be significant, including SU1 and PV1.

Table 9: Correlations in middle-size markets

	SU1	PV1	SU2	PV2	Negot-power
SU1	10000	-0628	-2776	-2619	-0543
	p= ---	p=720	p=106	p=129	p=757
PV1	-0628	10000	7224	7957	6164
	p=720	p= ---	**p=0.00**	**p=0.00**	**p=0.00**
SU2	-2776	7224	10000	8674	6703
	p=106	**p=0.00**	p= ---	**p=0.00**	**p=0.00**
PV2	-2619	7957	8674	10000	7277
	p=129	**p=0.00**	**p=0.00**	p= ---	**p=0.00**
Negot-power	-0543	6164	6703	7277	10000
	p=757	**p=0.00**	**p=0.00**	**p=0.00**	p= ---

Marked correlations are significant at p <05000
N=30 (Casewise deletion of missing data).

These findings seem to confirm the preceding hypothesis: hedonists need larger markets to find rewarding partners. Or, to state it differently, they are certainly more demanding, and less adaptive. But, on the other hand, when a wider social choice is available, this strategy proves as advantageous as the complementary one.

To sum up, the *comparison between the two modalities of partnership formation*, the goal-oriented and the gain-oriented shows both similarities and differences between them. As far as the similarities are concerned, it should be noted that, in large markets, all dimensions in both modalities are significantly correlated; in particular, both the goal-oriented agents and the gain-oriented agents with high negotiation-power find partners for interaction.

Table 10: Correlations in large-size markets

	SU1	*P.V.1*	*SU2*	*P.V.2*	*Negot-power*
SU1	1.0000	.6099	.5814	.3152	.6595
	p= ---	**p=0.00**	**p=0.00**	**p=0.03**	**p=0.00**
P.V.1	.6099	1.0000	.6680	.3738	.7737
	p=0.00	p=---	**p=0.00**	**p=0.01**	**p=0.00**
SU2	.5814	.6680	1.0000	.2892	.7618
	p=0.00	**p=0.00**	p= ---	**p=0.04**	**p=0.00**
P.V.2	.3152	.3738	.2892	1.0000	.3885
	p=0.03	**p=0.01**	**p=0.04**	p= ---	**p=0.00**
Negot-power	.6595	.7737	.7618	.3885	1.0000
	p=0.00	**p=0.00**	**p=0.00**	**p=0.07**	p= ---

Marked correlations are significant at $p < 0.05$
N=50 (Casewise deletion of missing data)

But our findings show interesting differences also. In particular, social utility of the goal type, in small and middle markets, is not correlated with any of the other dimensions considered, including the negotiation power. A given individual agent which is useful to satisfy others' goals may be depending upon them to a higher degree (it may need much more than what it can give). With the social utility of the gain type, instead, where the cost to benefit relation is necessarily considered, such a possibility is relatively less likely to occur: autilitaristically useful agent is one whose "price", so to speak, must be lower than what she has to offer. This seems to show that goal-oriented agents are somehow more demanding, more selective than gain-oriented ones. Utilitarian agents seem to content themselves more easily with what they find in the social world. Hedonists, instead, who are always looking for the agents who satisfy their goals need a "wider choice" in order find what they really want, what can really satisfy them.

8.6 Conclusions and Future Work

In this paper, a computational system, MICROdep, developed to form partnerships among agents with different goals and actions, was applied to explore two strategies of partnership formation: one (utilitarian) aimed at maximizing utility, and the other (hedonistic) aimed at satisfying one's goal(s).

Our findings seem to encourage the exploration through computer-simulation of different strategies of decision-making. So far, in our opinion, an essentially uni-dimensional view of this phenomenon has prevailed in decision theory, namely the idea that it is rational for self-interested agents to maximize their utility, whatever the content of such utility may happen to be. In our study, we have endeavored to show that self-interested agents can apply a somewhat different, although related, criterion for decision-making, namely goal-satisfaction. In particular, our initial findings seem to show that the two strategies vary inflexibility, namely, the goal-oriented is less flexible than the utility-oriented one. But one would wonder why: what is the use, the functionality of a goal-oriented strategy, since the latter appears to be more flexible? Of course, comparisons between these criteria, should be made on a variety of measures, including but not reduced to those analytically consistent with the utility principle (accumulation, capitalization, and the like). This is one of the objectives which MICROdep will be applied to in the future. One should check the stability of the two strategies in open markets, meaning in markets where newcomers may enter. Under such a condition, there may be a trade-off between flexibility and stability: the utility-based strategy, which is more flexible, produces less stable partnerships, since it is always possible to find out a more useful partner. While utility-maximization tends to be asymptotic, goal-satisfaction is not: there are limits which are no use to overcome. Obviously, the trade-off varies with hybrid categories: next-best hedonists will be less stable than pure hedonists, which will stand on the extremely stable and extremely rigid end of the continuum, while pure utilitarian will stand on the extremely flexible and extremely unstable one. Only further simulations with open social markets will allow this question to be answered.

References

Binmore, K. (1987): Modelling Rational Players: Part 1. Economics and Philosophy 3, 179-214.

Brandt, R. (1983): The Concept of Rational Action. Social Theory and Practice 9(2-3), 143-64.

Bratman, M. (1987): Intention, Plans, and Practical Reason. Harvard Univ. Press.

Castelfranchi, C (1996): Reasons: Belief Support and Goal Dynamics. Mathware & Soft Computing 3, 233-247.

Castelfranchi, C. (1997): Principles of Individual Social Action. In: Holmstrom-Hintikka, G. & Tuomela, R. (eds.): Contemporary Action Theory. Vol. 2: Social action. Kluwer, Dordrecht, 317-341.

Castelfranchi, C. & Conte, R. (1998): Limits of Economic Rationality Agents and MA Systems. Robotics and Autonomous Systems, Special Issue on Multi-Agent Rationality 24(3-4), 127-139.

Conte, R. (1998): Diversity in Rationality. A Multi-Agent Approach. In: Korta, K. (ed.): Logics, Language and Cognition, Kluwer, Dordrecht.

Conte, R. (1996): Foundations of Rational Interaction in Cognitive Agents: A Computational Approach. In: Liebrand, W. and Messick, D. (eds.): Frontiers in Social Dilemmas. Springer, Berlin.

Conte, R. & Castelfranchi, C. (1995): Cognitive and Social Action. UCL Press, London.

Conte, R. & Sichman, J. (1995): DEPNET: How to Benefit From Social Dependence. Journal of Mathematical Sociology 20(2-3), 161-177.

Conte, R., Veneziano, V., Castelfranchi, C. (1998): The Computer Simulation of Partnership Formation. Computational and Mathematical Organization Theory 4(4), 293-315.

Gmytrasiewicz, P .J. (1995): On Reasoning About Other Agents. In: DIMAS'95 - Proceedings of the First International Workshop on Decentralized Intelligent and Multiagent Systems- 22 Nov., Krakov, I/38-49.

Hargreaves-Heap, S. (1993): Post-modernity and New Conceptions Of Rationality In Economics. In: Gerrard, B. (ed.): The Economics Of Rationality. Routledge, London, 68-91.

Moser, P. K. (1990): Rationality in Action: General Introduction. In: Moser, P. K. (ed.):Rationality in Action. Contemporary Approaches. Cambridge, Cambridge University Press, 1-10.

Sichman, J. S., Conte, R., Castelfranchi, C. & Demazeau, Y. (1994): A Social Reasoning Mechanism Based on Dependence Networks. In: Cohn, A G. (ed.): Proceedings of the 11th European Conference on Artificial Intelligence. John Wiley & Sons, Baffins Lane, England, 188-192.

Part III

Game Theory and Cellular Automata

9. Cellular Automata as a Modelling Tool: Solidarity and Opinion Formation

Rainer Hegselmann, Andreas Flache, Volker Möller

Basic features of a *classical* cellular automaton (CA) are:

- Cells are arranged in a *regular D-dimensional* grid.
- Every cell adopts a state from a *finite* set of states.
- Time is *discrete*.
- Cells change their states according to *local* rules.
- The *same* transition rules apply to *all* cells.
- In each period cells are *updated* (simultaneously /sequentially).

Clearly, the CA framework is extremely simple. However, CA based approaches are increasingly used to model a variety of social dynamics (for some background cf. Hegselmann 1996, Hegselmann and Flache 1998). The most prominent reason is that CAs can be seen as *multi-agent systems based on locality with overlapping interaction structures.* In this perspective CAs are attractive as a modelling framework that may allow for a better understanding of *micro/macro relations.* The framework is suitable for addressing one of the most important problems in contemporary social sciences, economics included: what assumptions about human decision making are appropriate for theorising, explaining, or predicting social interactions and their dynamics? To endow cells in a CA with different decision making procedures allows to study *whether and to what degree different micro foundations matter for macro outcomes.* In one section of this paper we will demonstrate the usefulness of CA for that purpose. We develop models of solidarity networks which are based on *two different micro foundations.* In one version of the model the individuals are rational and anticipating beings, in another version we assume adaptive and "backward-looking" actors.

Many social scientists reject the CA approach on the grounds that the CA framework is far too simple, or –to put it mildly– far too idealised, to be an appropriate tool for modelling social processes. The assumptions of discrete time, regular grids, finite sets of states etc. may evoke *doubts about the framework itself.* One might argue that the CA approach is so overly simplistic that any analysis on this basis might be useless and futile. We do not share this concern. There are numerous applications of CAs that provided new and interesting insights into social dynamics (cf. Hegselmann 1996, 1998; Hegselmann and Flache 1998). However, there is *no* a priori guarantee that these results are robust against variation in the basic features of CAs, such as *discrete* time, *discrete* space, *finite* numbers of states, and *regular* grids. For this reason we analyse in the first section of this paper effects of the CA concept itself. For illustration, we use a model of the dynamics of an opinion formation process. For this model, we address consequences of the structural assumptions of the CA framework on model outcomes.

9.1 The CA Framework and Possible Framework Effects

Abelson (1964, cf. 1979) analysed the dynamics of opinion formation under mutual influence over time.[1] Abelson's model is *not* a CA based model. Its decisive features are analytically well understood. To analyse CA framework effects, we transform Abelson's model into an "orthodox" CA. Our concern here is *not* whether Abelson's model is an empirically appropriate description of opinion formation, or whether it is a fruitful analytical tool. *Instead* our concern is what happens if we force Abelson's model into a CA framework.

9.1.1 Abelson's Model of Opinion Formation

In Abelson's model every individual i holds an opinion u_i that is specified as a *real* number from the interval [-1,+1]. Opinion dynamics are based on the assumption that individual i's opinion u_i changes under the influence of j's opinion u_j. The rate of change, Δ, depends on the difference $(u_j - u_i)$ and is proportional to a factor y_{ij} that expresses j's degree of influence on i. More formally, the change of opinions is described by the differential equation

$$[1] \quad \dot{u}_i = \sum_{j \in N} y_{ij}(u_j - u_i),$$

where N is the set of actors. Figure 1 shows an example with three actors. The x-axis represents *continuous* time; the y-axis charts *continuous* opinions.

Figure 1 demonstrates that opinion dynamics converge to uniformity. In fact, it can be proven that under the regime of equation [1] any group will inevitably end up with *ubiquitous agreement*, if the group comprises at least one member that directly or indirectly (via others) imposes some degree of influence on all others ("compactness"). However, empirically ubiquitous agreement is not *the* typical outcome of opinion formation processes. This led Abelson to ask "... what on earth one must assume in order to generate the bimodal outcome of community

1 Abelson's model is to be seen in the tradition of models of opinion dynamics that started with French (1956) and Harary (1959).

cleavage studies" (153). In the following, we show that in some sense transforming Abelson's model into a CA based model may be sufficient to avoid ubiquitous agreement.[2] That implies that the modelling framework — CA or something else— may matter a lot.

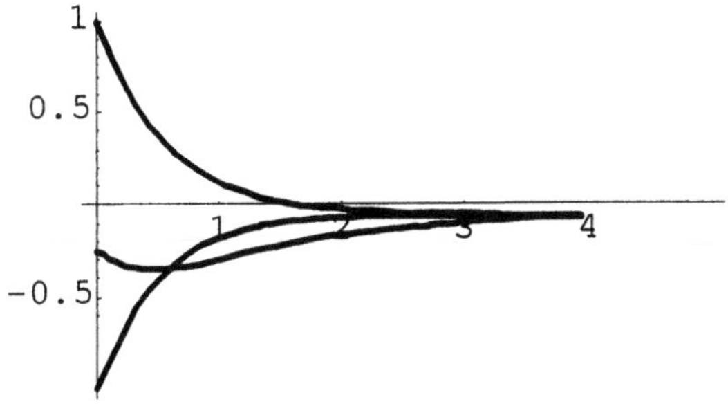

Figure 1: Continuous opinion dynamics.

9.1.2 Transforming Abelson's Model Into a CA

In the following we will transform Abelson's model into a CA based model that meets the CA requirements specified above. More in particular, this transformation involves *discretisation of space, time, and states.*

The *first* step of the transformation is straightforward. To discretise space, we assume a *2-dimensional rectangular grid*, where each cell is occupied by an individual that holds an opinion.

In the *second* step we make Abelson's model time discrete. The discrete time unit is 1. By transforming [1] we get

$$[2] \quad u_i(t+1) - u_i(t) = \sum_{j \in N} y_{ij}(u_j(t) - u_i(t)).$$

This equation can be rewritten as

$$[3] \quad u_i(t+1) = u_i(t)(1 - \sum_{j \in N} y_{ij}) + \sum_{j \in N} y_{ij} u_j(t).$$

[2] That does not mean that this is the only way to avoid the ubiquitous agreement outcome. Abelson himself mentions some alternatives.

For comparison of relative weights, we further assume that the sum of all influences of actors j on an actor i is normalised to 1, i.e. $\sum_{j \in N} y_{ij} = 1$. Equation [3] then simplifies to

$$[4] \quad u_i(t+1) = \sum_{j \in N} y_{ij} u_j(t).$$

In Abelson's original model opinions are elements of a *real valued interval.* This implies an *infinite* number of possible opinions. By contrast, in a classical CA the number of states is *finite* by definition — and opinions are the states of our cells. Accordingly, we transform in the *third* step Abelson's model to satisfy the requirement of a finite opinion space.

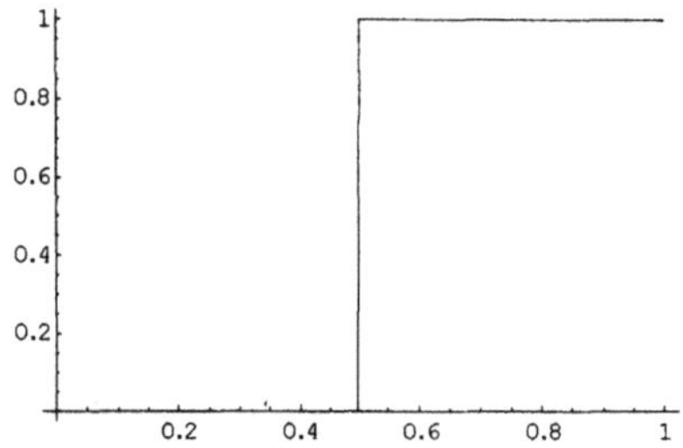

If x < 1/2, then y = 0
If 1/2 ≤ x, then y = 1

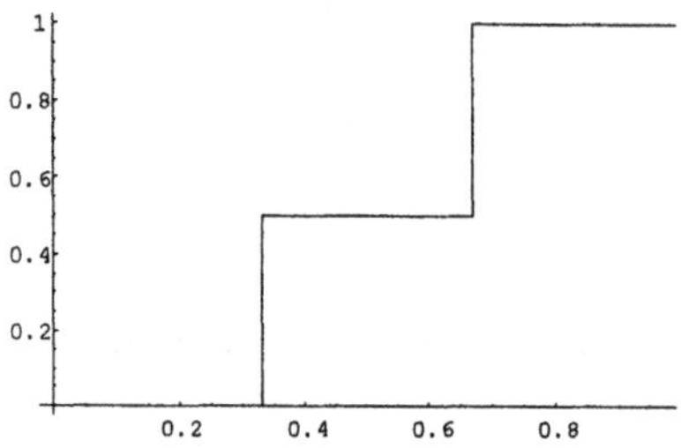

If x < 1/3, then y = 0
If 1/3 ≤ x < 2/3, then y = 0.5
If 2/3 ≤ x, then y = 1

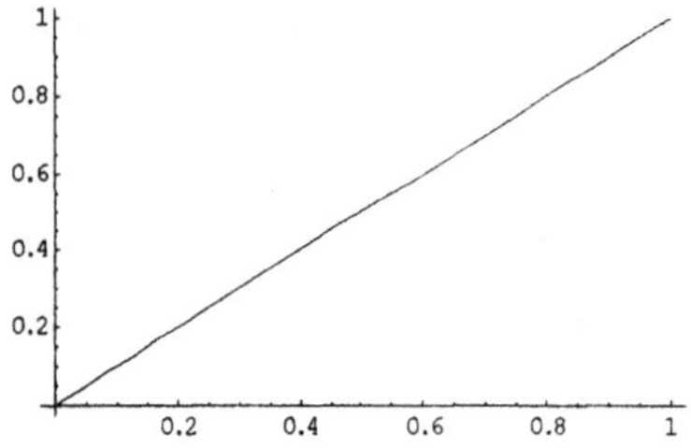

y=x, i.e. no discretisation

Figure 2: *Upper left*: two possible opinions, 0 and 1. *Upper right*: three possible opinions, 0, 0.5, and 1. *Left*: continuous opinions between 0 and 1.

The scale representing opinions depends on the substantive content of opinions and on measurement issues. In our context we obtain the finite number of opinions by a particular type of discretisation that is given by the step functions shown in Figure 2. Only the possible *values* of certain step functions constitute valid opinions. Those values are the input to which we apply the dynamics given by [4]. The result is an opinion value that in most cases is *not* in the set of valid opinions. The step function then transforms that value into an opinion $u_i(t+1)$ of the *finite number of valid opinions*. Subsequently, the resulting opinion is used as the input for the opinion dynamics of the next period and so forth.[3]

In short, the procedure to keep opinions discrete is as follows

$$[5] \quad \sum_{j \in N} y_{ij} u_j(t) \rightarrow \textit{step function} \rightarrow u_i(t+1).$$

Of course, the step functions in Figure 2 are not the only methods to obtain a finite number of possible opinions from Abelson's infinite set of real valued opinions.[4] However, we do not pursue alternative possibilities, because we focus on effects of the CA framework rather than on empirical adequacy.

In the *fourth* step we introduce *locality*, another decisive feature of CA. For this purpose, we confine the influence on a cell to its *neighbourhood*, i.e. a well defined area around a cell. There are different ways to define neighbourhoods. We employ the so called *von Neumann neighbourhood* (Figure 3), consisting of a centre cell *and* its next neighbours in the north, south, east, and west. (Note that a cell is *part* of its neighbourhood.). All actors i will be influenced by all their neighbours j (i included) by the *same* factor y_{ij} expressing j's degree of influence on i. Note that the sum $\sum_{j \in N} y_{ij}$ is assumed to equal 1. Let N_i be the set of neighbours of an actor i. $\# N_i$ denotes the number of neighbours in N_i. Under these assumptions [5] turns into

$$[6] \quad \frac{1}{\# N_i} \sum_{j \in N_i} u_j(t) \quad \rightarrow \quad \textit{step function} \quad \rightarrow \quad u_i(t+1).$$

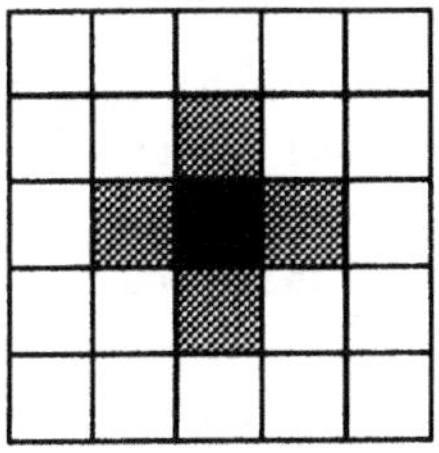

Figure 3: von Neumann Neighbourhood

3 As states we use a finite number n of real valued opinions. It would be easy to represent opinions by integer values. Nothing of what follows depends on that.

4 For the following it makes no difference whether the underlying real valued interval is [0,1], as we assume here, or the interval [-1,+1] as in Abelson's original model.

Figure 3 shows that a von Neumann neighbourhood comprises 5 individuals, i.e. $\#N_i = 5$. Equation [6] implies that in the *continuous case each actor adopts the average opinion of his neighbours,* while in the *discrete* case a step function is applied to that average opinion.[5] To recall, the continuous case corresponds to the lower "step" function in Figure 2, whereas the upper two step functions represent discrete cases.

Equation [6] shows the *fifth* step in transforming Abelson's model into a CA. The equation describes the *universal transition rule* that is applied to all cells of the 2-dimensional world that we constructed. The *sixth* and final step specifies the *updating procedure*. Updating procedures will be explicated below in the discussion of the corresponding simulation studies.

9.1.3 Effects of Discretisation of Time and Space

Imagine a time discrete and spatially transformed model, in which all actors update their *continuous* opinions on the basis of the average opinion in their von Neumann neighbourhood. All neighbours j of i have the same weight (i included!). This is the *first* remarkable observation we make: time discretisation and spatial transformation of the model *do not* change the outcome of ubiquitous agreement, if *continuous* opinions are assumed (randomly chosen and distributed). This result could even be obtained without simulation. In the following, we sketch a way to proof the result formally. We assume a cellular world of size $s \times s$, where every cell is occupied by one individual. Hence, the population size is $N = s \times s$. Every individual i $(i = 1,2,...,N)$ assigns the weight $1/\#N_i = 0.2$ to each of his 5 neighbours when calculating $u_i(t+1)$ as the (weighted) average of his neighbourhood's opinion in period t. The weights assigned by *all* individuals can be represented by a $N \times N$ matrix **W.** An entry y_{ij} expresses j's degree of influence on i or, conversely, the weight that i assigns to j's opinion. All *rows* of

[5] Work especially done by Latanè and Nowak suggests that individual differences in influence on others, influence on oneself, and preferences for particular opinions tend to suppress ubiquitous agreement. Instead, these factors seem to promote polarisation and clustering. Moreover, Latanè and Nowak claim that *at least one* of those factors is *necessary* to avoid ubiquitous agreement: "Clustering, consolidation, and continuing diversity were characteristic of all simulations under all conditions except when there were no individual differences in persuasiveness, no personal preferences, and no self-influence. Individual differences in persuasive ability, personal preferences, and self-influence are individually each unnecessary for the prevention of consolidation, but each does facilitate it, and at least one must be present" (66f). In the following we run the transformed Abelson model *without any* of the factors allegedly favouring "clustering, consolidation, and continuing diversity". We show that these phenomena nevertheless occur.

W hold five non-negative entries –including all diagonal entries y_{ii} – with value 0.2. All other entries of **W** are 0, and all row sums are equal to 1. Initial opinions of the individuals i $(i=1,2,...,N)$ can be characterised by a column matrix $\mathbf{U}=[u_1(t_o),u_2(t_0),\ldots,u_N(t_0)]$. The opinion dynamics is then given by the sequence **U**, **WU**, **W(WU)**, **W(W(WU))**, ... The associativity of matrix multiplication allows to rewrite this as **U**, $\mathbf{W}^1\mathbf{U}$, $\mathbf{W}^2\mathbf{U}$, $\mathbf{W}^3\mathbf{U}$, ..., $\mathbf{W}^t\mathbf{U}$, ... Moreover, note that it is trivially true that in matrix **W** there is *at least one chain of positive weights* from *each* individual *i to every other individual s*. More precisely, a chain of positive weights means that *i* assigns a positive weight to some *j*, *j* to some *k*, ..., and *r* to some *s*.[6] The reason that there is a chain of positive weights for every pairing *i* and *s* is the *overlap* between the von Neumann neighbourhoods in our cellular world. This applies regardless whether a finite checker board or a torus is considered. The positive chains are particularly important, because a theorem in Lehrer / Wagner (1981, 131) based on a theorem and proof in Doob (1953, 173) implies that under these conditions iterative matrix multiplication converges towards a matrix $\mathbf{W}^{\infty}$ with *identical rows*. Hence, the opinion dynamics ends up in ubiquitous agreement on the opinion vector $\mathbf{W}^{\infty}\mathbf{U}$. Note that this line of argumentation also applies for *Moore neighbourhoods* of arbitrary size. This reasoning implicitly assumes *simultaneous* updating – implied by the entries in **W**–, but it can easily be extended to *sequential updating*. For this purpose the entries of matrix **W** are to be computed on the basis of the assumption that updating will take place with equal probability $p_u = 1/N$. This implies that with probability p_u an equal weight of $1/\#N_i$ is given to all neighbours of *i* (*i* included). With probability $1-p_u$, *i* will *not* be updated, i.e. his opinion will not change. Formally, *i* assigns the weight 1 to y_{ii}. The *resulting* weights (i.e. weights, weighted by their probabilities) assigned to *i*'s neighbourhood are then $p_u/\#N_i$ for *i*'s neighbours $j, j \neq i$. For *i* himself, the weight is $1-p_u(1-1/\#N_i)$.[7] All other row entries are zero. Again all entries are ≥ 0 and all row sums equal one. After these steps the same reasoning applies as for simultaneous updating. Again, the outcome to be expected is ubiquitous agreement. As a general result we obtain:

---

[6] As a matter of fact there are lots of such chains from every individual to everybody else – though the number is finite.

[7] One can look at simultaneous updating as the case $p_u=1$.

Analytical observation

The expected outcome of continuous opinion change according to

$$u_i(t+1) = \frac{1}{\#N_I} \sum_{j \in N_I} u_j(t)$$

is ubiquitous agreement, regardless whether one assumes von Neumann or Moore neighbourhoods, simultaneous or sequential updating, a finite checker board or a torus.

Note that according to this observation *discretisation of time and space* in the original Abelson model does not affect the result of *ubiquitous agreement.*

9.1.4 Discrete Opinion Change

In a next step we introduce discrete opinions. We start with only two opinions, 0 and 1. Opinion dynamics are given by [6]. The upper left part of Figure 2 shows the corresponding step function. Figure 4 presents the result of a "typical" simulation run.

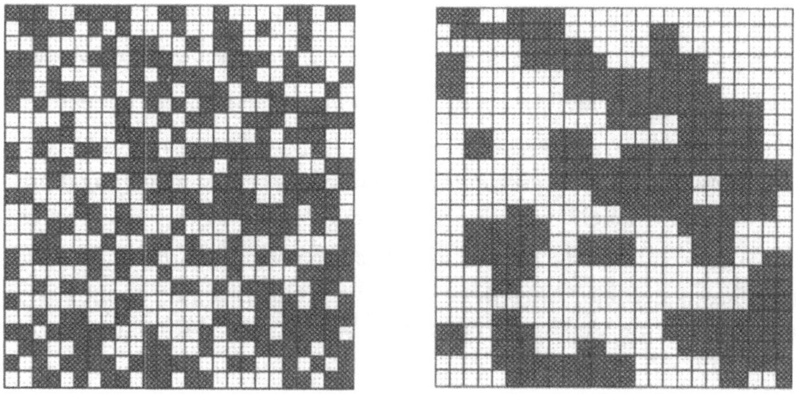

Figure 4: *Left*: initial distribution of opinion 0 (black) and opinion 1(white). *Right*: stable end configuration. Size of the world 25×25, checkerboard (*not* a torus), sequential updating, von Neumann neighbourhood.

Figure 4 shows that the opinion dynamics no longer converge on ubiquitous agreement. The polarised configuration in the right part of the figure *is stable.* To understand this result, consider the following. We distinguish three types of cells: Cells "living" *inside* the checkerboard with 5 neighbours; cells living *at the edges* but *not* in the corners, having 4 neighbours; and, finally, cells living in the 4 *corners* where they have 3 neighbours. Figure 5 describes the dynamic behaviour of the different types of cells. The cells are in one of two states, *opinion 0* or *opinion 1*. The numbers on the arrows indicate the condition under which the corresponding transition occurs. The transition takes place when one of the numbers indicated on the arrow is the number of neighbours having the same state than the focal cell. For example, the cell changes from state 0 to state 1, if one or two of its neighbours (including itself) are in state 0. Otherwise, the cell remains in state 0. All entries are calculated by applying the step function of Figure 2, upper left, to the dynamics governed by [6] in a von Neumann neighbourhood as given by Figure 3.

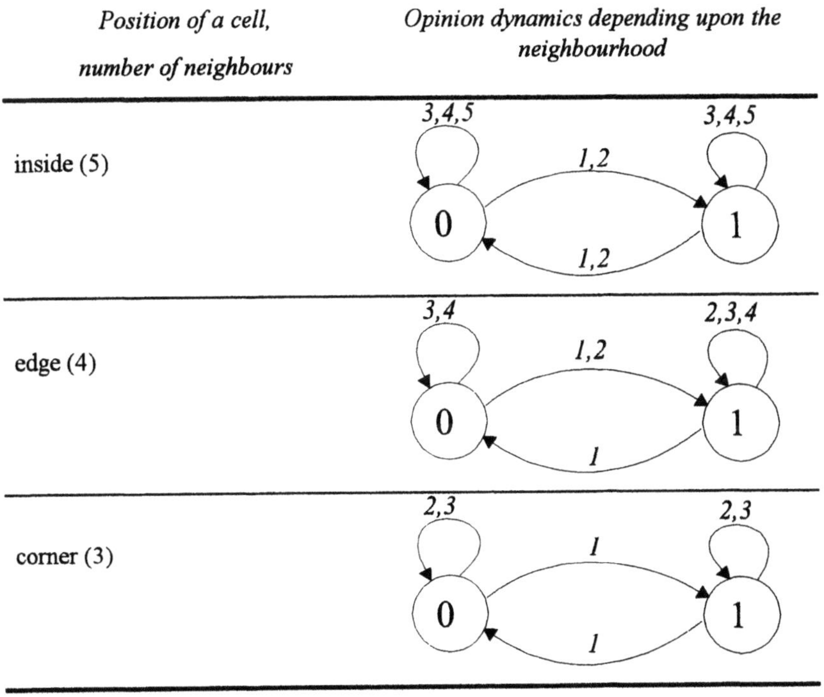

Figure 5: Dynamics of states.

Inspection of Figure 5 immediately shows that a configuration *is stable for ever*, if all *inner* and all *edge* cells have at least 3 neighbours being in their own state, and *all* corner cells have at least 2 of those neighbours. Again the result is *not* ubiquitous agreement but stable polarisation.

Figure 7 below demonstrates that this result can be *generalised.* The x-axes in Figure 7 indicates the number of different opinions in terms of discretisation steps: The first step of discretisation yields just 2 opinions, 0 and 1. In a second step we obtain 3 opinions: 0, 0.5, 1, and so forth up to 50 possible opinions. For each discretisation step think of the y-axes as divided into intervals corresponding to the step functions. Imagine that the y-axis, ranging from 0 to 1, is divided into 2 intervals for 2 opinions, into 3 for 3 opinions and so forth – though this is *not directly* visible in Figure 7. We colour these intervals to indicate the *frequency* of cells of a certain opinion after opinion dynamics converged. For reliability, we computed the *mean* frequency of 10 simulation runs per discretisation step. The simulations were run on a checkerboard (*not* a torus) with sequential updating, von Neumann neighbourhood, step functions according to Figure 2, and a dynamics according to [6].

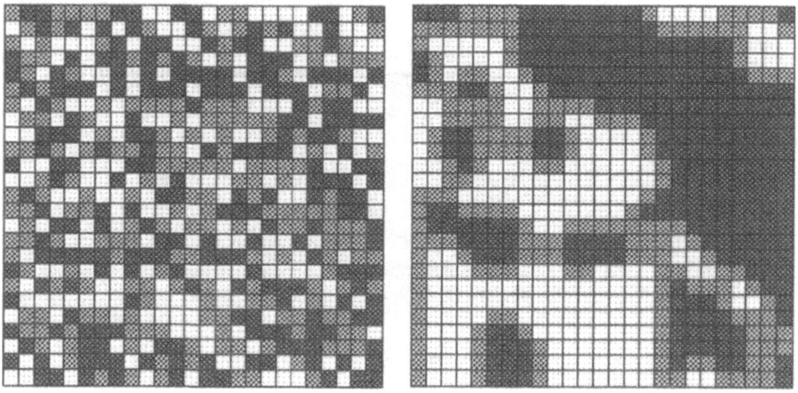

Figure 6: *Left*: initial distribution of opinion 0 (black), 0.5 (grey), and opinion 1(white). *Right*: stable end configuration. Size of the world 25 ×25, checkerboard (not a torus), sequential updating, von Neumann neighbourhood.

Figure 7 demonstrates that discretisation of opinions makes a considerable difference. While continuous opinion change generally ends up with ubiquitous agreement — given compactness — we obtain stable polarisation for discrete opinion change, a dramatically different result. Moreover, Figure 7 shows that the better we approximate the continuous case, the more the stable end configuration is centred on the middle of the opinion space.

9.1.5 Irregular Grids

Up to this point we used a *regular* grid as the underlying network structure for our cellular society and their opinion dynamics. More in particular, we employed a rectangular regular grid, which is quite common in the 2-dimensional case. We might also have used hexagonal or triangular structures, which are regular as well. In general, we denote grids as *regular* where all inner cells have the same number of neighbours, whatever our neighbourhood definition may be – von Neumann neighbourhood or a Moore neighbourhood of a given size. On a regular torus this even applies for the border cells. In the following we will show that relaxing the assumption of regularity matters in the following sense: *The typical outcome of opinion dynamics on a regular grid may no longer be typical if the same dynamics are simulated on an irregular grid.*

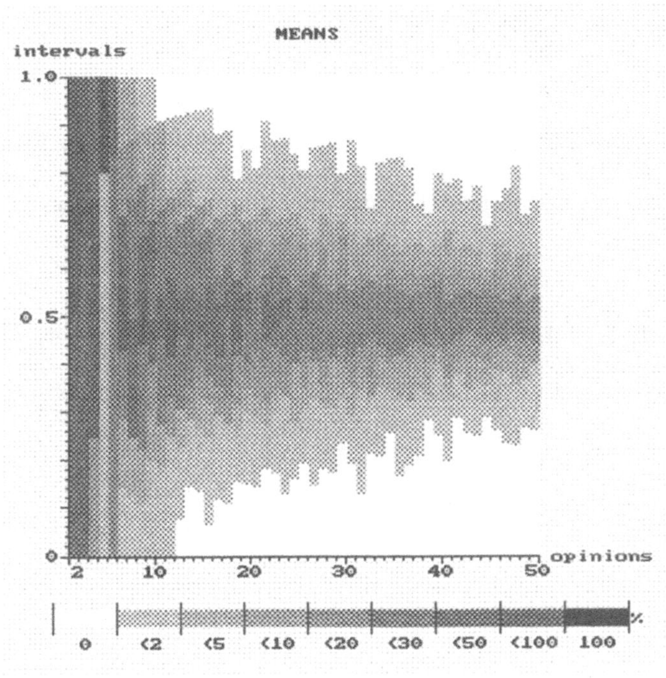

Figure 7: Discrete opinion dynamics with up to 50 possible opinions. Mean frequencies of opinions after convergence.

As an *example* we look at an opinion dynamics with 2 opinions. We use a slightly different updating rule that makes clustering a bit more difficult: We now assume that *one's own opinion does not count any longer*. Only the opinions of the neighbouring *others* are taken into account. Everything else (especially the step function) is kept constant. First we run the model on a *regular* grid. Subsequently, we use an *irregular* grid. The typical result on a regular grid is demonstrated by Figure 8. The simulation converges on global uniformity on opinion 1.

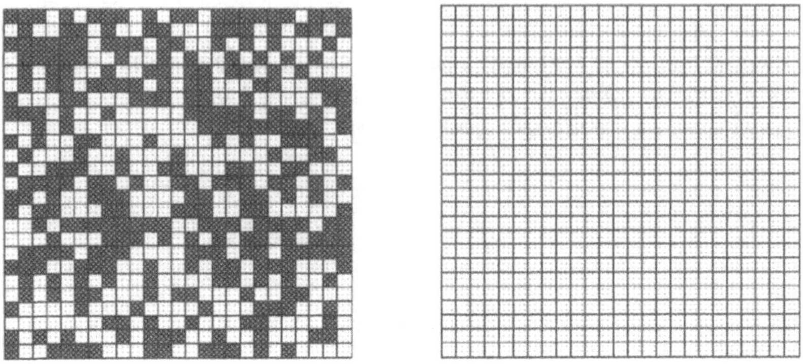

Figure 8: Left: initial distribution of opinion 0 (black) and opinion 1(white). Right: stable end configuration. Size of the world 25 ×25, checkerboard (not a torus), sequential updating, von Neumann neighbourhood. Difference to Figure 6: one's own opinion does not count.

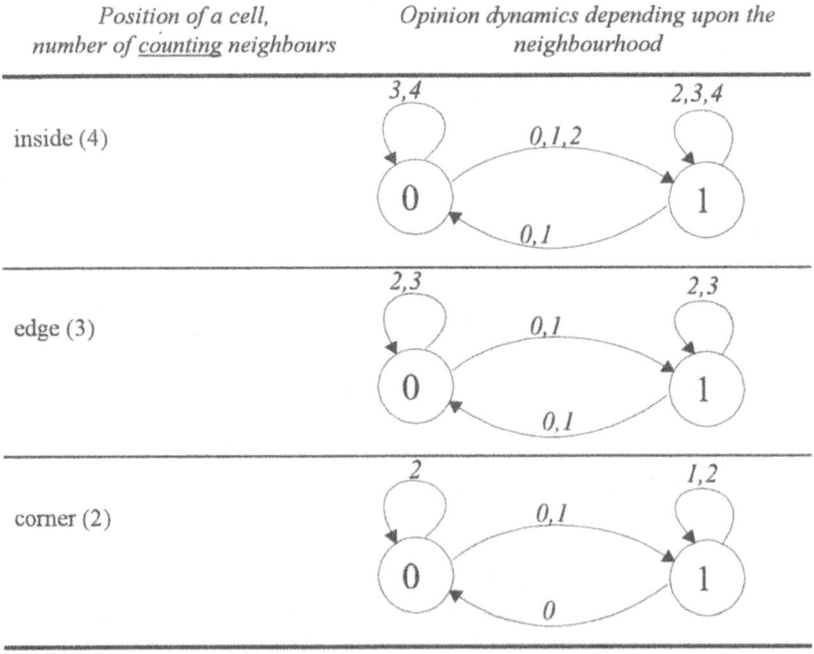

Figure 9: Dynamics of states. Only the opinions of others count.

To explain the result of Figure 8, we describe the cell dynamics in Figure 9 below. Figure 9 employs the transition graphs of Figure 5, with the exception of one difference. Again, the numbers on the arrows indicate the condition under which the transition corresponding to the arrow takes place. However, now the numbers indicate the number of neighbouring *others* that need to be in the same state as the focal cell.

Figure 9 shows that the crucial asymmetry in the opinion dynamics occurs when cells have two neighbours with opinion 0 and two neighbours with opinion 1. In this situation, a cell with opinion 0 will change to opinion 1, while a cell with opinion 1 would stay in its state. This is a consequence of the step function we employ. The upper left part of Figure 2 shows that this step function returns 1 for the argument 0.5. If the returned value were 0, the dynamics in Figure 8 would end up with ubiquitous agreement on opinion 0.

Closer inspection of Figure 9 shows that the opinion 0 does *not* necessarily die out. Opinion 0 can survive in clusters in which each *inside* cell has at least 3 neighbours of this opinion, each *edge* cell has at least 2, and each *corner* cell has 2 neighbours of opinion 0. Such configurations exist. Figure 10 shows an example. Note that the stripe could *not* survive if it would not touch the corner.

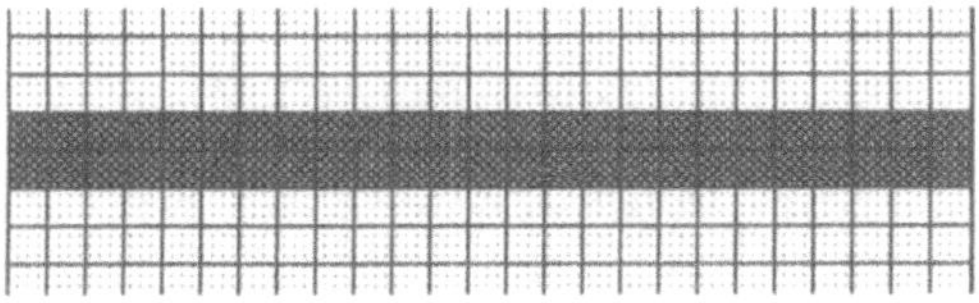

Figure 10: Stable configuration with surviving opinion 0.

Figure 10 shows that patterns exist in which opinion 0 is not doomed to die. However, it is extremely unlikely that an initial random distribution occurs that eventually entails one of those few patterns. Accordingly, the *typical* result of simulation runs under the given conditions is that of Figure 8, i.e. ubiquitous agreement. We now demonstrate that this will change if we run the *same* dynamics on *irregular* grids.

Figure 11 shows a *Voronoi-diagram,* the type of irregular grid that we use for our simulations. The basic idea for construction of a Voronoi diagram is the following: A number of points corresponding to the number of cells required are randomly distributed on a plane. Those points are called *generators*. The cell of a particular generator *i* is then defined as the area containing all points of the plane that are closer to *i* than to any other generator. The crosses in Figure 11 are the generators of the grid. The edges of the resulting polygons are points with equal distance to their neighbouring generators.[8]

Figure 12 shows a decisive feature of irregular grids: even the cells inside the grid have *different numbers of next neighbours*. Figure 13 shows how this feature affects the typical outcome of the opinion dynamics. (To recall, we assume that the updated opinion is computed on the basis of the value the step function returns for the average opinion of neighbouring *others*. One's own opinion does not count.)

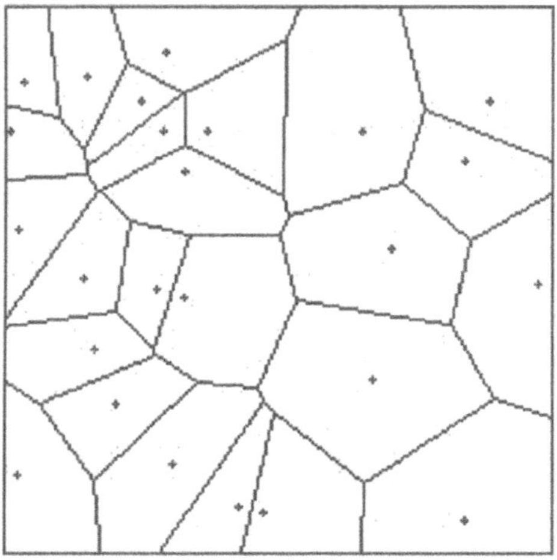

Figure 11: Irregular grid.

Replication of the simulations of Figure 8 and 13 shows that we normally obtain *ubiquitous agreement* on opinion 1 for regular grids, whereas it is fairly likely that *clusters of opinion 0 survive* on irregular grids. This effect is a consequence of the irregular grid structure. On a regular grid, a typical *convex* cluster has in almost all cases at least some cells at the border which have *inside* the cluster *less or only as many neighbours as they have outside*. In such a position a cell with opinion 0 will switch to opinion 1. That worsens the position of its like-minded neighbours. The cluster shrinks, and eventually vanishes. On irregular grids, however, conditions are different: there are much better chances for border cells to have more neighbours inside the cluster than outside — even with a convex cluster shape. The lower part of Figure 13 shows that all border cells of the surviving cluster have more neighbours inside the cluster than outside. While

[8] For the mathematical details see Okabe et al. (1992). Details of our implementation of Voronoi-diagrams and some sample data sets can be found in Möller (1997).

on regular grids such configurations are extremely rare, they are fairly likely on irregular grids. As a consequence the dynamics we analysed behave differently on regular and irregular grids.

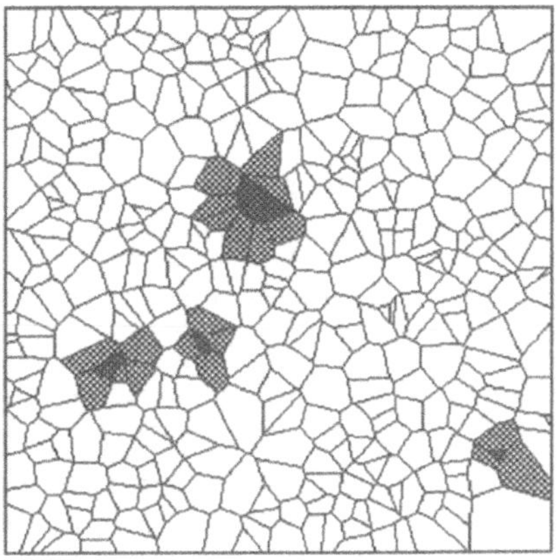

Figure 12: Cells with 3, 4, 5, and 8 neighbours.

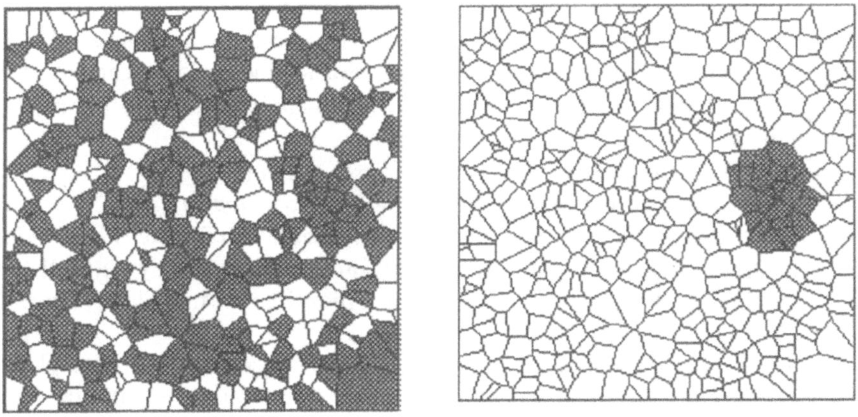

Figure 13: *left* - Initial random distribution. *right* - a stable and surviving cluster with opinion 0.

These analyses might suggest at least two methodological lessons. The *first* lesson regards discretisation: Making a continuous variable discrete may considerably change its dynamics. The mere number of discrete states may matter considerably. The details of the procedure for keeping states discrete, for instance the particular design of a step function, may have consequences. The *second* lesson regards grid structures: Regularity of grid structures may affect model outcomes. Results based on regular grids may no longer be obtained on irregular ones, even if all other conditions are equal.

This section demonstrated that some caution is in place when using the CA framework. However, the subsequent section will illustrate one of the merits of the approach. We show that CA provide an extremely useful framework for comparison of alternative micro foundations.

9.2 Rational vs. Adaptive Behaviour in Solidarity Networks: Analysing Effects of Micro Foundations in a CA Framework

In this section we use a CA to explore how sensitive model results respond to variation in the underlying micro foundation. Drawing on a substantive application from our previous work, we model the evolution of support networks among egoists who greatly differ in their neediness, are free to choose their partners, and will do that in opportunistic ways (cf. Flache and Hegselmann 1998). No central authority is involved. The question we address is how it shapes the structure of emerging solidarity networks whether we model egoistic actors as *rational* actors in a game theoretical sense or as *adaptive* actors. More precisely, we follow the standard approach in game theory and assume that rational actors a) choose their solidarity behaviour as if they maximise the utility of decision outcomes and b) in doing so aim to anticipate each others' (*rational*) decisions. By contrast, adaptive actors are modelled on the basis of a simple trial and error mechanism that assumes that actors a) are satisfying, i.e. they aim to find actions that are "good enough" rather than maximise utility and b) evaluate decisions on the basis of experience rather than anticipate decision effects[9].

Our central assumptions in modelling solidarity networks are:

- Support relationships are modelled by a *2-person support game*.
- The social space in which a network can emerge is a *2-dimensional cellular grid*. Actors can play the support game simultaneously with all their neighbours. Not all cells are occupied. From time to time players have the *chance for migration*.

Section 2.1 presents the support game. The basic assumptions about playing and migrating in the 2-dimensional world are explicated in (2.2), and the structure of rational and adaptive actors is sketched in (2.3) and (2.4), respectively. Results are presented and discussed in (2.5).

---

[9] For discussions of rational vs. adaptive micro foundations see Friedman 1953, Dawes and Thaler 1988, Macy 1990, Hechter 1992, Roth 1992, Macy 1993, Macy and Flache 1995, Flache and Macy 1996, Flache 1996.

9.2.1 The Support Game[10]

In the support game both players face simultaneously and independently the decision whether to provide help to their solidarity partner (C = Cooperation) or to not provide help (D = Defection). Cooperation implies to give help to the neighbour, while defection is not to help. Both players are characterised by the degree to which they need support, their *neediness* n ($0 < n < 1$). For simplicity, we assume that every actor living in the cellular world belongs to one of some *neediness classes*. The neediness of a member of class i is n_i. The more needy an actor Ego is the more he profits from receiving help from his neighbour Alter, and the less Ego is capable of providing help himself. High neediness of Alter limits the amount of help Ego can receive, while it increases the amount of help Ego will provide in case he follows the co-operative strategy with respect to Alter. Formally, we model i's gain from receiving help from j, G_{ij}, and i's loss from giving help to j, L_{ij}, as follows:

$$[7] \quad G_{ij} = n_i(1-n_j)B$$
$$L_{ij} = (1-n_i)n_jE$$

where the parameters B and E are positive constants that scale the benefit of receiving a unit of help (B) relative to the effort costs of producing a unit (E). Table 1 shows the payoffs of player i in the constituent support game. Following the standard in the game theoretical literature, we denote the payoffs corresponding with the outcomes CC, DC, CD and DD as **R**eward, **T**emptation, **S**ucker's payoff and **P**unishment, respectively.

Table 1: Payoffs of player i in the constituent support game

		Strategy player j C	D
strategy	C	$R_{ij} = G_{ij} - L_{ij}$	$S_{ij} = -L_{ij}$
player i	D	$T_{ij} = G_{ij}$	$P_{ij} = 0$

Table 1 demonstrates that the support game confronts egoistic actors with an incentive problem. Regardless whether an actor Ego receives help or not, in the constituent support game he prefers to defect himself. This incentive problem may create a *prisoner's dilemma situation*, because it might lead *both* players to withhold support even when both would prefer *mutual support*, *CC*, to *mutual*

[10] In this paper we deliberately keep presentation of the model short. We gave a detailed formal description in Flache and Hegselmann (1998) and Flache and Hegselmann (forthcoming).

defection, DD. The prisoner's dilemma only arises, however, when mutual support is attractive from *both* actors' point of view, i.e. $R_{ij} > P_{ij}$ and $R_{ji} > P_{ji}$. Whether this is the case depends on the neediness levels of the two players involved. The non-white area in Figure 15a visualises the range of pairings of neediness levels for which the game has a prisoner's dilemma (PD) structure. In the white regions of the figure, at least one player finds mutual support, CC, less attractive as compared to mutual defection, DD. Accordingly, in these regions mutual cooperation is impossible for rational actors. The figure shows that a prisoner's dilemma structure only arises when the neediness levels of the players involved are relatively similar. To summarise, the central features of the support game are

- actors prefer not to give help to providing help, all other things being equal;
- both actors prefer mutual support to mutual defection (i.e. the game has a *prisoner's dilemma* structure), if their neediness levels are sufficiently similar.

It is well known that even egoistic actors might attain a co-operative solution in a prisoner's dilemma when there is a "shadow of the future", i.e. when after every iteration the game is repeated with a certain continuation probability α. To allow for cooperation, we therefore assume that actors play this kind of repeated support game. In (2.4) and (2.5) below we discuss how, for each of the two models, players' neediness levels shape the minimum continuation probability required for cooperation.

9.2.2 Playing and Migrating in the Cellular World

For simplicity, we assume that individuals populating the cellular world belong to *nine different neediness classes* n_i with neediness 0.1, 0.2, ...0.9, respectively. They live at the surface of a *torus* so that the world can be represented as a checkerboard without borders. Individuals play support games with all neighbours in their *interaction neighbourhood.* The interaction neighbourhood is a von Neumann neighbourhood, i.e. neighbours are the neighbouring cells in the north, south, east and west. In each period every individual plays one iteration of every support game it is involved in. In addition, individuals receive an option to migrate with an *exogenously set probability q*. Migration is possible within a certain *migration window*, but *only to vacant destination cells.* Figure 14 shows interaction window and migration window. Migration options are evaluated and/or used in *sequential* order according to the results of a lottery. Subsequently, a new period starts. The migration window is a square of an odd size with the selected individual in the centre. It has the same size, for instance 5×5, for all individuals. Each actor, rational or adaptive, instantly *knows the neediness class of all his (real or potential) neighbours* and is informed about the payoffs of the support game.

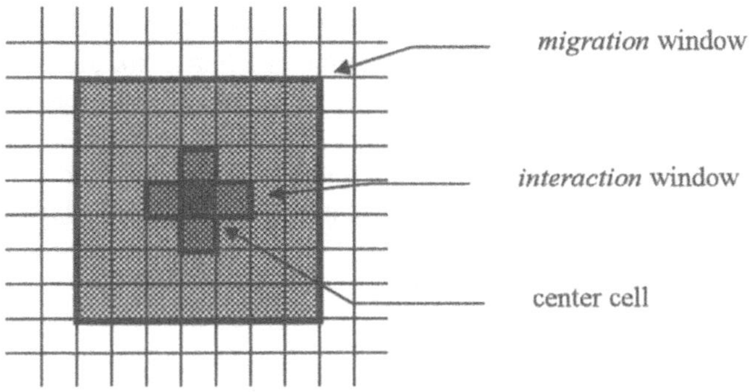

Figure 14: Basic windows.

9.2.3 A Model of (Boundedly) Rational Behaviour

In our model actors face two basic decisions. The *solidarity decision* is the decision whether and when to help a particular neighbour in the repeated support game. The *migration decision* addresses the question whether and where to migrate when a migration option is available. In the following we sketch how we model decision making of (boundedly) rational actors.

Solidarity Decision

In the constituent support game, rational actors will never provide help to their partner. In the repeated support game, however, rational decision making might result in perpetual mutual help. When both players follow a strategy of *conditional cooperation*, both of them *anticipate* that an attempt to deviate from co-operative behaviour inevitably evokes subsequent defection by the opponent. This in turn will endanger the prospects of mutual cooperation *in the future*. Accordingly, in a game where the probability for continuation is sufficiently large, the prospect of future losses might deter rational players from defection. The game theoretical literature (e.g. Friedman 1986, Taylor 1987), provided an analytical proof for the condition under which mutual conditional helping is an individually rational behaviour in a repeated support game. This result yields the following condition for the support game. Mutual conditional helping is individually rational when the probability α for continuation of the game is not lower than a certain threshold value α^*. The threshold is given by

[8] $$\alpha \geq \frac{T_{ij} - R_{ij}}{T_{ij} - P_{ij}} = \frac{L_{ij}}{G_{ij}} = \alpha_{ij}^* \qquad \text{for both players.}$$

Figure 15a shows how the combination of neediness levels n_i and n_j shapes this threshold continuation probability for rational actors. The figure illustrates for which pairings of neediness classes support relations can arise for rational actors. The horizontal axes represent neediness levels, i.e. n_i and n_j. The vertical axis indicates the threshold continuation probability. The white area where no value is plotted shows where the repeated support game is not a prisoner's dilemma. For this range of pairings mutual support is not profitable from at least one player's perspective.

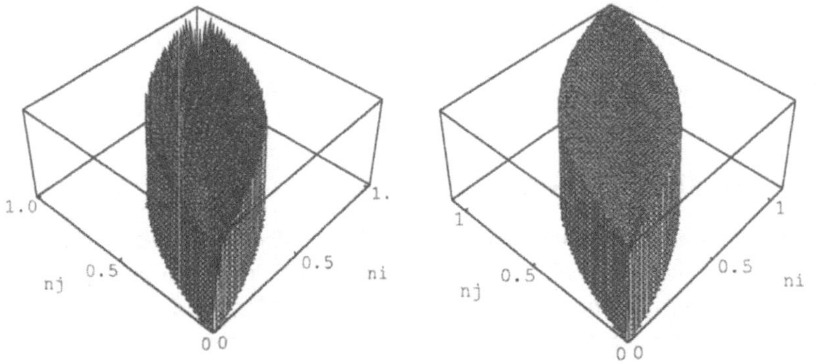

a) rational actors *b) adaptive actors*

Figure 15: Threshold continuation probability as a function of neediness levels n_i and n_j for the rational actor model (a) and the adaptive actor model (b). Vertical axis ranges from 0 to 1.

Figure 15a shows that cooperation between rational actors is only attainable when the neediness classes are *sufficiently similar*. Moreover, *middle neediness* classes have a larger range of potential partners than classes at the *extremes*. Finally, we find that the *more different* neediness classes are, the higher is the continuation probability required.

Migration Decision

It is straightforward to apply game theoretical results to the analysis of an isolated support game. However, it is extremely difficult to also model the interplay of support behaviour and migration behaviour in terms of perfectly rational decision making. The reason is that actors' beliefs about own and others' migration behaviour shape their estimate of the continuation probability α in a given relationship. This estimate – in turn – affects their anticipation of support decisions which again shape expected migration. We have not succeeded to resolve the problem of modelling perfectly rational decision making in this context. Instead, we assume that our rational actors are *bound to be rational.* Their solidarity decisions rely on an estimation of the continuation probability that is only a very rough and pessimistic guess. Moreover, in their migration decision actors are very short sighted. They anticipate only for the near future what will happen with the neighbourhood to which they move. Clearly, this decision making is far from fully rational. At the same time this does not affect our principle objective of comparison of rational and adaptive behaviour. As it will become apparent further below, even our boundedly rational actors are far more anticipating than their backward-looking counterparts. In the following we specify the assumptions of the rational actor model in more detail.

In our model, actors use a pessimistic proxy to assess the probability for continuation of the support game, α. More precisely, when a rational actor receives an option to migrate, he uses his estimate of others' rational behaviour to anticipate whether potential new neighbours would be willing to cooperate. "Pessimistic" means the individuals assume that every option will be used to leave. With q as the probability for getting a migration option, individuals estimate the continuation probability as $\alpha = (1-q)^2$. Given this estimate, everyone decides whether the condition for cooperation is met. Two neighbours i and j will support each other if α is at least as large as the corresponding threshold continuation probability.

Individuals use their pessimistic guess of α to assess the attractiveness of their own location and of potential new locations in the migration window. All individuals know that – relative to their own neediness class – they are in a *best social position* when they are surrounded by four individuals which are members of the best class willing to engage in support relationships with them. On the other hand, there are worst *social positions*: being surrounded by empty cells only or by only those individuals with whom support relations are not possible. We use a plausible heuristic to model risk preferences in migration behaviour. This is the heuristic "the worse off you are, the more you risk moves to bad positions in order to attain potential improvements in the future". More precisely, we assume that an individual is *satisfied* with a position that offers 50% *of the difference between the payoff at the best and the worst social position. Satisfied individuals* will migrate only if they can get a position which is at least as good as the position they have. *Dissatisfied individuals* will seize every migration option and accept even worse new locations, "hoping" that they find a better position by future moves accessible from the new location.

9.2.4 A Model of Adaptive Egoism in the Solidarity Game

The model of adaptive egoism in the solidarity game comprises two main components, the model of the solidarity decision and the model of the migration decision, respectively.

Solidarity Decision

We model solidarity decisions on the basis of a conventional Bush-Mosteller stochastic learning model (Bush and Mosteller 1955, see also Macy 1991). In this model, every actor *i* has in his "memory" one cooperation propensity for every neediness class represented in the population. The larger Ego's cooperation propensity with respect to Alter's neediness class, the larger is the probability that Ego will cooperate in a particular support game. Initially actors have no particular disposition for either cooperation or defection, hence in the first iteration they always cooperate with a probability of 50%. After every iteration of the game, Ego adapts his cooperation propensities according to the experiences he makes. The adaptation process proceeds in two steps. In the first step, Ego evaluates whether he is satisfied with the result of the preceding iteration of the game. For simplicity, we assume that actors are satisfied, when they attain 50% or more of the difference between the best and the worst payoff that is possible in the actual pairing of neediness classes. Otherwise they are dissatisfied. In the second step actors apply a *reinforcement learning rule*. The learning rule implies that the propensity to repeat co-operative behaviour increases, when *i* co-operated and experienced a satisfactory payoff, or when he defected and obtained a payoff falling below his expectation level. Conversely, *i* reduces his propensity to cooperate when he experienced a negative reinforcement for cooperation, or he defected and this defection was rewarded. The formal elaboration of the model guarantees that cooperation probabilities never exceed the valid range of [0,1].

The stochastic learning model generates a possibility for cooperation that is similar to the corresponding result for rational players. Again, mutual support may arise in a repeated game, but only when the game has a prisoner's dilemma structure and only when the probability of continuation, α, is sufficiently large. The mechanism by which mutual cooperation may come about is *spontaneous co-ordination on a mutually satisfactory outcome*. Our model implies that the outcome CC yields for both players a satisfactory payoff, if the support game has a prisoner's dilemma structure. Accordingly, in a prisoner's dilemma both players will increase their cooperation propensities, as soon as both of them co-operated by chance. A sufficient number of consecutive CC outcomes then drives both players to a cooperation probability of near 100%, stabilising a co-operative relationship. However, the less attractive at least one of the players finds mutual cooperation, the more iterations of the game are required to stabilise cooperation. Accordingly, for a particular pairing of neediness levels n_i and n_j a corresponding threshold continuation probability *for adaptive actors* is required to sufficiently stabilise mutual cooperation. Unlike the rational actor model, the stochastic learning model makes it extremely difficult to analytically compute this threshold continuation probability. Instead, we calculated a numerical approximation. Figure 15b shows how the combination of neediness levels shapes the estimated threshold continuation probability for adaptive actors. Axes in Figure 15b are defined as in Figure 15a.

A comparison of Figure 15a and Figure 15b demonstrates that the conditions for cooperation generated by both models are qualitatively similar. In both models a support relationship can only arise when the neediness classes are *sufficiently similar*. Moreover, both models imply that the continuation probability required for mutual cooperation increases, as neediness classes are *more different*. However, the figures also show that – ceteris paribus – adaptive actors need a higher continuation probability to attain cooperation. Broadly, this reflects that the learning mechanism is less efficient in "discovering" possibilities to gain from mutual support, as compared to rational actors' decision making.

Migration Decision

Like rational actors, adaptive players take their migration decisions by evaluating free social positions in their migration window relative to their present social position. Moreover, they follow the same heuristic for risk preferences. In contrast to rational actors, however, adaptive players are not capable of anticipating the future solidarity decisions of potential new neighbours. Instead, they estimate the attractiveness of a new location on the basis of experience they gathered in preceding interactions with members of the corresponding neediness classes. More precisely, whenever an adaptive actor plays an iteration of the support game, he uses the payoff attained to update his estimate of the attractiveness of the neediness class of his present opponent. The current estimate of the attractiveness of a particular class is a weighted mean of the payoffs attained in all preceding encounters with members of this class. The weights specify a decaying memory function, i.e. the more recent an encounter, the larger is the weight of the corresponding payoff.

9.2.5 Results and Discussion

In this section, we apply both simulation models to a scenario where the solidarity network is characterised by a relatively high degree of stability. We assume that actors receive on average a migration option only in every 20'th period, i.e. with a probability of 0.05 per period. This corresponds with a continuation probability of $\alpha = 0.903$. All individuals initially are randomly distributed in the cellular grid. Furthermore, we initialise the simulation as follows. We assume a von Neumann neighbourhood as interaction window. The migration window has a size of 5×5, with the given position in the centre. The world is a torus with 21×21 cells, populated by 9 neediness classes with 35 individuals per class. This results in a ratio of occupied cells to empty cells of approximately 3:1. Figures 16 and 17 below show the patterns of the solidarity network between rational actors and adaptive actors, respectively, as they typically emerge from the initial random configuration after 1000 simulation periods.

For interpretation of these figures notice the following:

- White cells are empty cells.
- Round cells are dissatisfied with their social position. Squared cells represent satisfied actors.
- Short white lines connecting two actors indicate bilateral cooperation.
- Different grey levels represent different neediness classes according to the legend at the bottom of the figures.

Figures 16 and 17 demonstrate that both models imply that *extensive solidarity networks* characterised by *segregation patterns* arise. Both networks are extensive in the sense that almost every member of the population has one or more support relationships with some other actor. However, the simulation also reveals clear differences in the segregation patterns. The network between rational actors may be characterised as an onion-like structure, where members with low neediness tend to form a "core", around which a "belt" of members of class 2 arises, which are surrounded by members of class 3 and so forth. Adaptive actors, by contrast, form a "class segregated" network in which clusters tend to arise that are homogenous in neediness levels, whereas only few relationships between members of different neediness classes emerge.

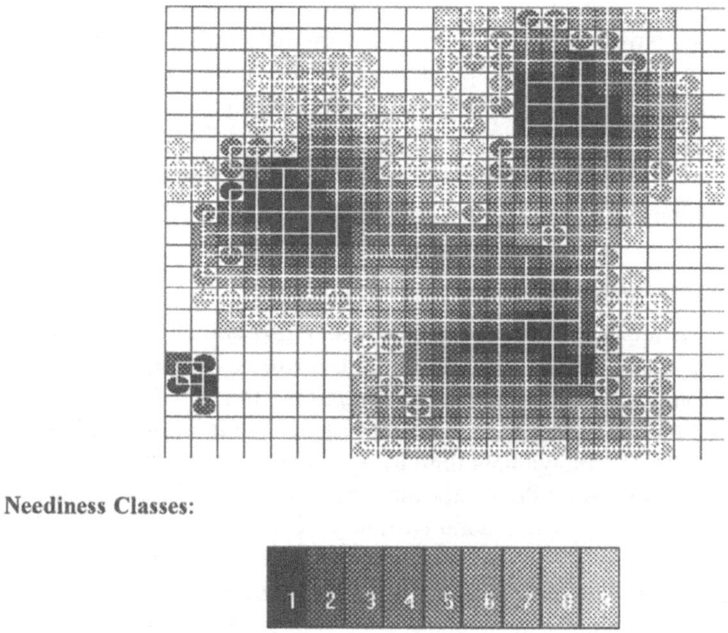

Figure 16: Onion-like network between rational egoists after 1000 simulation periods.

As Figure 15a indicates, the onion-like pattern of rational actors in Figure 16 results from the range of pairings that are possible with $\alpha = 0.903$. Members of class 1 can form support relations only with class 1 and class 2 members, but not with members of higher neediness classes. Moreover, they attain a larger payoff with their own kind due to the higher neediness of class 2. As a result, stable cores of mutual choices between members of class 1 can establish. These cores are only stable, however, when they are surrounded by members of the next attractive class 2. Members of class 2, in turn, remain only in positions where they have as many as possible partners of 1, 2 or the next attractive class 3, and so forth. At the end of the "hierarchy" members of class 9 remain only with class 8 and 9 to relate to, which forces them to stay at the periphery of the onion-like structure.

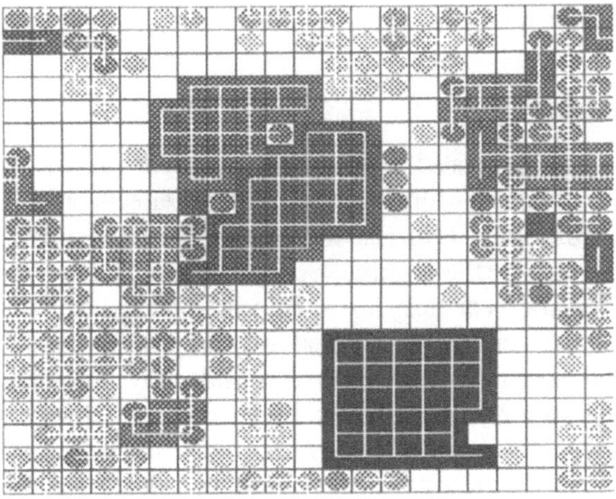

Neediness Classes:

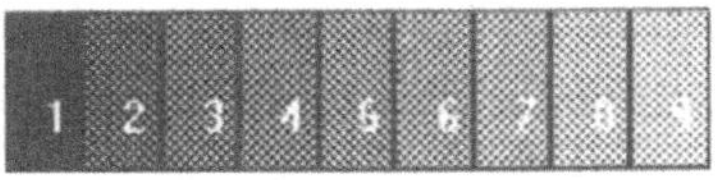

Figure 17: Class segregation network between adaptive egoists after 1000 simulation periods.

At first glance, the structural similarity of the conditions for cooperation for rational and adaptive actors seems to contradict with the obvious discrepancy between the segregation patterns of Figure 16 and 17. However, it is decisive for the network pattern that the threshold continuation probability for adaptive actors is somewhat larger than for rational actors. Figure 15b indicates that with $\alpha =$

0.903 adaptive actors succeed to cooperate only in those pairings with the lowest threshold continuation probability that is possible, i.e. *pairings with almost equal neediness*. Accordingly, adaptive egoists quickly learn that almost only partners of one's own neediness class are attractive, which drives them into the homogenous clusters shown by Figure 17. For rational egoists, however, a wider range of pairings was possible at this level of α, allowing for the formation of the onion-like pattern of Figure 16.

In this section we analysed solidarity networks using two different micro foundations. In one model, we assumed that actors are rational players in a game theoretical sense, in another model we applied a simple adaptive decision mechanism. Both simulations suggest that under the scenario we studied – relatively high stability of relationships – extensive solidarity networks characterised by segregation patterns may arise even between egoistic actors. Our study also demonstrated that using these two micro foundations allows us to assess how the predicted macro phenomenon may be affected by properties of individual decision making. In particular we found clearly different segregation patterns.

9.3 Conclusions

In this article we demonstrated dangers and merits of the CA framework. We used the example of opinion formation to illustrate effects built into and caused by the CA framework itself. It turned out that making a continuous opinion variable discrete may change the resulting opinion dynamics considerably. That result is a serious warning: One may be totally misled when modelling a variable as discrete that is in fact continuous in the real world process under study. We also showed that grid structures may matter considerably. Results obtained with regular grids may no longer be valid for irregular ones. These effects of the CA framework do not imply that studies using the framework necessarily fall victim to inherent artefacts. Our results do show, however, that careful consideration is required of modelling decisions upon appropriate grid structures and types of variables. Despite this danger, the CA framework is a useful tool to address a problem that is central to the modelling of social processes: the effects of variation in micro foundations on macro predictions. In the second part of this paper we used the framework to study and visualise how in the evolution of solidarity networks differences in micro behaviour may translate into clearly different network outcomes at the macro-level. Obviously CAs are both a dangerous *and* promising framework.

References

Abelson, R. P. (1964): Mathematical Models of the Distribution of Attitudes Under Controversy. In: Frederiksen, N. & Gulliken, H. (eds.): Contributions to mathematical psychology. Holt, Rinehart & Winstonm, New York, 142-160.

Abelson, R. P. (1979): Social Clusters and Opinion Clusters. In: Holland, P. W. & Leinhardt, S. (eds.): Perspectives on social network research. Academic Press, New York, 239-256.

Bush, R. R. & Mosteller, F. (1955): Stochastic models for learning. Wiley, New York.

Dawes, R. M., & Thaler, R. H. (1988): Anomalies: Cooperation. Journal of Economic Perspectives 2, 187-197.

Doob, J. L. (1953): Stochastic Processes. Wiley, New York.

Flache, A. (1996): The Double Edge of Networks: An analysis of the effect of informal networks on cooperation in social dilemmas. Thesis Publishers, Amsterdam.

Flache, A. & Macy, M. W. (1996): The Weakness of Strong Ties: Collective action failure in a highly cohesive group. Journal of Mathematical Sociology 21, 3-28.

Flache, A. & Hegselmann, R. (1998): Rational vs. Adaptive Egoism in Support Networks: How different micro foundations shape different macro hypotheses. In: Leinfellner, W. & Köhler, E. (eds.): Game Theory, Experience, Rationality. Foundations of Social Sciences, Economics and Ethics. In Honor of John C. Harsanyi [Yearbook of the Institute Vienna Circle 5]. Kluver, Dordrecht, 261-275.

Flache, A. & Hegselmann, R. (forthcoming): Rationality vs. Learning in the Evolution of Solidarity Networks: A Theoretical Comparison. In: Bonacich, P. & Carley, K. (eds.): Computational and Mathematical Organization Theory. Special Issue on Games and Networks.

French, J. R. P. jr. (1956): The Formal Theory of Social Power. Psychological Review 63, 181-194.

Friedman, M. (1953): Essays in positive economics. University of Chicago Press, Chicago.

Friedman, J. W. (1986): Game Theory with Applications to Economics, 2nd ed. 1991. Oxford UP, Cambridge.

Harary, F. (1959): A Criterion for Unanimity in French's Theory of Social Power. In: Cartwright, D. (ed.): Studies in Social Power. University of Michigan: Institute for Social Research, 168-182.

Hechter, M. (1992): The Insufficiency of Game Theory for the Solution of Real-World Collective Action Problems. Rationality and Society 4, 33-40.

Hegselmann, R. (1996): Cellular Automata in the Social Sciences - perspectives, restrictions, and artefacts. In: Hegselmann, R., Troitzsch, K. G. & Müller, U. (eds.): Modeling and Simulation in the Social Sciences from the Philosophy of Science Point of View (Theory and Decision Library). Kluver, Dordrecht, 209-234.

Hegselmann, R. (1998): Modeling Social Dynamics by Cellular Automata. In: Liebrand, W. B. G., Nowak, A. & Hegselmann, R. (eds.): Computer Modeling and the Study of Dynamical Social Processes. London, Sage, 37-64.

Hegselmann, R. & Flache, A. (1998): Understanding Complex Social Dynamics: A Plea For Cellular Automata Based Modelling. In: Journal of Artificial Societies and Social Simulation 1: http://www.soc.surrey.ac.uk/JASSS/1/3/1.html

Latané, B., Nowak, A. (1997): Self-Organizing Social Systems: Necessary and sufficient conditions for the emergence of clustering, consolidation and continuing diversity. In: Barnet, G. & Boster, F. (eds.): Progress in communication science: Persuasion, Ablex, Norwood, NJ, 43-74.

Lehrer, K., & Wagner, C. (1981): Rational Consensus in Science and Society. Reidel, Dordrecht.

Macy, M. W. (1991): Learning to Cooperate: Stochastic and tacit collusion in social exchange. American Journal of Sociology 97, 808-843.

Macy, M. W. & Flache, A. (1995): Beyond Rationality in Models of Choice. Annual Review of Sociology 21, 73-91.

Möller V. (1997): The irregular cellular automata.
page: http://www.uni-bayreuth.de/departments/philosophie/INHALT.HTM

Okabe A., Boots, B. & Sugihara, K. (1992): Spatial Tessellations - Concepts and applications of voronoi diagrams. Wiley & Sons, Chichester.

Roth, A. (1992): Game Theory as Part of Empirical Economics. In: Hey, J. D. (ed.): The Future of Economics. Blackwell, Cambridge, 107-114.

Taylor, M. (1987): The Possibility of Cooperation (revised edition of: Anarchy and cooperation, 1976). Wiley & Sons, London.

10. Evolution of Learning Rules in Space

Oliver Kirchkamp[1]

We discuss Monte Carlo studies of a cellular automaton that represents spatially distributed agents that interact with their neighbours in three ways: They play a repeated game, using a repeated game strategy, they learn repeated game strategies from their neighbours using a learning rule, and they learn learning rules. This threefold dynamics leads to learning rules whose implied behaviour is deterministic but also asymmetric in the sense that individual switching probabilities are more influenced by a learning player's own success rather than by the success of an observed neighbour. Furthermore stage game behaviour differs from the typical stage game behaviour found in the literature. In particular cooperation in prisoners' dilemmas disappears.

10.1 Introduction

In this chapter we relax one standard assumption that is common in spatial evolution: The literature on spatial evolution treats learning[2] rules as fixed over time, and restricts learning rules to be a variant of the type 'switch if better'. I.e. a learning agent copies another agent's strategy as long as the latter agent's payoff is higher than the first agent's payoff. A problem that arises in this context is that properties of the evolutionary dynamics depend on the underlying learning rules. Instead of choosing just another specific learning rule we will take a different approach according to which agents choose their own rules. This leads

---

[1] University of Mannheim, SFB 504, L 13, 15, D-68131 Mannheim, email: oliver@kirchkamp.de. I am grateful for financial support from the Deutsche Forschungsgemeinschaft, Sonderforschungsbereich 303 at the University of Bonn and Sonderforschungsbereich 504 at the University of Mannheim, as well to the hospitality of the European University Institute, Florence. I thank Karl Schlag and Avner Shaked for helpful and stimulating comments. Parts of the simulations were done with the help of the `meschach` library for matrix algebra. Some of the graphics were produced with the help of Mathematica and $\TeX$. gcc, emacs and friends were a great help while developing the software.

[2] Note that 'learning' denotes 'change of behavior' throughout the paper.

to another level of evolutionary dynamics. We want to find out which rules are chosen by the agents, in which way these rules relate to the commonly used 'switch if better' rule, and in which way agent's behaviour on the level of the stage game is influenced.

The literature on spatial evolution is quite young. This may be related to the fact that algebraic analysis of spatial evolution is often rather complex and has to be replaced by Monte Carlo studies, which become more feasible, given recent technological development.

Since Hotelling (1929) many spatial models have been specified and solved. Nevertheless spatial models which assume players' rationality to be limited have been neglected for a long time. Sakoda (1971) and Schelling (1971) were possibly among the first to present models of spatial evolution, later followed by Axelrod (1984, p. 158ff) and Nowak and May (1992, 1993), Bonnhoeffer, May and Nowak (1993) and Lindgren and Nordahl (1994). All of them have studied models where a population is represented as a *cellular automaton*. Players are represented as single cells which are in interaction with neighbouring cells and which are learning from neighbouring cells. Ellison (1993) and Eshel, Samuelson and Shaked (1996) have analytically derived properties of particular cellular automata.

Some of the above (Sakoda 1971, Schelling 1971, Ellison 1993) assume that players *optimise myopically.*[3] Others (Axelrod 1984, May, Nowak 1992, 1993, Bonnhoeffer, May and Nowak 1993, Lindgren and Nordahl 1994, Kirchkamp 1995, Eshel, Samuelson and Shaked 1996) assume that players *learn through imitation* of successful strategies. We will follow the latter approach.

Common assumptions of the literature that studies local evolution are *synchronous*[4] interaction and learning (Axelrod 1984, May, Nowak 1992, 1993, Bonnhoeffer, May and Nowak 1993, Lindgren and Nordahl 1994, Eshel, Samuelson, Shaked 1996) and the restriction that players *can not discriminate* among different neighbours, i.e. they either cooperate with all their neighbours or they do not cooperate with any neighbour at all (Nowak 1992, 1993, Eshel, Samuelson, Shaked 1996). In the current chapter we will relax these assumptions. We will study a model of a population where players' behaviour is *not synchronised* through an external clock and where players are able to *discriminate* between their neighbours.

---

[3] 'Myopic' means in this context that players assume that their opponents will behave in the next period as they did in the present one.

[4] With 'synchronization' we mean that it is *predetermined* whether in any given interval certain interactions or learning events take place. The most common specification is then that during each period *all* possible interactions and learning events take place. We call an event (like interaction or learning) not synchronized if for each possible interaction or learning step a random draw decides whether the interaction or learning step takes place.

We will concentrate on the evolutionary process itself. We discuss a model of evolution of the evolutionary rules, i.e. a model of evolution of learning rules. We analyse this model using Monte Carlo studies and compare the evolving learning rules and the induced population behaviour with rules and strategies studied in the literature.

Let us first go through some definitions of the term 'evolution': Economists use the expression 'evolution' to describe any kind of dynamic process. 'Evolutionary economics' in a Schumpeterian (Schumpeter, 1934) tradition analyses the dynamics of a process where individuals continuously find new knowledge, often in the form of new technologies. The process of *how* a new technology is found is often exogenous and not specified in detail. Evolutionary economics concentrates more on the effects newly developed technologies have on the evolution of growth of other technologies and of the economy as a whole.[5]

Economists also use the expression 'evolution' in the context of 'evolutionary game theory' (in contrast to 'evolutionary economics'). 'Evolutionary game theory' sometimes makes quite specific assumptions on individual behaviour and capabilities. Individual rationality is often limited. Popular restrictions are e.g. the assumption of *myopia*[6], *limited access to information*[7], *unreliable information*[8] and furthermore *inertia* that gives individuals only very rarely the opportunity to update their strategy.[9]

Some models in evolutionary game theory (which are often inspired by biology) exclude *optimising* behaviour. In these models individuals do not evaluate fictitious situations and then choose strategies based on this evaluation. Instead, successful strategies grow as long as they are successful while unsuccessful strategies vanish from the population.[10] The underlying story might (in biological models) model a *birth* process where users of successful strategies produce more offspring and a *death* process where users of unsuccessful strategies are more likely to die soon. In social contexts common motivations refer to successful strategies being more likely to be copied by other members of a society while users of unsuccessful strategies are more likely to abandon their strategy.

---

[5] See Richard Nelson (1995) for an exhaustive review of recent developments in evolutionary economics.

[6] Myopia in this context means that individuals optimise under the (wrong) assumption that they are the only ones who will change their behaviour.

[7] Limited access to information could, for instance, be reflected in subjects not being able to remember past actions or past payoffs at all.

[8] Unreliable information means that subjects most of the time remember past actions or payoffs correctly, but sometimes they remember incorrectly.

[9] Myiopia, limited and unreliable access to information, as well as inertia are, for instance, major aspects of models studied by Ellison (1993), Kandori, Mailath, Rob (1993) and Young (1993).

[10] See e.g. Maynard Smith and Price (1973) for static concepts and Taylor and Jonker (1978) and Zeeman (1981) for a dynamic model.

When discussing the evolutionary model in the remainder of the chapter we will use the latter interpretation: Players follow a specific rule that imitates strategies from other players based on the 'success' of these players.

Another important property of the model that will be presented is that it is *spatial*. Due to its intrinsic heterogeneity, a spatial model is harder to analyse than a non-spatial (homogeneous) one. However, there are two major reasons which justify the study of evolution in space. First, as shown by Axelrod's example, properties of a simplified non-spatial model often differ too much from properties of a spatial one, making it questionable to approximate a spatial with a non-spatial model. Second, space introduces heterogeneity in a very natural way. Such heterogeneity renders learning rules necessary and requires the study of their evolution.

In the recent literature on spatial population behaviour, cellular automata have become more and more popular. In the following paragraph we will briefly mention four categories that distinguish different cellular automata models: (1) agents may or may not migrate, (2) agents may or may not change strategies, (3) both migration and change of strategies may be driven by imitation of successful agents or by a more or less elaborate optimisation process, (4) the presence or absence of mutations.

The distinction between models with *migration* and models where *strategies change* may be exemplified by a number of previous studies. On the one hand, Sakoda (1971) and Schelling (1971) take players' states as fixed and introduce dynamics of the cellular automaton through movements of players, whereas Axelrod (1984), May, Nowak (1992, 1993), Ellison (1993), Eshel, Samuelson and Shaked (1996) take players' positions as fixed but allow players to change their states. Furthermore there are models where players are allowed to move *and* to change their states (Hegselmann 1994).

In the local models of Schelling, Sakoda and Ellison players are capable to *optimise* at least myopically. A different approach is taken by Axelrod (Axelrod 1984, p. 158), Nowak and May (1992), Kirchkamp (1995) and Eshel, Samuelson and Shaked (1996) who consider a model where players learn through *imitation*.

Axelrod (1984, p. 158) analyses a cellular automaton where initially all cells are filled with players that use various strategies. In each period, players gain payoffs from a tournament against all their neighbours. Between periods, players copy successful strategies from their neighbourhood. This process gives rise to complex patterns of different strategies. Axelrod found that most of the surviving strategies in such spatial environment were successful in a non-spatial context as well. However, some of the *locally* surviving strategies show only intermediate success in Axelrod's *global* model. These are strategies that get along well with themselves but are only moderately successful against other strategies. They can be successful in the spatial model because they typically form clusters where they only play against themselves. Lindgren and Nordahl (1994) study an extension of Axelrod's model where they introduce mutations of the repeated game strategies.

While Axelrod and also Lindgren and Nordahl study a population with a large number of possible strategies, Nowak and May (1992, 1993) and Bonnhoeffer, Nowak, and May (1993) analyse a population playing a prisoners' dilemma where players are restricted to strategies that may not distinguish among neighbours and that have no memory. The potential to learn a completely new strategy may, however, compensate for this restriction. They study various initial configurations and analyse the patterns of behaviour that evolve with synchronous interaction.

Eshel, Samuelson and Shaked (1996) derive analytically the behaviour of a particular cellular automaton. The price they have to pay for the elegance of the analytical result is the restriction to a small set of parameters. They have to assume a particular topology and neighbourhood structure. They find that there is a range of prisoners' dilemmas where increasing the gains from cooperation reduces the amount of players who actually cooperate. Since it is hard to find more than a purely mathematical intuition for this result, it might be regarded as an artefact that results from the discrete structure of the model.

In the following we will focus on the *learning rules* that players use to change their strategy. We drop the assumption that learning rules are given exogenously and analyse evolution of strategies together with evolution of learning rules. Thus, players will be characterised by a repeated game strategy and a learning rule. Players will use their individual learning rule to revise their strategy. They will use a replicator dynamics to update their learning rule.

We find that in the long run learning rules deviate in two respects from the learning rules studied in the literature. They are not deterministic in the sense that many learning decisions are stochastic events. Further, surviving learning rules are asymmetric in the sense that switching probabilities are more sensitive to changes in the learning player's payoff than to changes in the observed neighbour's payoff.

We show that players' strategies which emerge under endogenous evolution learning rules differ from those emerging under exogenous rules. In contrast to the fixed learning rules studied by Axelrod (1994) and Nowak, May (1992, 1993) we find that cooperation in prisoners' dilemmas and coordination on equilibria which are not risk dominant in the sense of Harsanyi and Selten (1988) vanishes with our endogenous learning rules.

10.2 The Model

In the following we consider a model where agents are located on a network, like a torus, or a big checkerboard whose edges are pasted together. The torus has a size 21×21. Each cell of the torus is occupied by exactly one agent. Properties of each agent in the first round of a Monte Carlo run will be assigned randomly. Interaction among agents is threefold: Agents play games against their immediate neighbours, learn repeated game strategies from their neighbours and learn learning rules.

The underlying stage game can be described as follows:

Table 1: The stage game payoff matrix

		Player II: C	Player II: D
Player I	C	g, g	h, -1
	D	-1, h	0, 0

Parameters g and h will be chosen randomly from (-1.1,1.1) for g and from (-2.2,2.2) for h. The space of games can be represented as follows:

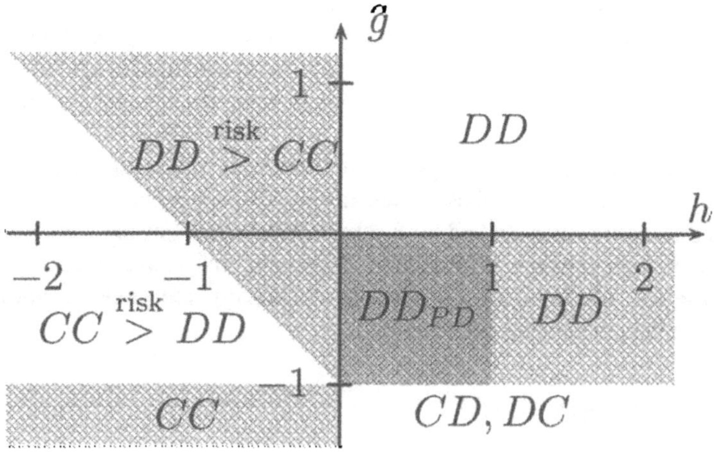

Figure 1: The space of games

The range $-1<g<1$ and $-2<h<2$ includes both prisoners' dilemmas and coordination games. All games with $-1<g<0$ and $0<h<1$ are prisoners' dilemmas

(DD_{PD} in the above figure), and all games with $g>-1$ and $h<0$ are coordination games. Coordination Games are denoted with $CC \overset{risk}{>} DD$ or $DD \overset{risk}{>} CC$. In the case described above, coordination games have two Nash-equilibria in pure strategies. One, where both players choose C, and another one where both players choose D. The learning process that we will describe later on will select, depending on the values g and h, mostly one of these equilibria. One of the criteria that are used in theory to select among equilibria is *pareto-dominance*. Following this criterion means selecting the equilibrium where payoffs are higher, i.e. both players choose C as long as $g>0$ and D otherwise. Another criterion is *risk-dominance* which takes into account that the opponent might deviate from the equilibrium strategy and tries to avoid losses that such a deviation may yield. The area where CC risk dominates DD are marked $CC \overset{risk}{>} DD$, and those where DD risk dominates CC are marked $DD \overset{risk}{>} CC$. The concepts of Pareto dominance and risk dominance are developed in detail in Harsanyi and Selten (1988).

Each agent will be characterised by three properties: A repeated game strategy that is used against all neighbours, states of this repeated game strategy for each neighbour, and a learning rule that is used to update the repeated game strategy.

For the simulations discussed below we have to make specific assumptions, such as a particular learning speed, a given interaction frequency, etc. How sensitive are the results of this simulation to specific assumptions? In Kirchkamp (1998) we show within a different context that results which are obtained with the following very simple and specific assumptions can be extended to a broader range. The purpose of the analysis that follows is to give a simple example of a simulation model that, despite its simplicity, shows several interesting properties.

For learning and interaction we will assume Moore-neighbourhoods of radius one[11]. To avoid artefacts, timing will be modelled asynchronously, i.e. actions, like strategic interaction among two agents, learning of an agent, as well as updates of an agent's learning rule will be determined randomly and independently for each agent (for each pair of agents). Interactions for a given pair of agents will take place with probability 1/2, learning for each agent will take place with probability 1/24. Repeated game strategies are restricted to Moore-machines with less than three states — thus strategies like tit-for-tat, grim, tat-for-tit, defect always, cooperate always are included.[12]

---

[11] I.e. a player will interact with (and learn from) the eight players whose horizontal and vertical distance from him or her does not exceed 1.

[12] Moore-machines are finite automata that are described as sets of states. Each state defines a stage game strategy and a transition function that takes the opponent's action as an argument and maps back into the set of states of this automaton. The transition functions prescribe with which state to proceed depending on the stage game behavior of the opponent. With the definition of the initial state that defines with which state to begin before we observed our opponent's action the definition is complete. For 2×2 games there are 2 automata with only one state. The stage game strategy is one of the 2 possible stage

We restrict ourselves here to simple learning rules that use only information on two players' payoffs: The payoff of the learning player itself and the payoff of a single player that is chosen randomly at each learning step from the learning neighbourhood.

Thus our learning rules have to choose only one of at most two strategies. This makes learning rules easy to describe and easy to parameterise. Taking more neighbours into account would complicate the learning rules and would possibly make the results too hard to interpret.

In our Monte Carlo study we consider among all behavioural rules only the class of learning rules where the probability of switching to an observed strategy is a truncated linear function of a player's own payoff (sown) and the payoff of the sample player (ssamp). These rules will be called 'linear switching rules'. Of course, we cannot require an overall linear learning rule that forces players to switch with probability $a_0 + a_1 s_{own} + a_2 s_{samp}$ to the observed strategy since values of this expression need not be in [0,1]. We therefore associate negative values of the above expression to 'not switching' and values greater than 1 to 'switching for sure'.

Note that what may look like a technical complication is actually a strength of the model. The rules that we describe above can, with appropriately chosen parameters, model a fully linear and always probabilistic rule. But they can also, with other parameters, model a completely deterministic rule. Finally, this specification also captures several interjacent rules. One of the questions that we want to answer with the help of this model is whether the evolutionary process will select probabilistic, deterministic or mixed rules.

Note further that this specification also covers rules that are sensitive to relative performance of strategies. Rules that are based on aspiration levels, as well as rules which are based on a concept of 'switch-if-better', can be described as a special case.

Evolution of learning rules could occur on a local level similar to the modelling of evolution of repeated game strategies. However, evolution of learning rules is not very frequent, and therefore may imply a larger effort, or, in other words, a wider scope than the local environment that was used for strategy updates. For simplicity we consider the extreme case of global evolution of learning rules — replication of learning rules is based on global fitness of a learning rule. To define evolution of learning rules we assume a process that approximates

---

game strategies. The transition function always maps back to this state. For 2×2 games there are 26 automata with two states. This set of two-state automata covers a broad and interesting range. One example is tit-for-tat. In the initial state tit-for-tat cooperates. The transition function maps back to this state as long as the opponent cooperates as well. Otherwise the transition function maps to the second state with the associated strategy 'defect'. A similar transition function maps back to the first state, provided that the opponent cooperates.

standard replicator dynamics when the population is large. We select rules that are going to be replaced (to 'die') proportional to their non-success and select rules that are going to be replicated (to be 'born') proportional to their success. In each period a random draw decides how many learning rules will replicate for the current period. Probabilities are chosen such that, on average, each player changes the learning rule after 240 periods. This is reasonably fast but still much slower than the average lifetime of a strategy (24 periods).

Thus in a world with 21 learning rules, at each period a random draw would decide with probability 21/240 whether a learning rule is going to be replicated. Depending on how many learning rules are to replicate, sequentially pairs of rules are selected: One rule to replicate, another to be replaced. A rule that is going to be replicated is selected proportional to its success while the learning rule which is to be replaced is selected proportional to its non-success.

In the next period all previous users of a replaced rule will use a slightly perturbed version of a replicated rule, i.e. a random linear combination of the 'replicated rule' and a random rule that will be very close to the 'replicated rule'.[13]

There will also be mutations within the replication of the repeated game strategies. With a small probability (of 1/100) a player that actually learns to switch to a certain repeated game strategy will not pick this strategy but a random strategy instead.

A problem arises due to the fact that positions on a heterogeneous network are different. Sometimes advantageous learning rules might be surrounded by unfavourable environments while disadvantageous learning rules might benefit from its beneficial surroundings. This effect slows down the evolutionary dynamics since the average payoff of a learning rule is perturbed by the noise of this 'neighbourhood effect'. While this might be an interesting effect to study, it is not in the focus of our analysis. We have learned from other simulations that this effect vanishes with higher dimensionality. As a short cut and to speed up Monte Carlo runs we aggregate payoffs of several players. At every point in time there are exactly n different learning rules, each of them used by one of n players. Players with their learning rules are distributed randomly on a square grid. Fitness of a learning rule is determined as the average payoff of all its representatives. Another way to speed up the Monte Carlo runs is what we call migrations of learning rules. Whenever a learning rule is replicated, a random draw decides with probability p_m whether the users of the newborn rule change places with randomly chosen members of the population.[14] The migrating players continue to use strategies and states of their predecessors in their new position.

---

[13] The linear combination puts weight x^l on the 'replicated rule' and $1-x^l$ on the random rule. In our simulations $l=1/10$ and x is distributed equally between 0 and 1.

[14] If not stated otherwise learning rules migrate with probability $p_m=1$.

Each Monte Carlo run starts with the selection of a game. We chose payoffs g and h from uniform and independent distributions in (-1.1, +1.1) and (-2.2, +2.2) respectively. It takes normally 2000 to 5000 rounds to reach a stable distribution for learning parameters. The results displayed below are averages over the last 5000 rounds of runs that each lasted 10 000 rounds.

In order to cover a wide range of parameters and initial values we have normally carried out at least 800 different runs each starting with different configurations of repeated game strategies, learning parameters and payoffs.

In terms of CPU usage one simple Monte Carlo run[15] uses about two days exclusive use of an RS6000 machine with a 7013/530 processor.

10.3 Results

In the following we will first describe properties of the endogenous learning rules and then discuss stage game behaviour that is implied by these rules.

The switching probabilities, averaged over 800 repetitions, are displayed in Figure 2:

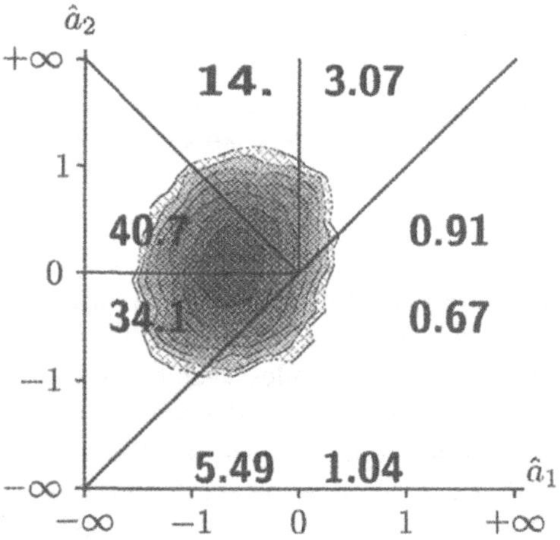

Figure 2: Distribution of learning rules.

[15] A simple Monte Carlo run means 800 simulations lasting for 10000 rounds each where learning and interaction neighbourhoods have radius 1 and the grid has size 21 × 21.

Since we can not display a distribution over the three-dimensional space (a_0, a_1, a_2) we show here only the projection into (a_1, a_2). The figure was derived from a table of frequencies with 30 × 30 cells and is simultaneously a density plot and a table of relative frequencies: Different densities of the distribution are represented by different shades of grey. The highest density is represented by the darkest grey.[16]

The picture also contains a table of relative frequencies. The percentages within each of the eight sectors represent the amount of players that use a learning rule with parameters in the respective range.

We first note that most of the endogenous rules are 'reasonable' in the sense that $a_1 + a_2 < 0$ in only 8.1% of all cases. However, rules are not symmetric in the sense that players are much more sensitive to changes in their own payoff than to changes in the sampled neighbours' payoffs. This asymmetry seems to be in contrast with observations from the literature on optimal learning rules (Boergers 1995, Schlag 1996). Boergers and Schlag find that when players are *infinitely patient,* the optimal learning rule ascribes equal weights to both the player's own payoff as well as to the observed payoff.

Since our players do not optimise it is meaningless to call them more (or less) *patient*. However there is an *implicit patience* in the dynamics: A player's fitness influences the speed of the evolutionary process even in finite time. A learning rule that quickly reaches a *good*, although *not optimal*, score outperforms a *slow* but *optimal* rule. Thus, evolution within finite time favours quick, possibly sub-optimal, learning rules. Now, why are these rules asymmetric? A successful neighbour might be successful only with the help of some other neighbours who could be too far away from the learning agent to be relevant for the agent's own payoff. Therefore an agent might be more successful, at least in the short run, when the agent cares more about his or her own payoff than about a neighbour's payoff.

Let us next compare the *stage game behaviour* under endogenously evolving learning and under fixed learning rules. Figure 3 shows the stage game behaviour for the rule 'switch if better'. 'Switch if better' can be approximated by the linear switching rule described earlier when $a_0 = 0$ and $a_2 = -a_1$ close to infinity. Hence whenever the sampled payoff is larger than the player's own payoff, the sampled

[16] Densities are derived from a table of frequencies with a grid of size 30 × 30. We actually map logs of densities into different shades of gray. The interval between the log of the highest density and the log of 1% of the highest density is split into seven ranges of even width. Densities with logs in the same interval have the same shade of gray. Thus, the white area represents densities smaller than 1% of the maximal density while areas with darker shades of gray represent densities larger than 1.9%, 3.7%, 7.2%, 14%, 27% and 52% of the maximal density.

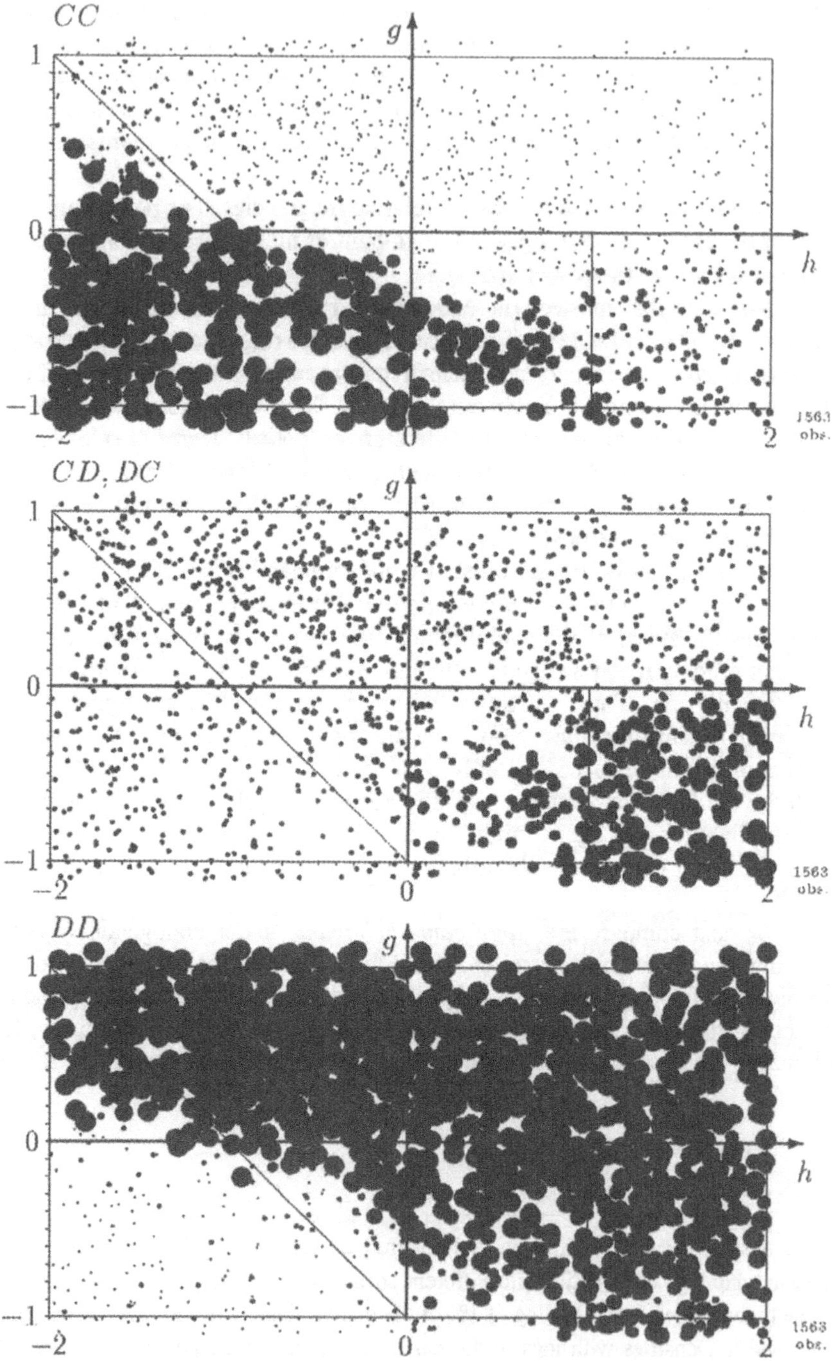

Figure 3: Stage game behaviour with 'switch if better'.

strategy is copied. Otherwise the agent keeps his or her own strategy.[17] In our framework this is a close approximation of the 'switch if better' rules from the literature.

Figure 3 displays relative frequencies of pairs of stage game strategies (*CC*, *CD*, *DC*) for various games (*g*, *h*), given the same parameters (size of the network, size of learning and interaction neighbourhoods, frequency of strategy updates, mutations for repeated game strategies etc.) as for the model described above. The only difference here is that learning rule is fixed from the beginning as 'switch if better' and is no longer subject to an evolutionary process.

In figure 1 we have introduced a graphic representation of different symmetric 2×2 games using two parameters g and h. We use this representation to describe stage game behaviour for different games. Each run of the Monte Carlo study refers to a particular game with payoffs g and h. Such a run leads in the long run to certain relative frequencies of pairs of stage game strategies. In the above figure we display relative frequencies of pairs of stage game strategies as they were realised during the Monte Carlo study. Each possible pair of strategies (*CC*, *CD*, *DD*) is represented in one diagram. For example, thick circles in the *CC*-diagram mean that *CC* was played rather often whereas small circles or no circles at all means that the respective strategy was played rarely or not at all.

The results in Figure 3 show that in contrast to predictions of normative game theory, the fixed learning rule 'switch if better' leads to cooperation for a large range of prisoners' dilemma games.

Most of the cooperation is found for g values close to -1 and h values close to 0, i.e. in a region where mutual cooperation is particularly efficient and not very risky.

The region $h>1$ and $-1<g<0$ is not a prisoners' dilemma. In this region alternating between *CD* and *DC* yields higher payoff than the Nash Equilibrium *DD*. Note that the socially efficient *CD* and *DC* is also found by 'switch if better'.

Let us next turn to coordination games. A coordination game has two equilibria, one where both players play *CC* and one where both players play *DD*. The recent literature on evolution in the presence of small mutations (see Kandori, Mailath, Rob 1993, Young 1993) argues that evolution tends to favour the risk dominant equilibrium. In figure 3 we can indeed separate the area of coordination games where mostly *CC* is played from the area where mostly *DD* is played. Following the results of Kandori, Mailath, Rob and Young we should expect these areas to be separated by the diagonal that divides *CC*-risk dominant equilibria from *DD*-risk dominant equilibria (see figure 2). For the 'switch if better' learning rule we find that an intermediate case between Pareto dominance and risk dominance is selected. Hence, at least in a spatial context, both for

---

[17] Other deterministic rules that sample all players in the neighborhood and then copy the best player (a rule used by Nowak and May) or copy the strategy with highest average payoff yield very similar results on the level of stage game behavior.

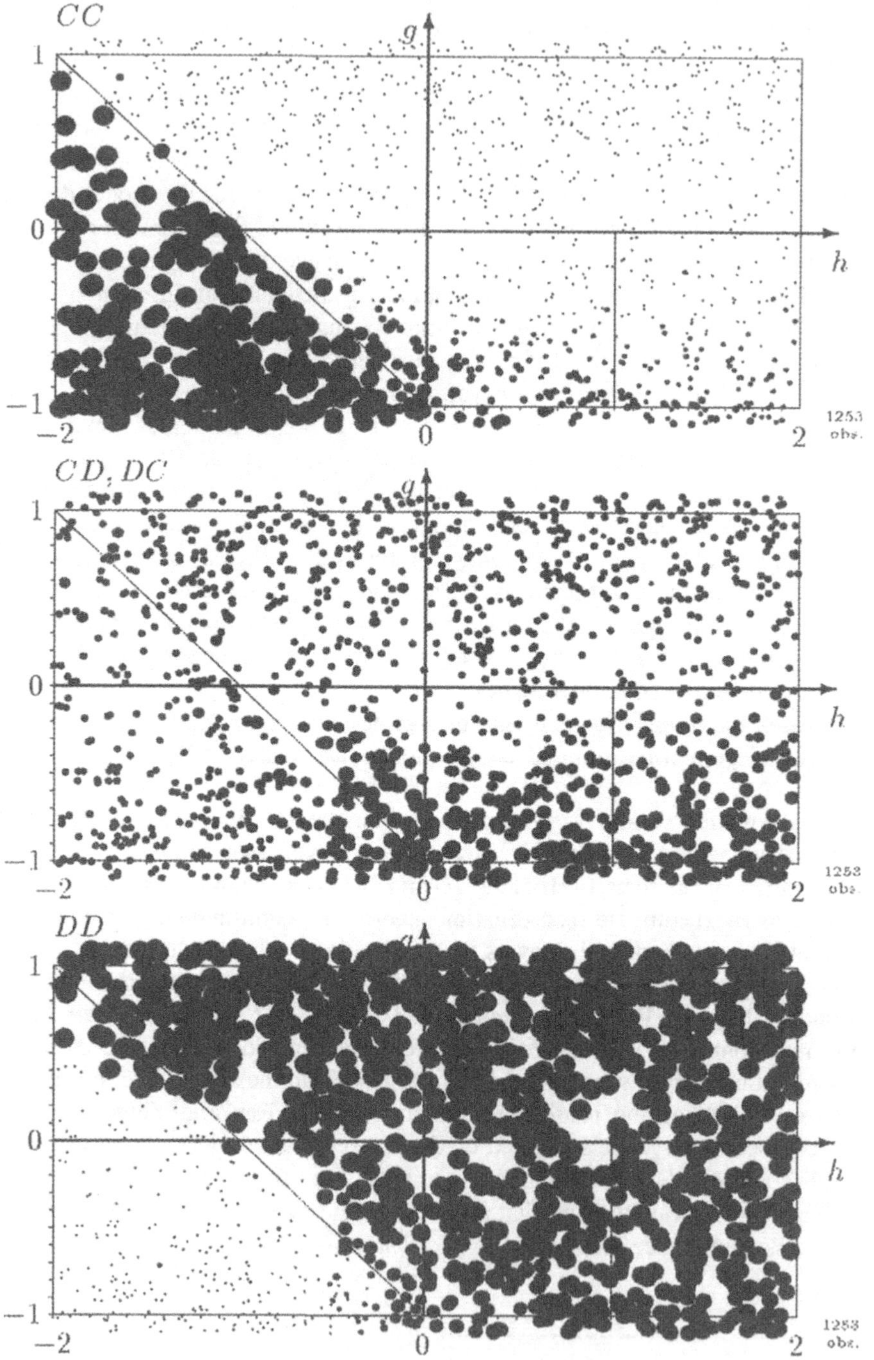

Figure 4: Stage game behaviour with endogenous learning rules.

prisoners' dilemma games and for coordination games the learning rule 'switch if better' favours Pareto dominance.

Next, let us turn to endogenous evolution of learning rules. Relative frequencies of pairs of stage game strategies are displayed in figure 4.

Here we find almost no cooperation for prisoners' dilemmas. We further find in coordination games that the selection of pairs of strategies is mostly determined by the criteria of risk dominance.

Apparently it is the pre-specification of the 'switch if better' rule that generates cooperation in prisoners' dilemmas. As soon as evolution chooses its own learning rule, players shift away from deterministic rules, and thus abandon cooperation.

10.4 Conclusions

In this chapter we addressed two questions: First we considered whether learning rules of the type 'switch if better', which are used in the literature on spatial evolution, can be justified as the outcome of an evolutionary selection dynamics.

Our findings indicate that the dynamics that we analysed here select rules which are different from the ones commonly studied in the literature. In particular, endogenous learning rules tend to be less sensitive to the neighbours' success than to the player's own success. This contrasts with findings from the literature (Boergers 1995, Schlag 1996) showing that 'optimal learning rules' are symmetric in the sense that players are equally sensitive to the observed neighbours payoffs and to their own payoff. Our Monte Carlo study suggests that this symmetry of optimal learning rules results from the assumptions of 'infinite patience' and 'infinite time horizon'. Assuming a finite time horizon forces players to learn 'fast' which, in turn, leads to asymmetry.

Second, we asked whether endogenous learning rules imply the stage game behaviour found in the literature. Here we found that important properties of stage game behaviour, like cooperation for some prisoners' dilemmas and the lack of coordination on risk dominant equilibria may disappear with endogenous learning rules.

Notwithstanding, we face a conceptual problem: While we did not specify a fixed learning rule, we made assumptions concerning the underlying meta-learning process, i.e. the evolutionary dynamics that selects learning rules. Changing these assumptions may affect results, as it actually did, at least on the level of stage games. Kirchkamp (1996) studied a more sophisticated meta-learning process in which agents smooth success of all learning rules from their neighbourhood. This was done by using a quadratic fit to choose the maximum of the smoothed function. While this specification seems to differ significantly from the model used here, properties of learning rules seem to be fairly similar in both models. In both models we obtain a 'suspicion' effect according to which learning rules are more sensitive to an agent's own payoff than to an observed neighbour's payoff. So far, the simple model presented here seems to be consistent with the more complex one suggested by Kirchkamp (1996). However, the implied stage game behaviour is different. Agents now cooperate in at least some prisoner's dilemmas and also behave more cooperatively in some coordination games.

One might get the impression that endogenising learning rules only shifts the question which learning rule to specify to the level of meta-learning rules. However, we could escape this endless circle if changing rules on a certain meta-level has little or no impact. Our study is a step in this direction. Changing learning rules has substantial impact on players' behaviour. Changing the dynamics that selects learning rules seems to have less impact. With different rules we still find common features of the implied learning rules (like 'suspicion').

Having said that, learning rules from the literature on spatial evolution (since they do not share this feature) become more questionable. It may still be possible to justify 'switch if better' rules for various reasons, e.g. their simplicity or the fact that deterministic rules are more plausible. However studying endogenous learning rules may help us to look for new fixed rules and view the present ones differently.

References

Ashlock, D., Stanley, E. A. & Tesfatsion, L. (1994): Iterated Prisoner's Dilemma with Choice and Refusal of Partners, In: Langton, C. G. (ed.): Artificial Life III, no. XVII. In: SFI Studies in the Sciences of Complexity, Addison-Wesley, 131 – 167.

Axelrod, R. (1984): The evolution of cooperation. Basic Books, New York.

Binmore, K., & Samuelson, L. (1994): Muddling Through: Noisy Equilibrium Selection. Discussion Paper B–275, SFB 303, Rheinische Friedrich Wilhelms, Universität Bonn.

Boergers, T. & Sarin, R. (1995): Naive Reinforcement Learning with Endogenous Aspirations. Second International Conference on Economic Theory: Learning in Games. Universidad Carlos III de Madrid.

Bonhoeffer, S., May, R. M. & Nowak, M. A. (1993): More Spatial Games. International Journal of Bifurcation and Chaos 4, 33–56.

Ellison, G. (1993): Learning, Local Interaction and Coordination. Econometrica, 61, 1047–1071.

Eshel, I., Samuelson, L. & Shaked, A. (1996): Altruists, Egoists and Hooligans in a Local Interaction Model. Tel Aviv University and University of Bonn.

Glance, N. S., & Huberman, B. A. (1993): Evolutionary Games and Computer Simulations. Proceedings of the National Academy of Sciences 90, 7716–7718.

Harsanyi, J. C., & Selten, R. (1988): A General Theory of Equilibrium Selection in Games. The MIT Press, Cambridge, Massachusetts.
Hegselmann, R. (1994): Zur Selbstorganisation von Solidarnetzwerken unter Ungleichen. In: Homann, K. (ed.): Wirtschaftsethische Perspektiven I, no. 228/I. In Schriften des Vereins für Socialpolitik, Gesellschaft für Wirtschafts- und Sozialwissenschaften, Neue Folge, Duncker & Humblot, Berlin, 105–129.
Hotelling, H. (1929): Stability in Competition. Economic Journal 39, 41–57.
Kandori, M., Mailath, G. G. & Rob, R. (1993): Learning, Mutation and Long Run Equilibria in Games. Econometrica 61, 29–56.
Kirchkamp, O. & Schlag, K. H. (1995): Endogenous Learning Rules in Social Networks. Rheinische Friedrich Wilhelms Universität Bonn, Mimeo.
Kirchkamp, O. (1995): Spatial Evolution of Automata in the Prisoners' Dilemma. Discussion Paper B–330, SFB 303, Rheinische Friedrich Wilhelms Universität Bonn.
Kirchkamp, O. (1996): Endogenous Evolution of Learning Rules. Discussion Paper B–379, SFB 303, Rheinische Friedrich Wilhelms Universität Bonn.
Kirchkamp, O. (1998): Simultaneous Evolution of Learning Rules and Strategies. EUI Working Paper ECO 98/31, European University Institute Florence.
Lindgreen, K. & Nordahl, M. G. (1994): Evolutionary Dynamics of Spatial Games. Physica D 75, 292–309.
May, R. M. & Nowak, M. A. (1992): Evolutionary Games and Spatial Chaos. Nature 359, 826–829.
May, R. M. & Nowak, M. A. (1993): The Spatial Dilemmas of Evolution. International Journal of Bifurcation and Chaos 3, 35–78.
Maynard Smith, J. & Price, G. R. (1973): The Logic of Animal Conflict. Nature 273, 15–18.
Nelson, R. (1995): Recent Evolutionary Theorizing about Economic Change. Journal of Economic Literature 33, 48–90.
Sakoda, J. M. (1971): The Checkerboard Model of Social Interaction. Journal of Mathematical Sociology 1, 119–132.
Schelling, T. (1971): Dynamic Models of Segregation. Journal of Mathematical Sociology 1, 143–186.
Schlag, K. (1996): Why Imitate and If So, How? A Bounded Rational Approach to Multi-Armed Bandits. Discussion Paper B–361, SFB 303, Rheinische Friedrich Wilhelms Universität Bonn.
Schumpeter, J. (1934): The Theory of Economic Development. Harvard University Press, Cambridge.
Taylor, P. D. & Jonker, L. B. (1978): Evolutionary Stable Strategies and Game Dynamics. Math. Biosc. 40, 145–156.
Weibull, J. W. (1995): Evolutionary Game Theory. MIT Press, Cambridge, Massachusetts.
Young, H. P. (1993): The Evolution of Conventions. Econometrica 61, 57–84.
Zeeman, E. (1981): Dynamics of the Evolution of Animal Conflicts. Journal of Theoretical Biology 89, 249–270.

11. Maintaining Diversity: Simulating the Role of Nonlinearity and Discreteness in Dynamic Social Impact[1]

Bibb Latané and Hiroaki Morio[2]

Dynamic social impact theory (DSIT) uses Latané's (1981) individual-level theory of social impact to explain and predict the emergence of group-level phenomena as consequences of social influence among interacting people. An extensive program of social simulation has resulted in numerous recent articles and chapters reporting the emergence of several group-level phenomena as very general consequences of social influence among spatially distributed individuals (see the symposium in the fall 1996 *Journal of Communication* for an introduction and review of dynamic social impact theory, and Latané, 1996a; b; c; d; 1997 for discussions of the assumptions and implications of computer simulation of social processes).

Three robust phenomena discovered by these simulations include: *consolidation,* or a reduction in diversity as minorities are exposed to more adverse influence than majorities; *clustering,* as neighbors come to resemble one another more than strangers; and *continuing diversity* as clustering allows minorities to survive by forming local majorities. These phenomena hold for simulations using a variety of simple and complex influence networks or social spaces, populations ranging from 9 to 196,000 actors, and a large assortment of computing platforms and simulation procedures.

The existence of these phenomena is not restricted to virtual reality, but extends to empirical tests conducted among real people discussing a wide variety of issues in electronic discussion groups (Latané & Bourgeois, 1996; Latané & L'Herrou, 1996) and in lecture halls (Harton, Green, Jackson & Latané, 1998). The phenomena have interesting implications for understanding the large-scale emergence of social representations (Huguet & Latané, 1996; 1998) and even the evolution of culture (Schaller & Latané, 1996).

Consolidation and clustering, like the air we breathe, are so pervasive as often to escape notice. In a seminal study, Festinger, Schachter, and Back (1950) demonstrated the existence of and developed a theoretical basis for these phenomena (a.k.a. pressures to uniformity in groups) in a postwar MIT housing project. However, recent interest, as shown by textbook treatments such as Myers (1994),

[1] We thank Ramzi Suleiman for his encouragement and helpful comments.
[2] Department of Psychology, Florida Atlantic University

has concentrated on two opposing processes identified by Moscovici—minority influence and group polarization (Moscovici, 1985; Moscovici & Zavalloni, 1969)—usually assumed to moderate but not negate consolidation, which continues to be taken for granted.

By contrast, continuing diversity, like the flight of the bumblebee, has seemed theoretically improbable if empirically inevitable. In fact, in 1964 the distinguished social psychologist Robert Abelson presented a mathematical proof that for a wide range of social influence functions (basically whenever opinion discrepancy leads to a linear, incremental tendency for individuals to change their opinion towards greater agreement), members of fully connected influence networks should inevitably come to full agreement[3]. That is, in the absence of competing psychological or social processes, linear social influence in which change is proportional to the degree of pressure should inexorably lead to ultimate ubiquitous agreement. Abelson was quick to recognize that in this case, mathematical proof was not a good guide to real-world outcomes, and has since discovered and/or discussed a number of processes which might inhibit the rush to agreement.

Nowak, Szamrej, and Latané (1990) showed that Abelson's proof does not apply under conditions of nonlinear or non-incremental change in fully connected but spatially distributed networks (in which not everyone is connected directly to everyone else). They created computer simulations of social impact in a population of 400 people varying in persuasive strength and arranged in a 20X20 two dimensional spatial matrix. People were assumed to maintain their initial categorical pro or con position unless the strength, immediacy, and number of people advocating the opposite position exceeded the strength, immediacy, and number supporting them. Under this nonlinear change rule, although the minority was reduced in size, the emergence of spatial clustering led to the maintenance of continuing diversity.

Latané and Nowak (1997) employed SITSIM, a simulation program written in Pascal and designed to enable the systematic factorial variation of a large variety of theoretical assumptions, boundary conditions, and parameters, to determine the robustness and sensitivity of these phenomena (see Latané, Nowak, & Liu, 1994 and Nowak & Latané, 1994 for more details on SITSIM simulations). According to Latané and Nowak (1997), for social systems to self-organize into clusters with continuing diversity, individuals must be spatially distributed with decreasing social influence with increasing distance (to enable the formation of clusters), and attitude change must be nonlinear (to keep people at the borders

---

[3] A "fully connected" social network is one in which there is at least one influence chain or path connecting any two individuals. It is not necessary that everyone have the opportunity for direct influence on everyone else, just that there exists some path by which indirect influence may be spread between any two people. In other words, a fully connected network cannot be divided into two or more non-communicating factions.

from compromising their attitudes into uniformity).

Recently, however, Hegselmann, Flache, and Möller (1997) have attacked this view, putting forth the claim that nonlinearity is not necessary and that it is sufficient to maintain continuing diversity if opinions are discrete rather than continuous. They created cellular automata in which for each successive time period, individuals adopt the average of the prior opinions of their four nearest neighbors, ignoring their own previous opinion and those of the rest of the population (technically, this is known as a von Neumann neighborhood). Then they systematically varied the precision with which the attitudes could be coded, from continuous to categorical with 41, 14, 6, 3, and 2 discrete categories. Their results were complex, but clearly showed that clustering and continuing diversity could emerge in a variety of discrete systems. They concluded that the reason for continuing diversity is not the nonlinearity of attitude change, but rather that "the dividing line between clustering or not lies between continuous and discrete systems."[4]

Although Hegselmann et al. proved that multi-category discrete cellular models produce clustering and continuing diversity, their evidence is based on a nearest neighbor cellular automaton, and is not directly applicable to the continuous distance models adopted by Latané and Nowak (1997) — in fact, it extends the generality of the SITSIM results. Hegselmann et al's cellular automata and SITSIM are similar in that both treat time as discrete, systematically updating individual attitudes through a series of iterations according to simple rules. However, they differ in that: 1) cellular automata usually assume that cells can take a finite (usually small) number of states, whereas SITSIM can be set to allow a variety of change rules varying from dichotomous flip-flops to linear change (i.e., an infinite number of states), and 2) cellular automata usually employ strictly local rules in which cells are influenced only by their nearest neighbors, whereas SITSIM can vary the rules from local to global, with each individual typically affected by all the others in inverse proportion to some function of their distance.

As Hegselmann, et al. have shown, cellular automata can be modified to simulate continuous attitudes (as usual, finding results consistent with Abelson's, 1964, proof). The continuous space of DSIT can also be modified to approximate a cellular automaton by setting the distance exponent sufficiently large that only nearest neighbors can have measurable influence. In fact, Latané and Nowak (1997) report that "under these conditions, clustering and consolidation were reduced to relatively minor (although still statistically significant) effects" (p.

---

[4] More recently, the Hegselmann group (this volume) seems to have backed off this claim, which was a central point in their 1997 presentation. Their new analyses, received too late for detailed consideration in this chapter, continue to provide extensive but unacknowledged support for the fact, first reported by Nowak, Szamrej, & Latané, 1990, that opinion dynamics behave quite differently when Abelson's assumption of linear change is relaxed.

62).

The primary purpose of the simulations in the present chapter is to investigate the necessary and sufficient conditions for the maintenance of continuing diversity in a cellular automaton framework. By introducing different discrete and nonlinear continuous attitude functions, the simulations will allow us to explore the relationships between the nature of attitudes and emergent group-level phenomena resulting from social influence in a variety of social spaces. In every case, the models are fully-connected and linear averaging should lead to ultimate universal agreement. Although the models in this study differ somewhat from Hegselmann et al.'s cellular automata, they enable a fair test of the relative importance of nonlinearity and discreteness in the maintenance of continuing diversity.

A second purpose of this chapter is to demonstrate the potential advantages of using commercially available spreadsheet software on personal computers to construct simulations. The simulations reported here were created by Microsoft Excel97, a powerful spreadsheet program running in the Windows environment. A major advantage of spreadsheet programs over traditional computer languages such as C++ or Pascal is that spreadsheets are more familiar to lay people, allowing everyone to examine, try out, and modify simulations. Such manipulation does not require special training or professional expertise in computer programming, making computer simulation, like the vernacular Bible, accessible to all. Such simulations may also be less subject to human error than simulations written in conventional programming languages. The spreadsheets used for the simulations in this study can be downloaded from the web at http://www.seafrolic.org/.

11.1 Spreadsheet Models of Dynamic Social Impact

Figure 1 reproduces a spreadsheet designed to simulate opinions in a group of 24 people exchanging messages over 50 rounds of communication. Each column in Figure 1 represents one of the 24 people, with rows representing different attributes or different points in time.

11.1.1 Social Space or Network Geometry

Row 4 identifies the column letter or ID of each of the 24 persons, with the entries in Rows 5-8 representing the IDs of the four people to and from whom they send and receive messages. Thus, Person A talks to W, X, B, and C, Person B to X, A, C, and D; etc. This particular geometry or social space can be conceived as if the 24 people were sitting around a circular table with a large flower piece in the center. Each person is able to talk to the two people on either side, and the whole one-dimensional array wraps around to form a ribbon (Latané & L'Herrou, 1996). One could switch to other network configurations or geometries simply by changing the IDs in Row 5-8.

11.1.2 Opinions

Opinions are coded as numbers between 0 and 1, representing the range of possible choices on each round of discussion. Rows 10 and 64 show the initial pre-discussion opinions and each following row contains opinions at the subsequent round. The conditional formatting feature of EXCEL97 is used to make the color of cells with values less than 1/3 yellow, cells with intermediate values green, and cells with values larger than 2/3 blue (in a black and white reproduction, these colors show up as different shades of gray, graduated from lighter to darker.)

	A	B	C	D	E	F	G	H	I	J	K	L	M	N	O	P	Q	R	S	T	U	V	W	X
1	Figure 1. Spreadsheet simulations of dynamic social impact																							
2																								
3	*Addresses of group members and their neighbors*																							
4	a	b	c	d	e	f	g	h	i	j	k	l	m	n	o	p	q	r	s	t	u	v	w	x
5	w	x	a	b	c	d	e	f	g	h	i	j	k	l	m	n	o	p	q	r	s	t	u	v
6	x	a	b	c	d	e	f	g	h	i	j	k	l	m	n	o	p	q	r	s	t	u	v	w
7	b	c	d	e	f	g	h	i	j	k	l	m	n	o	p	q	r	s	t	u	v	w	x	a
8	c	d	e	f	g	h	i	j	k	l	m	n	o	p	q	r	s	t	u	v	w	x	a	b
9																								
10	*Linear change: Attitude $_{(t+1)}$ = neighborhood average $_{(t)}$*																							
11	1.0	.0	.0	1.0	.0	.0	.0	1.0	.0	.0	.0	1.0	1.0	1.0	1.0	1.0	.0	.0	.0	1.0	.0	.0	1.0	.0
12	.4	.4	.4	.2	.2	.4	.2	.2	.2	.4	.4	.6	.8	1.0	.8	.6	.4	.4	.2	.2	.4	.4	.4	.4
13	.4	.4	.3	.3	.3	.2	.2	.3	.3	.4	.5	.6	.7	.8	.7	.6	.5	.4	.3	.3	.3	.4	.4	.4
14	.4	.4	.3	.3	.3	.3	.3	.3	.3	.4	.5	.6	.7	.7	.7	.6	.5	.4	.4	.3	.3	.4	.4	.4
15	.4	.4	.3	.3	.3	.3	.3	.3	.4	.4	.5	.6	.6	.6	.6	.6	.5	.4	.4	.4	.4	.4	.4	.4
16	.4	.3	.3	.3	.3	.3	.3	.3	.4	.4	.5	.6	.6	.6	.5	.6	.5	.5	.4	.4	.4	.4	.4	.4
17	.4	.3	.3	.3	.3	.3	.3	.3	.4	.4	.5	.5	.6	.6	.6	.5	.5	.5	.4	.4	.4	.4	.4	.4
18	.3	.3	.3	.3	.3	.3	.3	.4	.4	.4	.5	.5	.5	.6	.6	.5	.5	.5	.4	.4	.4	.4	.4	.4
61	.4	.4	.4	.4	.4	.4	.4	.4	.4	.4	.4	.4	.4	.4	.4	.4	.4	.4	.4	.4	.4	.4	.4	.4
62																								
63	*Dichotomous attitudes: Attitude $_{(t+1)}$ = 1 if neighborhood average $_{(t)}$ >.5, 0 else*																							
64	1.0	.0	.0	1.0	.0	.0	.0	1.0	.0	.0	.0	1.0	1.0	1.0	1.0	1.0	.0	.0	.0	1.0	.0	.0	1.0	.0
65	.0	.0	.0	.0	.0	.0	.0	.0	.0	.0	.0	1.0	1.0	1.0	1.0	1.0	.0	.0	.0	.0	.0	.0	.0	.0
66	.0	.0	.0	.0	.0	.0	.0	.0	.0	.0	.0	1.0	1.0	1.0	1.0	1.0	.0	.0	.0	.0	.0	.0	.0	.0
67	.0	.0	.0	.0	.0	.0	.0	.0	.0	.0	.0	1.0	1.0	1.0	1.0	1.0	.0	.0	.0	.0	.0	.0	.0	.0
68	.0	.0	.0	.0	.0	.0	.0	.0	.0	.0	.0	1.0	1.0	1.0	1.0	1.0	.0	.0	.0	.0	.0	.0	.0	.0
69	.0	.0	.0	.0	.0	.0	.0	.0	.0	.0	.0	1.0	1.0	1.0	1.0	1.0	.0	.0	.0	.0	.0	.0	.0	.0
70	.0	.0	.0	.0	.0	.0	.0	.0	.0	.0	.0	1.0	1.0	1.0	1.0	1.0	.0	.0	.0	.0	.0	.0	.0	.0
71	.0	.0	.0	.0	.0	.0	.0	.0	.0	.0	.0	1.0	1.0	1.0	1.0	1.0	.0	.0	.0	.0	.0	.0	.0	.0
114	.0	.0	.0	.0	.0	.0	.0	.0	.0	.0	.0	1.0	1.0	1.0	1.0	1.0	.0	.0	.0	.0	.0	.0	.0	.0

Figure 1: Spreadsheet simulations of dynamic social impact

Each cell in Row 10 contains the same formula: =ROUND(RAND(),0), which returns a random number rounded off to either 0 or 1. Thus, initial opinions are random and extreme. The formula for Row 64 refers only to the corresponding value in Row 10, making initial opinions identical for the two models.

11.1.3 Linear Averaging Model

For subsequent rounds under the linear change model, each individual's opinions are calculated as the average of their own opinions and those of everybody connected to them on the previous round. Thus, each cell in Rows 11 to 61 of Column A contains an identical formula designed to calculate this average (to conserve space in the display, many of these rows are hidden). The formula is written in abstract fashion using the INDIRECT() and ROW() functions so that it can be copied directly to the other columns and rows, reducing the possibility of clerical errors. The INDIRECT() function returns the value in the cell specified in parentheses, and the ROW() function returns the row number of the current cell[5].

To show how simple the simulation, the formula is reproduced here in its entirety:

```
=AVERAGE(INDIRECT(A$4&(ROW()-1)),
         INDIRECT(A$5&(ROW()-1)),
         INDIRECT(A$6&(ROW()-1)),
         INDIRECT(A$7&(ROW()-1)),
         INDIRECT(A$8&(ROW()-1)))
```

Cells A4 through A8 contain the neighbors' addresses. If we take cell A12 as an example, A4 through A8 contains A, W, X, B, and C, and INDIRECT(A$8&(ROW()-1) returns C12, resulting in the formula =AVERAGE(A12, W12, X12, B12, C12).

11.1.4 Dichotomous Attitudes Model

Under the nonlinear catastrophic change model, a discrete threshold function is used to transform the averages calculated by the linear model by using EXCEL's IF function (the formula IF (Average>.5,1,0) returns 1 if Average is greater than .5, 0 otherwise). The results of such a model are shown in rows 64-114 (again, many of the rows are hidden to conserve space).

Thus, Figure 1 contains two sets of results, each representing people following a different model. One set is shown from Row 11 to Row 61, and the other, based on the same initial values, is from Row 64 to Row 114 (Row 19-60 and Row 72-113 are hidden because of the limited space). The sheet is constructed so that the two models evolve independently — that is, after the initial round, opinions in one model can have no direct effect on opinions in the other. Each replication of fifty iterations of a simulation yields indices of clustering, consolidation, and continuing diversity. To insure that results are stable across different distributions of initial attitudes, all simulations employed 500 replications.

---

[5] EXCEL97 has two notations for addressing cells. Address elements with a dollar sign ($), such as the row numbers showing neighbor's addresses, are absolute: they will not be altered when the formula is copied. Address elements without a dollar sign, such as the column letters identifying each person, are relative: they will be automatically converted to the appropriate value when copied. This automated system for copying formulas considerably reduces the probability of human error in writing the simulation.

11.1.5 Clustering

An immediately obvious feature of Figure 1 is clustering, the propensity for group members to become more similar to their neighbors than to those at a greater distance. Clustering should occur whenever people are more affected by their neighbors than by those who are further away in space. It can be measured by computing the ratio of the average variance among neighbors to the total variance. The statistical significance of such a ratio can be approximated by the F distribution with 5 and 23 degrees of freedom. Thus, ratios larger than 1.2 can be considered as significant. As to be expected with the strong degree of local influence, clustering was universal, increasing during the initial course of every single replication for both the linear and nonlinear models. Since clustering is so pervasive, it will not be reported in the analyses to follow.

11.1.6 Consolidation

As Figure 1 also shows, linear averaging causes attitudes, all either 0 or 1 at the initial round (Row 10), quickly to converge toward the mean of the initial distribution. Every single one of 500 replications of this simulation with the linear averaging model showed a reduction in the variance among attitudes from the first to the fiftieth round. Consolidation is also prevalent in the nonlinear model, but not so ubiquitous, with reduced variance occurring on only 80% of the 500 replications. In this case, the reduction in variance reflects a decrease in the size of the minority. In the remainder of this chapter we report the percentage of the 500 replications for which consolidation was significant — i.e., the variance on the final round was less than the 5th percentile of the first round distribution.

Continuing diversity is the ability of minorities to survive rather than being eliminated. With categorical models, continuing diversity is achieved if people continue to be distributed across two or more categories. With the linear change model, however, although variance will approach zero as trials approach infinity, tiny amounts of variation will remain for many trials. Therefore, we coded uniformity whenever the range of final values was less than one seventh the initial range, or .14 (this somewhat arbitrary number was chosen to reflect human limitations on the perception of differences in opinion).

In the linear model, as expected from Abelson's proof and many subsequent simulations, diversity did not survive even once in 500 replications. By contrast, it was maintained over 92% of the time with the nonlinear model, failing only when the initial minority was too small and spread out to organize itself into clusters.

Table 1. The prevalence of consolidation, clustering, and continuing diversity.

	Linear averaging model	Dichotomous attitudes
Clustering	100%	100%
Consolidation	100%	80%
Continuing diversity	0%	92%

11.1.7 Plan of Analysis

Figure 1 and Table 1 pose the basic question we address in this chapter — why is there a difference in continuing diversity between the two models of attitude change? Latané and Nowak (1997) claim it is due to nonlinearity, Hegselmann et al. (1997) claim it results from discreteness. In the remainder of this chapter, we will report four series of simulations. First, we will replicate Hegselmann et al.'s results by looking at the number of categories, systematically varying this between 2 and 100. Then, we will examine the effect of varying the size of the middle category in a three category system. Third, we will compare results from discrete and linear models with results from two continuous but nonlinear functions. Finally, we will examine the effect of the degree of nonlinearity in continuous functions.

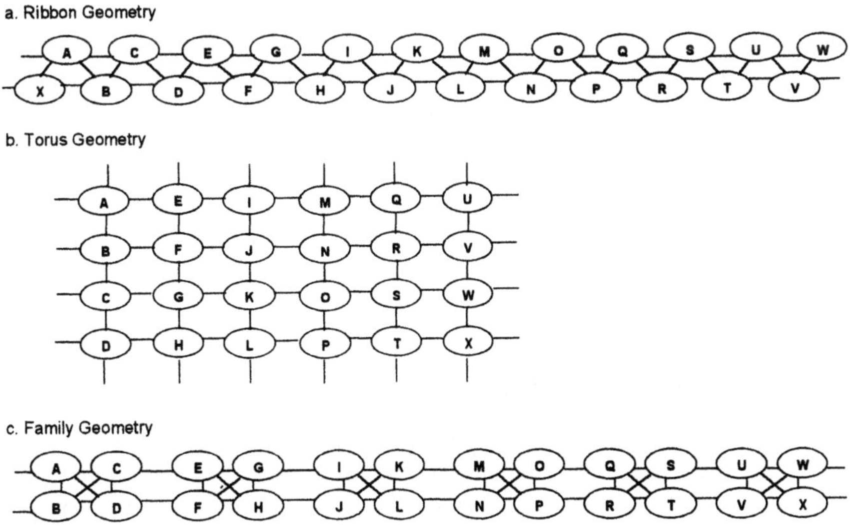

Figure 2: Three geometries of social space

11.1.8 Social Space

Each of these simulations will be conducted with three different geometries (see Figure 2 for diagrams). First is the simple linear geometry, as illustrated in Figure 1. To make the results more comparable to those reported by Nowack, Szamrej, & Latané, (1990) and Hegselmann, et al., (1997), we will add a Torus Geometry which can be conceived as people evenly distributed in a two-dimensional space communicating with their four nearest neighbors to the north, south, east and west. Finally, we will also include results from simulations conducted with a third geometry designed to represent an even richer, more realistic social space. The Family geometry can be regarded as consisting of six four-person families living on a circular cul-de-sac. The members of each family (e.g., father, mother, son, and daughter) speak regularly to each other and to one outside friend from the neighboring family. This hierarchical geometry can also be taken to represent workgroups in an organization.

Simulation 1: Varying The Number of Discrete Categories

Attitudes are often measured by Likert-type scales in which people are asked to choose a number between, say, 1 and 7 to represent their attitude. Simulation 1 was designed to test whether consolidation, clustering, and continuing diversity are sensitive to the number of different categories in such an opinion space. The simulation compared discrete models ranging from 2 to 100 categories with a continuous model, which can be regarded as having an infinite number of categories.

To accomplish this, we created a function designed to divide the opinion space into a specified number of equal intervals and to change individual's attitudes according to which interval contained their neighborhood average. For example, to produce three categories, individuals with neighborhood averages from 0 to .33 were given a new attitude of 0, those with a neighborhood average of .34 to .66 were given a new attitude of .5, and those with a neighborhood average of .67 to 1 were given a new attitude of 1.0[6].

6 With the discrete models, averages of neighboring attitudes are sorted into a specified number of categories using a simple threshold function in which x represents the neighborhood average, n the number of categories, and TRUNC the integer portion of their product: Attitude = TRUNC(x * n) / (n - 1); If Attitude > 1 Then Attitude = 1; End Function.

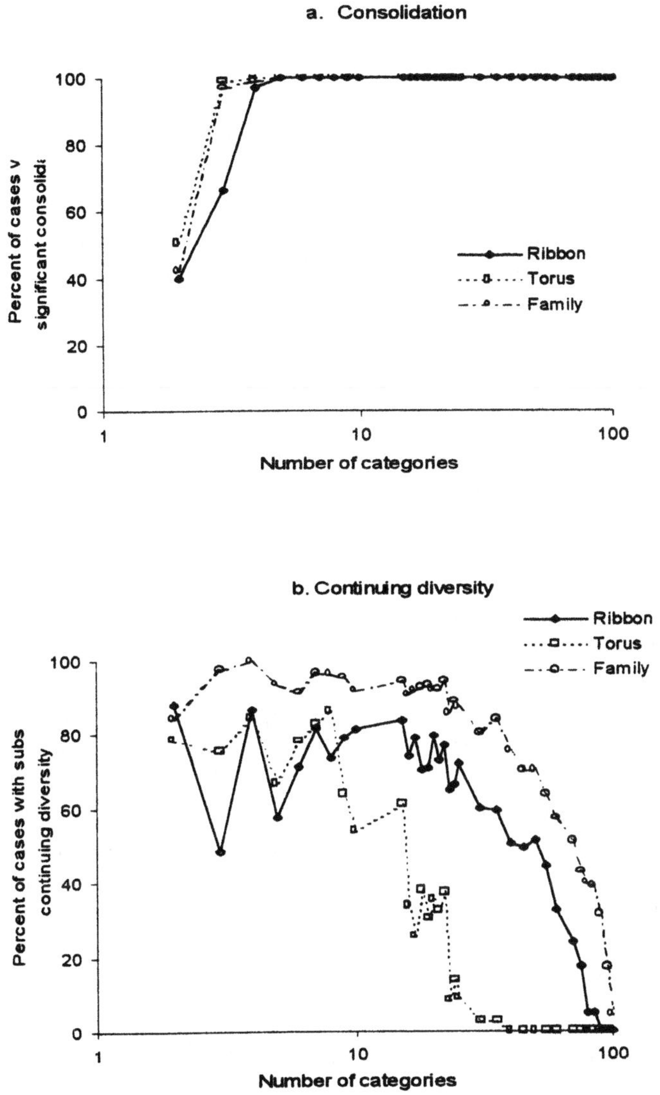

Figure 3a,b: Effect of number of categories

Because clustering was pervasive in every simulation, regardless of the spatial geometry, the number of categories, or whether attitudes were linear or nonlinear, continuous or discrete, for brevity we will not report it further. Figure 3 plots consolidation and continuing diversity as a function of the number of categories for each social space or geometry.

11.2.1 Consolidation

In a simulation world in which people try to accommodate their views to those of their neighbors, consolidation is the rule rather than the exception, and this becomes more pronounced as the number of categories increases. Consolidation was significant about half the time in dichotomous two-category systems, almost always with four or more categories. Especially for the three-category systems, there tended to be somewhat less consolidation for the Ribbon geometry, in which people have neighbors on only two sides and therefore are not so exposed to adverse influence.

11.2.2 Continuing Diversity

Continuing diversity was substantial with small and intermediate numbers of categories, but essentially disappeared as the number of categories began to approximate a linear system. Continuing diversity was most prevalent with the Family geometry, presumably because once a minority opinion becomes entrenched in a local group, each group member can be exposed to at most one source of outside influence, and no single outside source can affect more than one group member. In the Ribbon and Torus, by contrast, minority members on the borders of clusters are exposed to at least as many people with differences of opinion as supporters.

Curiously, the Ribbon geometry exhibits a sharply reduced incidence of continuing diversity with three categories. This seems to result from a two-step process whereby most people initially convert to a neutral attitude only to become vulnerable to influence from blocs of people with extreme values. Such blocs can move along the uni-dimensional Ribbon as a wave, sweeping the entire circle of social space. This phenomenon is echoed in reduced form in the five and seven-category systems, but does not appear in the two-category model because people always have extreme values, or in the two-dimensional Torus because influence is less focused.

11.3 Simulation 2: Varying the Size of the Middle Interval

In addition to showing the strongest effects of geometry, the three-category model is interesting because it reproduces the form of typical attitude surveys, in which people can go on record as being pro, con, or undecided. Our second set of simulations varies the size of the middle interval in a three-category model from 0% to 100% of the total. For example, unlike Simulation 1's 33% middle interval, the 20% middle interval had cutting points at 40% and 60%, the 50% middle interval at 25% and 75%, etc. In the latter case, if the neighborhood average of self and neighbors opinions was between .25 and .75, the new attitude became .5;

otherwise 0 or 1[7].

A middle interval of 20% or less produced results identical to those with the dichotomous function (Figure 4). Consolidation increased as the size of the middle interval increased, in stepwise fashion, with the Ribbon again being most vulnerable to complete consolidation with intermediate middle intervals.

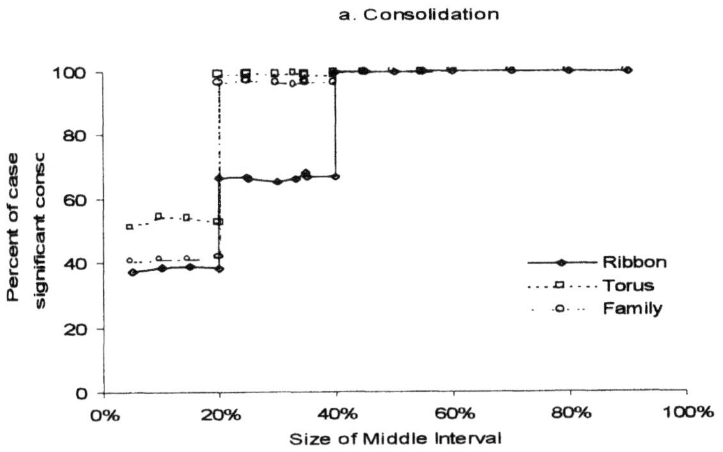

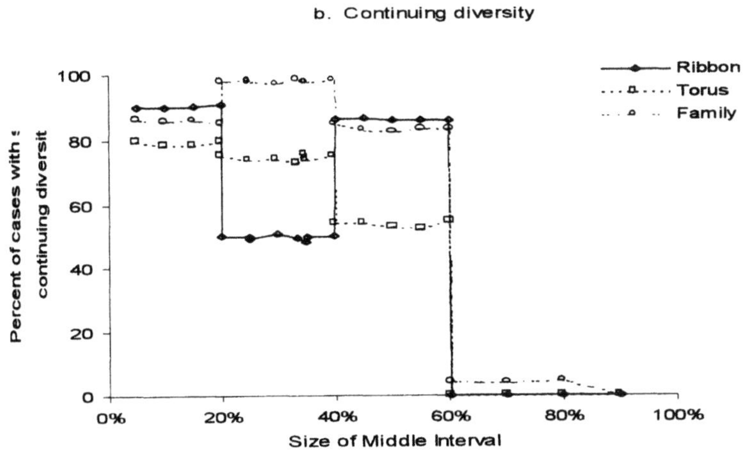

Figure 4a,b: Effect of size of middle interval

Middle intervals up to 60% demonstrated considerable continuing diversity for

---

[7] The Excel97 function to create three-category models with flexible length of middle interval was: L = 1 / (y + 2); H = (y + 1) / (y + 2); If x <= L Then Attitude = 0; Else; If x < H Then Attitude = 0.5; Else: Attitude = 1; End If; End Function, with x representing the previous neighborhood average and y the proportion of the middle interval to the total.

all geometries. As in Simulation 1, the Ribbon geometry produced a surprising dip in continuing diversity with moderate sized middle intervals (between 20 and 40%). The Family geometry showed the opposite tendency: greater continuing diversity with moderate middle intervals.

For the Ribbon and Torus, a middle interval greater than 60% (with cutting points at .2 and .8) requires unanimous agreement on an extreme position to keep attitudes from converging to the neutral value of .5. With these cutting points, no single fence sitter can survive without social support, any two fence sitters will convert everyone in their neighborhood to neutrality, and this process will propagate itself throughout the network. In the Family geometry, the reduced exposure of family members to the rest of the network means that extremist groups can coexist with neutral groups with middle intervals as large as 80%, but only if they become established very early.

The size of the middle interval can be taken as a way of representing how hesitant people are to express firm opinions. If nobody is willing to express an extreme opinion on a topic, nobody can find social support for their own extreme position, and opinions should converge to neutrality.

In general, the neutral middle in a three-category system serves as an attractor, sucking up people with extreme positions and rarely releasing them. If this region of comfortable fence sitting becomes too large, consolidation proceeds unchecked and diversity is always eliminated as it becomes impossible to maintain an extreme position unless everyone else in the group does also.

Clearly, however, continuing diversity is maintained with a wide range of middle intervals. These simulations show that the predictions of DSIT are very robust to a variety of discrete categorical models, even under the relatively unfavorable conditions of a limited cellular automaton model. Thus, as discovered by Hegselmann, et al., consolidation and continuing diversity are not limited to bipolar systems. On the other hand, with enough categories, the results become practically identical to those obtained with linear averaging.

11.4 Simulation 3: Testing Continuous Nonlinear Functions

Hegselmann et al. have correctly pointed out that discreteness beyond two categories leads to continuing diversity. But such systems are also nonlinear. The key point at issue between Hegselmann et al. (1997) and Latané & Nowak (1997) is whether nonlinearity or discreteness is the critical condition for continuing diversity. In other words, would any departure from linear opinion change produce continuing diversity (as suggested by Latané & Nowak) or does it require step functions (as implied by Hegselmann et al.)? We compared linear and discrete functions with a nonlinear continuous attitude change function to answer this question. A sine function, relatively flat at both ends and steep in the middle, can be seen as intermediate between a linear and a dichotomous function (Figure 5). If nonlinearity rather than discreteness is critical, continuing diversity should be maintainable even with continuous models such as this.

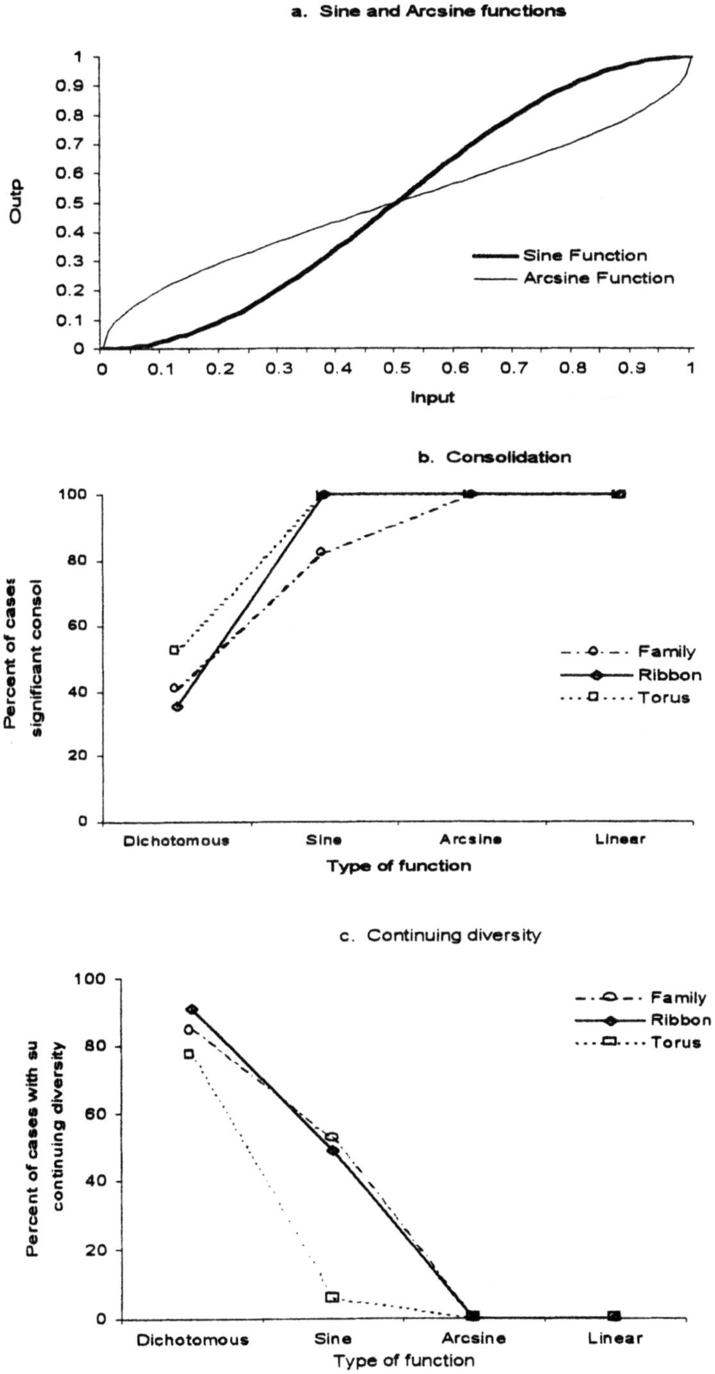

Figure 5: Effect of different forms of nonlinearity

But perhaps not all forms of nonlinearity should have the same effect. Of particular interest is the inverse sine or arcsine function, which is twisted in the opposite way from the sine function — steep at the ends and flat in the middle[8].

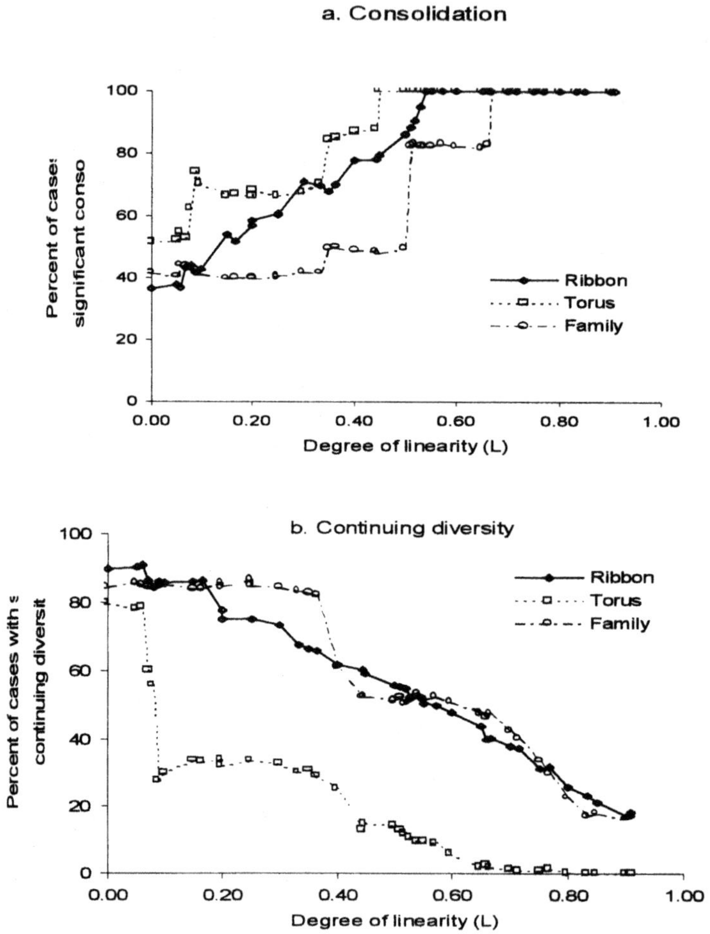

Figure 6a,b: Effect of degree of linearity

---

[8] The EXCEL97 formulae were: Attitude =SIN(x*PI()-PI()/2)/2+0.5; and Attitude =(ASIN(2*x)+PI()/2)/PI(), with x representing the previous neighborhood average.

11.4.1 Consolidation

Consolidation was more characteristic under both forms of the continuous function than with discrete categories, presumably because the attractor regions are not so well defined.

11.4.2 Continuing Diversity

Continuing diversity depended on the nature of both nonlinearity and spatial geometry. With the Ribbon and Family geometries, the sine function produced continuing diversity more than half of the time, supporting the findings of Latané and Nowak (1997) that a discrete categorical model is not a necessary condition. However, little continuing diversity was seen with the Torus geometry, demonstrating the vulnerability of Torus inhabitants to social influence. As discussed by Latané and Nowak (1997), in the Torus, the borders of minority clusters are always convex and tend to erode from the corners. As they discovered, this tendency can be reduced if individual differences in strength are introduced, allowing strong individuals to anchor the borders.

The arcsine function, although also nonlinear, produced 100% consolidation and never any continuing diversity. That is, it led to just as much ubiquitous agreement as the linear model. Why did the results differ so between sine and arcsine functions? Although both are nonlinear and continuous, they differ in the number and location of attractor regions — the arcsine function has one attractor in the middle, into which everyone tends to gravitate. By contrast, the sine function has two attractors, one at each end, and because the slope is steepest in the middle, people at the extremes are protected from change.

11.5 Simulation 4: Varying the Degree of Nonlinearity

How much linearity is needed to destroy continuing diversity? To answer this question, Simulation 4 introduces another way of creating a nonlinear but continuous function in which the degree of linearity (L) can be varied[8]. When L is equal to one, the function is strictly linear, and as L becomes smaller, the middle part of the function becomes steeper and the function less linear, with a concave upward slope from 0 to .5 and a concave downward slope from .5 to 1[9].

As a general tendency, there was more consolidation with increases in the linearity of the model, with the Family geometry again producing less consolidation than the other two geometries.

---

[9] The EXCEL97 function with x representing the previous neighborhood average and L the linearity parameter was: If x < 0.5 Then; Attitude = ((2 * x) ^ (1 / L)) / 2; Else: Attitude = ((((x - 1) * (-2)) ^ (1 / L)) / 2) * (-1) + 1; End If; End Function.

With the Family or the Ribbon geometry, continuing diversity was inversely proportional to the degree of linearity, but was still substantial even when the linearity index exceeded 0.8. Consistent with earlier simulations, the Torus geometry is more susceptible to the negative effects of linearity, but produced detectable degrees of continuing diversity with linearity indices up to at least .6.

Thus, continuing diversity was characteristic of this continuous model throughout most of its range, disappearing only when it approached pure linearity. Only a slight departure from linearity is sufficient to produce continuing diversity.

11.6 Comments and Conclusions:

11.6.1 Nonlinearity (Including Discreteness) of Opinion Space Is A Critical Determinant of The Dynamics of Group Interaction.

- As Abelson (1964) proved, linear attitude change functions always lead to ubiquitous agreement in fully connected networks.
- However, consistent with Latané and Nowak (1997), slight departures from linearity yield considerable degrees of continuing diversity, even in these relatively impoverished cellular automaton models with no individual differences in strength.
- It appears that, contrary to Hegselmann, et al. (1997), nonlinearity rather than discreteness is the critical condition. The existence of discrete categories may often produce continuing diversity, but it is not necessary, as continuous nonlinear functions can have the same effect.
- Nonlinearity may be the critical issue, but the nonlinearity must be such as to promote at least two different attractor regions, or flat areas.
- A "catastrophe" theory of attitudes (Latané and Nowak, 1994; Harton & Latané, 1997; Liu & Latané, 1998) suggests that as attitudes become more involving, they also become more extreme or categorical and less linear, leading people to be less likely to shift incrementally. Our simulation results suggest that as the importance of issues or the involvement of people increases, consolidation and clustering should lead to a greater likelihood of continuing diversity.
- An interesting question, not explored here, is what happens if discrete categories are not ordered? That is, what if rather than choosing among pro, neutral, and con, the choice is chicken, beef, or fish?

11.6.2 How Attitudes Are Communicated May Be As Important As How They Are Represented In An Individual Mind.

- The categorical models in this study can be interpreted in two ways: 1) People may have limited interest in or ability for representing attitudes precisely and store them only as categories. 2) People may store their attitudes as raw values on a continuous scale, but express them by choosing from a finite number of categorical options. Both possibilities would lead to similar results in the simulations reported here, because neighborhood averages are primarily determined by expressed rather than private attitudes.

- The number of categories can be considered as representing the number of points in a Likert-type attitude scale. Likert scales typically employ 5 to 11 options, and the present results suggest that the actual number of categories may not be too important, at least with respect to group dynamics. Despite some fluctuation in the Ribbon geometry, similar numbers of categories in the present simulations produced similar degrees of consolidation, clustering, and continuing diversity.
- Likert scaling may be based on the questionable assumption of equal category sizes (Mitchell & Jolley, 1992), but the dynamics predicted by DSIT held even when intervals were not equal, suggesting that social norms against extremity may not have much effect on consolidation, clustering, and continuing diversity.
- Whether or not to allow people to express a neutral or undecided position may have important consequences for opinion dynamics, as there were substantial increases in consolidation as the number of categories increased beyond two.

11.6.3 Distance Matters.

- Decreasing influence with distance is necessary to create neighborhood clusters which give members of a global minority the protection of being in a local majority. The role of space is thus crucial to the dynamics of group attitude change.
- Unlike the spatial models analyzed by Latané and Nowak (1997), cellular automaton models do not easily accommodate continuous variations in distance, but rather employ a discrete distance decay function with a single, all-or-none threshold for influence. These models can also be seen as embodying an overly simplistic network model, a regular set of reciprocal pairwise connections among some but not all members of a group.
- Nevertheless, consolidation, clustering, and continuing diversity occur both in cellular automata representing nearest neighbor networks, and continuous distance spatial models.
- It would be interesting to develop mixed models in which people are members of networks superimposed on an underlying spatial distribution. Each individual could have both a pair of addresses corresponding to their latitude and longitude on the face of the earth, and a set of relationships expressed in a sociometric matrix for those people who have come to know one another through family, school, or workplace ties. Distance could be calculated in two ways — the difference between spatial locations, or the network distance. Social impact would be calculated by weighting people's influence according to the lesser of these two forms of immediacy.

- Latané and Nowak (1997) considered only a regular Torus. The present results suggest that in addition to individual differences, irregularities in the structure of social space may also help to protect minority clusters from consolidation. The Ribbon geometry makes people vulnerable to adverse influence from only two, rather than four sides, and the Family geometry allows small enclaves of minority members to hole up in partial seclusion from the outside world. These results are provocative with respect to the possible preservation of cultural or linguistic diversity along river banks (such as the Amazon river system) and in mountain valleys (such as West Virginia and Switzerland).

11.6.4 Individual Differences Make A Difference.

- Cellular automaton models tend to have no (Hegselmann, et al., 1997) or not enough (this chapter) self influence. A great deal of evidence shows that the most important determinant of one's time t+1 opinion is one's own time t opinion. Further, simulations show that such self-influence can contribute to the maintenance of continuing diversity.
- Cellular automaton models, including those in this chapter, usually do not allow individual differences in strength, or peoples' ability to persuade others and resist persuasion themselves. Individual differences in strength promote continuing diversity by allowing strong personalities to anchor the borders of minority islands, preventing them from eroding into the majority sea. Individual difference could be added to an EXCEL97 simulation model by creating a row showing each individual's persuasive strength (initialized as a random value between 0 and 1), and modifying the formula to use these numbers as weights when calculating averages.

11.6.5 Cellular Automaton Models, Including Those Used By Hegselmann et al. And In The Present Chapter, Are Too Deterministic.

- This is reflected in the step functions relating the size of the middle interval to the resulting dynamics.
- Adding a random element to the change formula (i.e., increasing the 'temperature') would solve the problem, while actually increasing the tendency of such social systems to self-organize, although it may lead to increased consolidation (Latané & Nowak, 1997).

11.6.6 Consolidation, Clustering, And Continuing Diversity Are Pervasive Features of Social Influence Settings Where People Try to Accommodate Their Opinions to One Another.

- Consolidation comes about from the fact that the minority is by definition at a numerical disadvantage.
- Clustering is characteristic whenever some influence pairings are more likely than others (this is another way of saying there is a spatial distribu-

tion of people),
- Continuing diversity is common because clusters shield minorities from consolidation.

11.6.7 Spreadsheets Are Useful Tools For Social Simulation.

- Spreadsheets may not be as flexible or powerful as conventional programming languages, but their advantages are apparent: less time required to learn, clarity of logical structures of simulations, and visibility of rules.
- As spreadsheet programs and the personal computers they run on become more and more powerful, the limitations of spreadsheets will continue to disappear.
- Simple spreadsheets can reveal complex dynamics — and it doesn't take a computer programmer, a priest, or even the Pope to understand how.

References

Abelson, R. P. (1964): Mathematical models of the distribution of attitudes under controversy. In: Frederiksen, N. & Gulliksen, H. (eds.): Contributions to Mathematical Psychology Holt, Rinehart & Winston, New York, 147-160).

Festinger, L., Schachter, S. & Back, K. (1950): Social Pressures in Informal Groups: A Study of Human Factors in Housing. Stanford University Press, Stanford, CA.

Harton, H. C. & Latané, B. (1997): Thought- and information-induced polarization: The mediating role of involvement in making attitudes extreme. Journal of Social Behavior and Personality 12, 271-299.

Harton, H. C., Green, L. R., Jackson, C. & Latané, B. (1998): Discussion leads to consolidation, clustering, and correlation — but not necessary the correct answer: A classroom demonstration of dynamic social impact. Teaching of Psychology 25, 31-35.

Hegselmann, R., Flache, A. & Möller, V. (1997): Solidarity and Social Impact in Cellular Worlds: Results and Sensitivity Analyses. Paper presented at Dagstuhl Seminar on Social Science Microsimulation: Tools for modeling, parameter estimation, and sensitivity analysis. Schloss Dagstuhl, Germany, (May).

Huguet, P. & Latané, B. (1996): Social representations as dynamic social impact. Journal of Communication 46, 57-63.

Huguet, P., Latané, B. & Bourgeois, M. (1998): The emergence of a social representation of human rights via interpersonal communication: Empirical evidence for the convergence of two theories. European Journal of Social Psychology, 28, 831-846.

Latané, B. (1981): The psychology of social impact. American Psychologist 36, 343-365.

Latané, B. (1996a): Dynamic social impact: The creation of culture by communication. Journal of Communication 46, 13-25.

Latané, B. (1996b): Dynamic social impact: Robust predictions from simple theory. In: Hegselmann, R., Müller , U. & Troitzsch, K. (eds.): Modeling and Simulation in the Social Sciences from the Philosophy of Science Point of View. Kluwer Theory and Decision Library, Dordrecht, 287-310.

Latané, B. (1996c): The emergence of clustering and correlation from social interaction. In: Hegselmann, R. & Peitgen, H. O. (eds.): Modelle Sozialer Dynamiken: Ordnung, Chaos, und Komplexität. Hölder-Pichler-Tempsky, Vienna, 79-104.

Latané, B. (1996d): Strength from weakness: The fate of opinion minorities in spatially distributed groups. In: Witte, E. & Davis, J. (eds.): Understanding Group Behaviors: Consensual Action by Small Groups. LEA, Hillsdale, NJ, 193-219.

Latané, B. (1997): Dynamic social impact: The societal consequences of human interaction. In: McGarty, C. & Haslam, A. (eds.): The Message of Social Psychology: Perspectives on Mind and Society. Blackwell, Oxford, England, 200-220.

Latané, B. & Bourgeois, M. J. (1996): Experimental evidence for dynamic social impact: The emergence of subcultures in electronic groups. Journal of Communication 46, 35-47.

Latané, B. & L'Herrou, T. (1996): Social clustering in the Conformity Game: Dynamic social impact in electronic groups. Journal of Personality and Social Psychology 70, 1218-1230.

Latané, B. & Nowak, A. (1994): Attitudes as catastrophes: From dimensions to categories with increasing involvement. In: Vallacher, R. & Nowak, A. (eds.): Dynamical Systems in Social Psychology. Academic Press, New York, 219-249.

Latané B. & Nowak, A. (1997): Self-organizing social systems: Necessary and sufficient conditions for the emergence of consolidation, clustering, and continuing diversity. In: Barnett, G. & Boster, F. (eds.): Progress in Communication Sciences: Persuasion, v. 13. Ablex, Norwood, NJ, 43-74.

Latané, B., Nowak, A. & Liu, J. (1994): Measuring emergent social phenomena: Dynamism, polarization, and clustering as order parameters in dynamic social systems. Behavioral Science 39, 1-24.

Liu, J. H. & Latané, B. (1998): The catastrophic link between the importance and extremity of political attitudes. Political Behavior, 20, 105-126.

Mitchell, M. & Jolley, J. (1992): Research Design Explained. Harcourt Brace Jovanovich College Publishers, Fort Worth.

Moscovici, S. (1985): Social influence and conformity. In: Lindzey, G. & Aronson, E. (eds.): The Handbook of Social Psychology, 3rd ed. Erlbaum, Hillsdale, NJ.

Moscovici, S. & Zavalloni, M. (1969): The group as a polarizer of attitudes. Journal of Personality and Social Psychology 12, 124-135.

Myers, D. G. (1994): Exploring Social Psychology. McGraw-Hill, New York.

Nowak, A. & Latané, B. (1994): Simulating the emergence of social order from individual behavior. In: Gilbert, N. & Doran, J. (eds.): Simulating Societies: The Computer Simulation of Social Phenomena University College London Press, London, 63-84.

Nowak, A., Szamrej, J. & Latané, B. (1990): From private attitude to public opinion: A dynamic theory of social impact. Psychological Review 97, 362-376.

Schaller, M. & Latané, B. (1996): Dynamic social impact and the evolution of social representations: A natural history of group stereotypes. Journal of Communication 46, 64-72.

12. Representation Methods and the Emergence of Inter-Group Cooperation

Ilan Fischer[1] & Ramzi Suleiman[2]

We report a simulation study which explores the effects of different representation methods and election frequencies on the emergence of mutual cooperation between two conflicting groups. We assume that the inter-group conflict can be modeled as an iterated Prisoner's Dilemma game played by the groups' representatives. In addition, we assume that the performance of each representative influences her constituents and that this, in turn, affects her prospects to be reelected. At the end of a constituency period, new elections are called for, and their results determine whether the delegate remains in her position or is replaced by another representative. Our main objectives were: (1) to investigate the effects of three representation methods: Random representation, Mean representation, and Minimal Winning Coalition representation, on the evolution of cooperation between the groups; and (2) to investigate the effect of the frequency of elections in the two groups on the evolving inter-group relations. Outcomes of 1080 simulations yield the following main results: (1) there exists an election frequency which optimizes the likelihood that the inter-group process converges to mutual cooperation; (2) similar cooperation characteristics evolve under the Minimal Wining Coalition and the Random representation methods; and (3) the Mean representation method results in low levels of inter-group cooperation.

When should a government step down and return power to its people? This issue was debated in several democratic parliaments, but was never consensually resolved. In practice, constituency periods span from one year, to five, six, and even seven years. Obviously, the length of election periods are of no intrinsic value, unless seen in the context of optimal constituency periods given some defined objectives. The inspection of debates over this issue in various parliaments reveals two opposing arguments: the first favors short constituency periods on the bases of expected high level of commitment, and strong delegate-electorate dependency. The author of the fifty second issue of The Federalist described fre-

[1] Ben Gurion University, Department of Behavioral Sciences,
Email: ifischer@bgumail.bgu.ac.il

[2] University of Haifa, Department of Psychology, Email: suleiman@psy.haifa.ac.il

quent elections as: "Unquestionably the only policy by which the dependence between the people and their representatives can be effectually secured" (Hamilton & Madison, 1788). A few years later, the same ideas reappeared in England when Jeremy Bentham prescribed annual elections to keep parliament members in close contact with their constituents and to insure that they did not have sufficient time to develop distinct interests of their own in their capacity as politicians (Bentham, 1843). James Mill in his "Essay on Government" wrote that "the smaller the period of time during which any man retains his capacity as representative, as compared with the time in which he is simply a member of the community, the more difficult it will be to compensate the sacrifice of the interests of longer period by the profits of misgovernment during the shorter" (Mill, 1828).

The second argument claims that longer constituency periods are more favorable because they guarantee a reasonable degree of independence for the delegate and allow her to develop long term policies. Under longer periods, delegates are protected from being too closely tied to the interests and opinions of their constituents. A clear example of such arguments was voiced in the case of the Septennial act of 1716 which extended the life of the House of Commons to seven years. In debating against this act, the Tories urged for more frequent elections, but the Whig leaders insisted on the need that parliament members have a large measure of independence. Similar arguments were, in fact, the reason why since the end of the French revolution, all proposals to hold frequent elections to the national legislatures in European countries were rejected (Birch, 1971).

An intriguing conclusion was drawn by Amacher & Boyes (1978) who claimed that since both arguments have opposing implications for the political system, their juxtaposition should lead to the theoretical prediction that there exists an optimal election frequency. Amacher & Boyes distinguished between 'decision costs' associated with direct elections expenditures, and 'external costs' composed of the expected costs individuals would endure as the result of an elected official behaving in an independent way which contradicts the desires of her constituents; The authors claim that by choosing this optimal frequency, election costs can be minimized.

While the study of Amacher & Boyes (1978) investigated the issue of election frequency from an intra-group perspective, two recent simulation studies (Suleiman & Fischer, 1996; Fischer & Suleiman, 1997), introduced a new approach to the question by investigating the influence of election frequencies in two interacting groups on the emergence of cooperation between the two groups. Following standard game theoretic approaches to inter-group conflicts (c.f. Schelling, 1960; Brams, 1975) the groups' interaction was modeled as a 2-person PD game, but contrary to the standard approach, the two groups were not represented by two unitary players who represent their groups for the whole duration of the game. Rather, it was assumed that each group is comprised of a *population* of individuals, and that each individual is characterized by a certain probability that he or she will cooperate with an out-group member. The elected delegates represented their groups in the ongoing interaction for a predetermined constituency period, after which elections are held, and their results determine whether each delegate is reelected, or replaced by a new delegate. Between two successive elections, the elected delegates play the repeated PD game and their corresponding strategies are defined by their probabilities of cooperation. Following each interaction, each representative affects the cooperativeness of each in-group member, according to a social influence function that is directly proportional to the representative payoff and inversely proportional to the social distance (on the cooperativeness continuum) between the representative and the target member. As a result, the distribution of cooperativeness in each group can change after each interaction; this, in turn, affects the choice of the representative for the next encounter.

In our two previous studies, the groups' delegates were elected according to a Minimal Winning Coalition (MWC) representation method. Under this procedure, a "minimal coalition"[3] that contains 51% or more group members is formed in each group, and its mean is delegated to represent the group between two successive elections. Under this representation method, our main findings suggest that: (1) the dynamics of the inter-group conflict evolve into a few, well defined, patterns; and (2) for most election frequencies, mutual defection was not an enduring pattern while mutual cooperation evolved as an enduring one, and (3) there exists an election frequency which optimizes the likelihood that the inter-group process converges to mutual cooperation. In the present study we continue to focus on the emergence of cooperation between two groups.

The main purpose of the present study is to test the robustness of our previous results under different representation methods. In addition to the MWC method, we investigate the evolution of cooperation under two other representation methods: (1) a Mean Representation (MR) under which the elected group delegate represents the average opinions in her group regarding the readiness to cooperate with the out-group delegate; (2) a Random Representation (RR) under which the group delegate is chosen randomly. The Mean representation method can be seen as an extreme democratic procedure, because all constituents' attitudes toward inter-group cooperation are accounted for in choosing a "mean delegate". In contrast, the Random representation method stands on the other extreme, in that it delegates a representative based on a random selection. The comparison of simulation results under the three representation methods is aimed at achieving the following objectives: (1) to study the effects of different election procedures on the relations between two conflicting groups, and on the prospects for the evolution of cooperation between the groups; and (2) to investigate the effect of the *frequency* of elections in the groups on the evolving inter-group relations; especially on the emergence of mutual cooperation between the groups.

---

[3] For a detailed description of the MWC representation method, see Suleiman & Fischer (1996) or Fischer & Suleiman (1997).

The following section provides a detailed description of the inter-group simulation. The section which follows summarizes the main results of this study, and the final section concludes with a brief discussion.

12.1 The Simulation

The basic features of the simulation can be described as follows:

(1) The inter-group conflict is modeled by an iterated two-person PD game played by elected delegates of the two groups. The specific PD payoff matrix in each game is shown in Table 1.

Table 1: Prisoner's Dilemma Payoff Matrix

		Delegate B: C	Delegate B: D
Delegate A	C	5 , 5	-10 , 10
	D	10 , -10	-5 , -5

(2) The simulation begins by defining two group sizes, n and m. Each group member is characterized by an initial propensity, or probability p, that she will cooperate in a PD game with an out-group member. Group members with p = 0 are those who will never cooperate, while those with p = 1 will always cooperate. To obtain a reasonable degree of flexibility in defining the distributions of cooperation in the two groups, we used a beta function for defining these distributions. This function is sufficiently rich to describe most random variables having a restricted range (Derman et al., 1973). It can assume a wide variety of shapes, including uniform, symmetric, positively skewed, negatively skewed, and U-shaped functions. In formal notation, the betadensity function for the probability of cooperation, p, can be written as:

$$f(p) = \begin{cases} p^u (1-p)^v / B(u,v) & \text{if } 0 \leq p \leq 1 \\ 0 & \text{Otherwise} \end{cases} \qquad (1)$$

where u and v are the parameters of the beta distribution ($u \geq 0$, $v \geq 0$), and B(u,v) is the Beta function (see e.g. Johnson & Kotz, 1969).

The results reported here were obtained for symmetrical distributions with $u = v = 5$ for the two groups. For these values, the distribution is bell-shaped, and resembles a normal distribution in the range (0, 1). In all the simulations, continuous beta functions were approximated by discrete functions with equal intervals of $\Delta = 0.05$.(3) Following the construction of each group, the simulation defines the group's delegate, according to one of three representation methods:

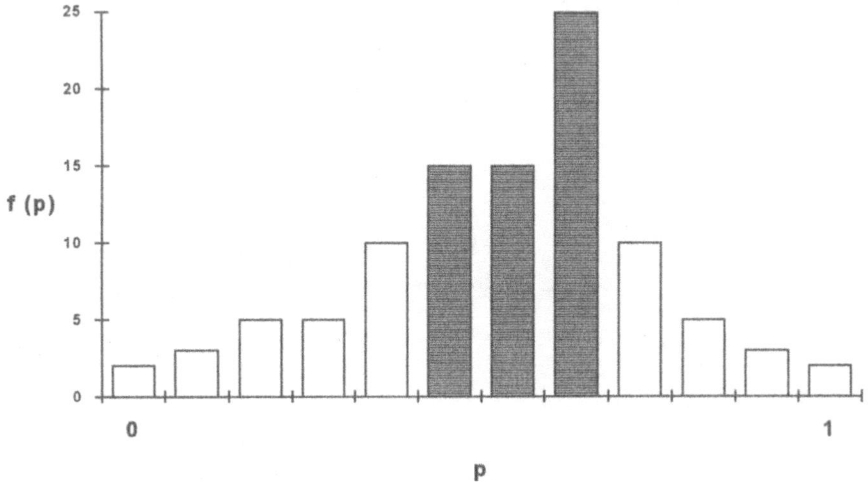

Figure 1: The minimal winning coalition for a hypothetical distribution of cooperation.

(a) The "Minimal Winning Coalition" (MWC) method: Under this representation method, the simulation starts by locating all possible "non-fragmented"[4] groups which contain 51% or more group members. Of all the possible subgroups satisfying this condition, the simulation selects the most compact one on the cooperativeness continuum, and the mean of this MWC is chosen to represent the entire group in the PD game played against the second group representative. To illustrate this procedure, consider a population with the cooperation distribution portrayed in Figure 1.

To locate the group delegate, the simulation starts by finding all possible non-fragmented subgroups containing 51% or more of the popula-

[4] A non-fragmented group is one that is defined over a continuous segment of the cooperativeness range.

tion. The simulation proceeds by selecting the most compact one (i.e., the one with the minimal range on the p continuum). This is the dark shaded subgroup in Figure 1. This subgroup is sought as the minimal winning coalition in democratic elections held in this population, and its mean is sought as the elected group delegate. This delegate represents the entire group for the duration of a constituency period. After this period, new "elections" are held according to the procedure described above and a new delegate is "elected".

(b) The “Mean Representation” (MR) method: Under this condition, the simulation computes the mean attitude (probability of cooperation) of the entire population, and this mean is then delegated to represent the group for the entire duration between two successive elections.

(c) The “Random Representation” (RR) method: Under this condition, the simulation chooses a random delegate who represents the population for the entire duration between two successive elections[5]. Note that this method does not necessarily represent an anarchistic situation. It can also imply that the criteria for electing a delegate are orthogonal to the population's attitude toward inter-group cooperation. Such criteria can be related to the delegate's position regarding intra-group issues (economic, educational, etc.).

(4) The time intervals between successive elections (viewed as a measure of democracy) are defined separately for each group.

(5) After being elected, the delegates engage in an iterative PD game with each other. At every stage of the game, the strategy of each delegate is probabilistic and is defined by her probability of cooperation. Following each interaction, the simulation records each delegate's payoffs, as well as her cumulative gain.

(6) Following each interaction, each group delegate influences the propensities for cooperation of all her group members. This process can be considered as an ongoing social learning process (Axelrod, 1984) in which group members learn and adapt their strategies depending on their delegate's performance. We assume that the social influence of a delegate d on any group member i is inversely proportional to the social distance (measured on the cooperation continuum) between d and i. Moreover, we assume that this influence is proportional to the delegate's payoff. To express this relationship in formal notation, denote the probabilities that an individual i and the delegate d will cooperate at period t by $p_i(t)$ and $p_d(t)$, respectively. Also denote the outcome for delegate d at period t by $O_d(t)$. The social influence of d on i can be expressed by:

---

5 Only “populated” attitudes were considered in this selection procedure.

$$p_i(t+1) = p_i(t) + K\,O_d(t)/(p_d(t)-p_i(t)) \quad (2)$$

where $p_i(t + 1)$ is the probability that the individual i will cooperate at period t+1, and K is a constant. In the reported simulations, the values for the outcome $O_d(t)$ are given by the matrix in Table 1. Accordingly, a delegate d can either win 10 or 5 units, or lose 10 or 5 units. The numerical value for K in the reported simulations was K = 0.1.

In essence, the social influence function in Equation 2 resembles recent formulations of Latane's "social impact function" (Latane, 1981; Nowak, et al., 1990). Nevertheless, it has a number of unique properties that differentiate it from Latane's function in several important ways: (1) distances here are social and not spatial, because they are defined on the continuum which maps the individuals' propensities to cooperate; (2) group members are influenced by the outcomes of their delegate (the source), not by the strength of his or her attitude; (3) the type of social learning defined in Equation 2 is probabilistic and not deterministic as in Latane's model; and (4) social influence is sought as a continuous or an incremental process in which delegates gradually affect their constituents' cooperativeness, whereas in Latane's theory the change of attitudes is a binary, all-or-none process.

Note that Equation 2 has a singular point when $p_d(t) - p_i(t) = 0$, which means that the delegate's social influence on a group member with $p_i(t) = p_d(t)$ is not defined. To account for such cases, the probability $p_i(t + 1)$ at the singular point is defined as:

$$\begin{aligned} &\text{if } O_d(t) \geq 0 \text{ then } && p_i(t+1) = p_i(t) \\ &\text{if } O_d(t) < 0 \text{ then } && p_i(t+1) = p_i(t) + m\,K\,O_d(t) \quad (3) \end{aligned}$$

Where m is a random parameter which assumes the values ± 1 with equal probabilities.

According to this definition, for a non-negative delegate outcome ($O_d(t) \geq 0$), the initial propensity of group member i to cooperate is not changed (i.e., it remains equal to the probability of cooperation of the delegate). On the other hand, if delegate d loses ($O_d(t) < 0$), then the probability that i will cooperate at period t+1 will either increase or decrease, depending on the realization of the random parameter m.

To illustrate the dynamics of social influence generated by the model, consider the distribution in Figure 2 which shows a hypothetical population at some period t.

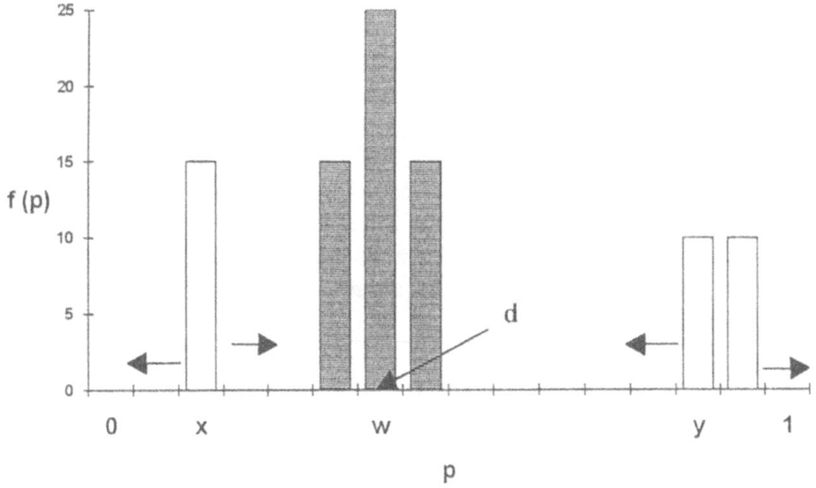

Figure 2: The social influence process in a hypothetical population.

If delegate d wins at period t, then she will "pull" all other group members toward her position (i.e. closer to her probability of cooperation). The "pull" force will be strongest on members in cluster w. In addition, it will be stronger on members in cluster x, whose positions are closer to the position of the delegate, than on members in cluster y whose positions are more distant from the delegate's position. Conversely, if the delegate loses, then all group members will be "pushed" away from her position, and the "push" force will be inversely proportional to the proximity between their positions and the delegate's position.

Following each interaction between the delegates, the simulation updates the cooperation propensities of all other group members according to the social influence function described above. Each delegate remains in position until the next elections, regardless of the changes in the distribution of cooperation that take place in her group. When elections are due, the representation method together with the most recent distribution of cooperation determines who becomes the new group's delegate for the next constituency period.

While Equation 2 describes the form of social influence in almost all cases, special treatment was needed for the two boundary situations in which group members shift toward all-cooperate or all-defect, while their delegate lags behind them and incurs losses. Suppose that members in group A shift to the right and concentrate very close to the $p = 1$ point, while their delegate, who is still in position, is located somewhere very close, but still to the left (having a lower p value). In the case in which the delegate of group B is much less cooperative, group A's delegate will lose continuously, and all group members having higher p values will be "pushed" away from her towards the far right end (i.e., toward the all-cooperate position). When the next elections are due, the new delegate will be an all-cooperate one and will lose continuously. Still, a majority of her group members will have higher p values , and thus will continue to be "pushed" to the right end of the p continuum. Because these members can not be "pushed" to the right further, they may be "stuck" at the right end of the p continuum. The same situation, though less frequently, might occur at the all-defect end. To avoid such counter-intuitive situations, we modified the social influence rule for these extreme situations, such that group members "pushed" away from a losing delegate will move inward, towards the center of the p range.

Operationally, the extreme boundaries were defined as the 2.5% ends on both sides of the p continuum. The simulation treatment of the upper boundary condition can be formalized as follows:

for $p_i(t) \geq .975$, $p_d(t) \geq .975$, and $O_d(t) < 0$:

$$\text{If } p_i(t) > p_d(t) \text{ then } p_i(t+1) = p_d(t) + K\, O_d(t) \qquad (4\text{-a})$$

Similarly, the lower boundary condition can be formalized as follows:

for $p_i(t) \leq .025$, $p_d(t) \leq .025$, and $O_d(t) < 0$:

$$\text{If } p_i(t) < p_d(t) \text{ then } p_i(t+1) = p_d(t) - K\, O_d(t) \qquad (4\text{-b})$$

For all other situations $p_i(t+1)$ is determined according to one of the influence functions stated before (i.e., Equations 2 or 3).

12.2 Results

As stated earlier, the main objectives of the reported simulations were: (1) to investigate the effects of periodic elections of group representatives, under three different representation methods, on the evolution of inter-group relations between two conflicting groups; and (2) to investigate the effect of the frequency of elections in the two groups on the evolution of cooperation between the groups. The simulations were run on a PC. The results reported here were obtained for two equal groups with size $n = m = 300$. The initial propensities for cooperation in both groups were identical, with symmetrical beta distributions with parameters

$u = v = 5$. To test the effect of the election period on the evolution of inter-group conflict, for each of the three election procedures (MWC, MR,& RR), we ran thirty simulations under each one of twelve election conditions. These conditions were: (1, 1), (10, 10), and (1000, 1000) up to (10000, 10000) with intervals of 1000, where each pair indicates the number of iterations between successive elections in the first and second group. The total number of simulations was 1080 (3 election procedures x 12 election conditions x 30 simulations for each election condition), and each simulation was run for 80,000 iterations. For every iteration, the computer program generated (and displayed) the distributions of cooperation for the two groups, their delegates' probabilities of cooperation, and their cumulative payoffs. In reporting the results, however, we focus on the first two variables and omit the reporting of the payoffs, which is of no value for understanding the evolution of cooperation between the two groups.

To study the emerging patterns of inter-group dynamics, we plotted the probabilities of cooperation of the groups' delegates as functions of the number of iterations. The investigation of these graphs revealed that for some simulations, the inter-group dynamics followed certain well defined "pure" patterns. In most other simulations, the emerging patterns exhibited some alternating mixes of these "pure" patterns. Typically, a certain "pure" pattern prevailed for a relatively long period (counted in iterations), after which the dynamics shifted to another "pure" pattern, and so on. The emerging pattern was chaotic only for very few simulations. For the reported simulations (each of which consisted of 80,000 periods), an "enduring" pattern was defined as one that stabilized for at least 5000 iterations.

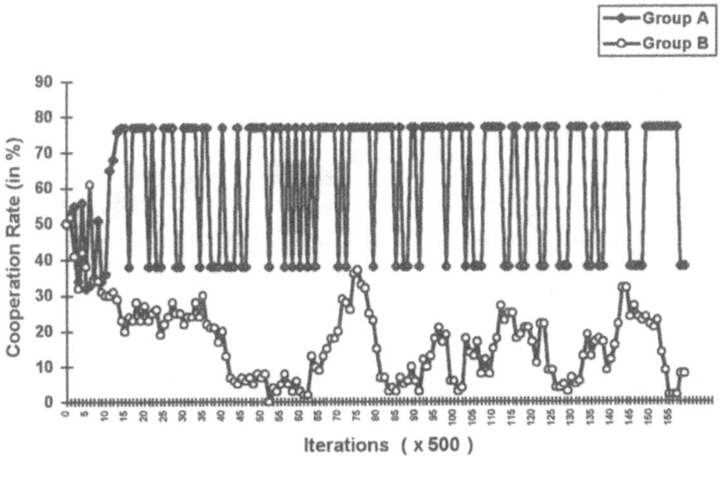

Figure 3: The delegates' rates of cooperation for a typical Defection vs. Oscillation pattern.

Before focusing on the results concerning the evolution of the mutual cooperation under the various conditions, it is useful to present the main patterns obtained in the simulations[6]. These patterns were:

[6] For a detailed description of the evolution of these patterns (under the MWC representation method) see Sulieman & Fischer (1996) and Fischer & Sulieman (1997).

(a) **A Defection vs. Oscillation (D-O) pattern**: A typical example of this pattern obtained under the (10,10) elections condition is shown in Figure 3.1.

Figure 3 depicts the delegates' probabilities for cooperation as a function of the iterations. It clearly shows that after some transitional phase, a relatively stable steady state emerges in which group A's delegate continuously oscillates between high and low levels of cooperation, while group B's delegate converges slowly toward a very low level of cooperation.

(b) **A Defection vs. 50% Cooperation (D-50%C) pattern**: This pattern was obtained more frequently under conditions of short election periods. Figure 4 depicts a typical example for this pattern, in which after a short transitional period the cooperation rate of group A's delegate declines rapidly while group B's delegate stabilizes at a 50% cooperation rate.

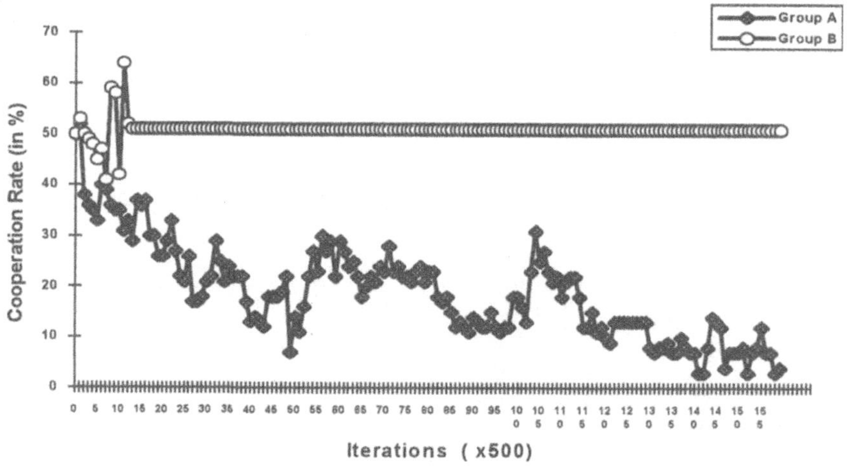

Figure 4: The delegates' rates of cooperation for a typical Defection - 50% Cooperation pattern.

The evolution of the intra-group process in this pattern was qualitatively similar to the one obtained for the Oscillation vs. Defection pattern[7]. The only difference here is that group B evolves into two subgroups: all-cooperators and all-defectors, each containing exactly 50% of the population which is less than the minimum 51% needed to form a winning coalition. In this situation, no single group can form a winning coalition. The only possible coalition is one that encompasses the entire group, with a delegate having a 50% cooperation rate.

---

[7] For a detailed description of this process see Fischer & Suleiman (1996).

(c) **The Cooperation - Cooperation (C-C) pattern:** This pattern is the main focus of the current work. In this pattern, the two groups converge to mutual co-operation as shown in the result portrayed in Figure 5.1.

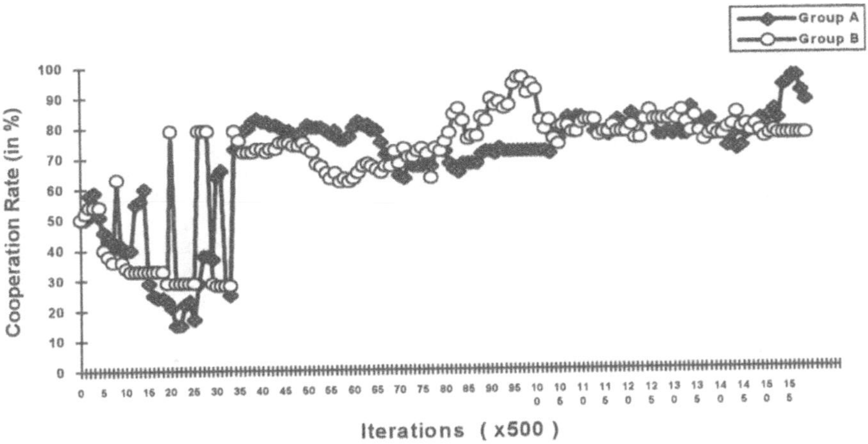

Figure 5.1: The delegates' rates of cooperation for a Cooperation-Cooperation pattern.

The maximum level of mutual cooperation obtained in this result is quite impressive (about 80% for both delegates). In most other results the delegates' rates of cooperation were lower. For classification purposes, we decided to classify a pattern as a C-C pattern if it exhibited a clear and enduring pattern (for at least 5000 iterations) of mutual cooperation, and if the rate of cooperation of both delegates did not drop below the 51% threshold.

The corresponding intra-group dynamics for the result in Figure 5.1 is depicted in Figure 5.2. The vertical axis shows the frequency distributions, and the horizontal axis shows the probabilities of cooperation given in equal intervals of 0.05 (denoted in the graphs by 0, 1, ..., 19).

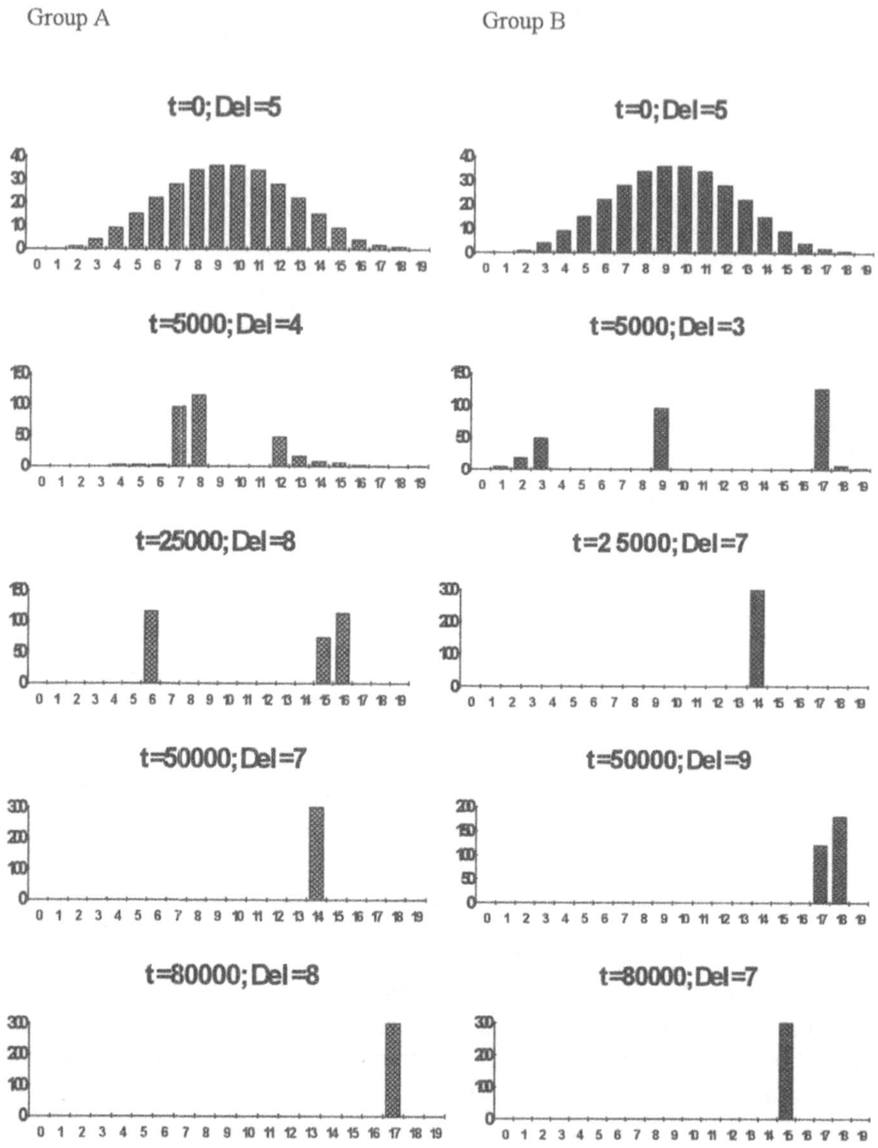

Figure 5.2 The intra-group processes for a typical Cooperation - Cooperation pattern.

The figure shows that after some transitional period the two groups move towards high cooperation rates. When such a situation is reached, both delegates win (in expected values) and mutual cooperation is maintained.

In addition to the "pure" patterns described above, two additional patterns emerged and were sustained for sufficiently long durations (more that 5000 iterations) to justify their classification as enduring patterns. These are the **Cooperation vs. Defection** and the **Mutual Defection** patterns. Figure 6 depicts a simulation result that demonstrates a Cooperation vs. Defection pattern . The figure shows that for two relatively long phases of the evolution (marked by P1 & P2 in the figure), group A's delegate is 100% cooperative, while group B's delegate is very uncooperative (with cooperation rates of 22% and 2% during phases P1 and P2 respectively). An example for a Mutual Defection pattern is demonstrated in Figure 7. The figure clearly shows that during two adjacent phases (marked by P1 & P2) the levels of cooperation of the two groups' delegates are quite low. This D-D interaction is more salient in the second phase (P2) during which the rates of cooperation of the two delegates drops to very low levels.

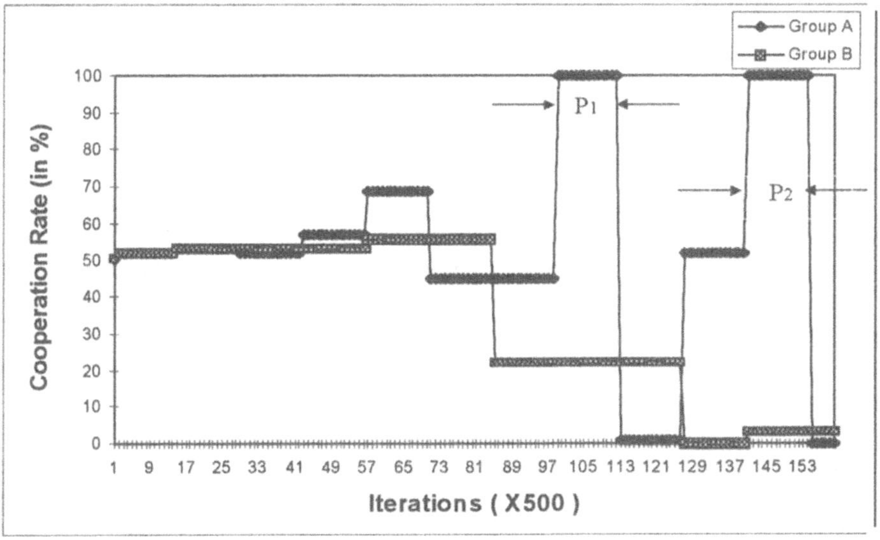

Figure 6: The delegates' rates of cooperation for a simulation result with Cooperation vs. Defection phases.

As noted earlier, some simulation results were predominantly comprised of one of the "pure" patterns described above, but most others exhibited an alternating mix of "pure" patterns, as well as some transitional chaotic patterns. Figures 8 demonstrates some typical pure and mixed simulation results.

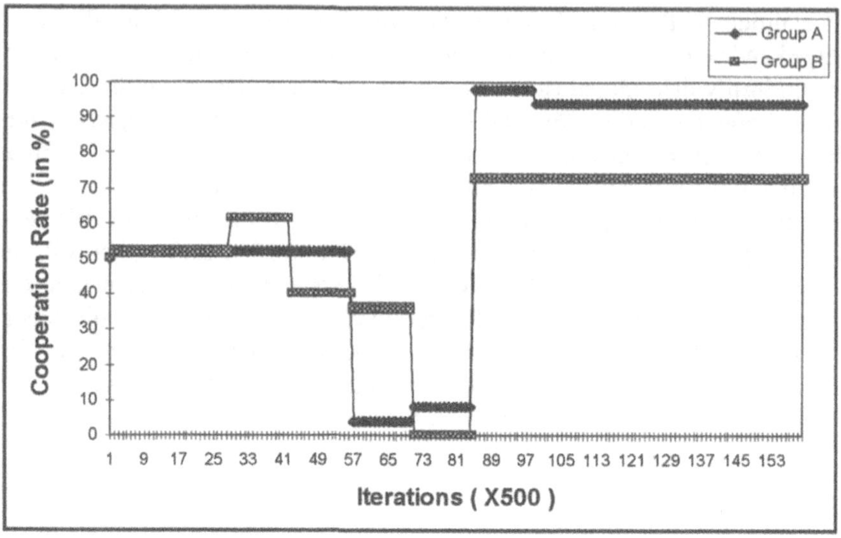

Figure 7: The delegates' rates of cooperation for a simulation result with Mutual Defection phases.

To investigate the effect of election frequency on the evolution of mutual cooperation, we sampled each of the 1080 simulation results at eight sampling points separated by equal intervals of 10 thousand iterations. The sampling points were 10, 20, ... 70, 80 thousand iterations. For each of the three representation methods, and each election condition, we calculated the relative frequency of the mutual cooperation pattern. At each sampling point, a pattern was classified as a C-C pattern only if the dynamics characterizing this pattern was stable for at least 2500 iterations on each side of the sampling point.

Figures 9-a to 9-c depict the relative frequencies of the occurrence of mutual cooperation (in %), for each of the three election methods, and each election condition.

The emerging pattern of mutual cooperation for the MWC (Fig. 9-a) demonstrates the following main features: (a) for all election conditions, the levels of mutual cooperation first decrease, then increase monotonically and stabilize on a certain cooperation rate that differs from one election condition to another; (b) the highest mutual cooperation rates (of about 85%) are achieved under the 5000 elections period condition; and (c) lower rates of cooperation are reached for frequencies above and below the 5000 period.

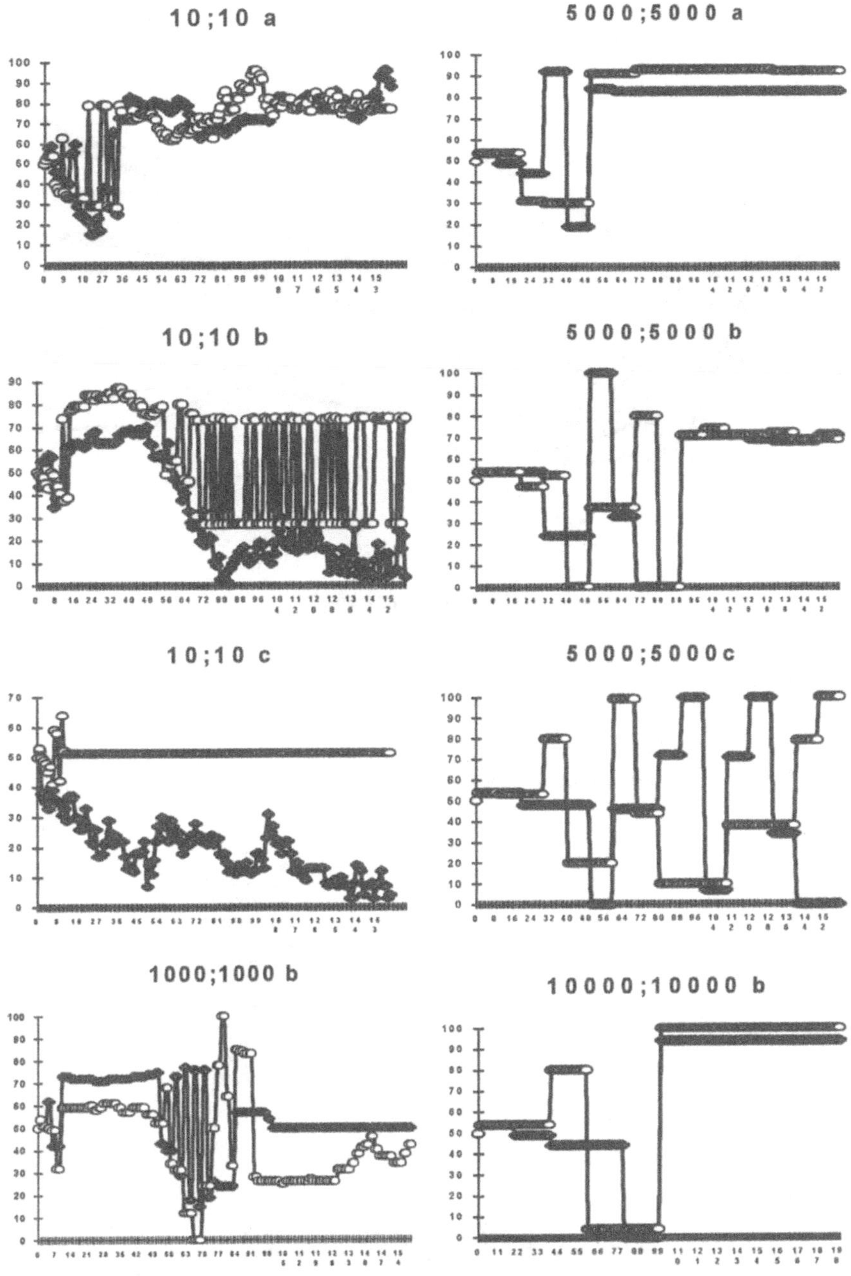

Figure 8: Typical simulation results for some elections frequencies.

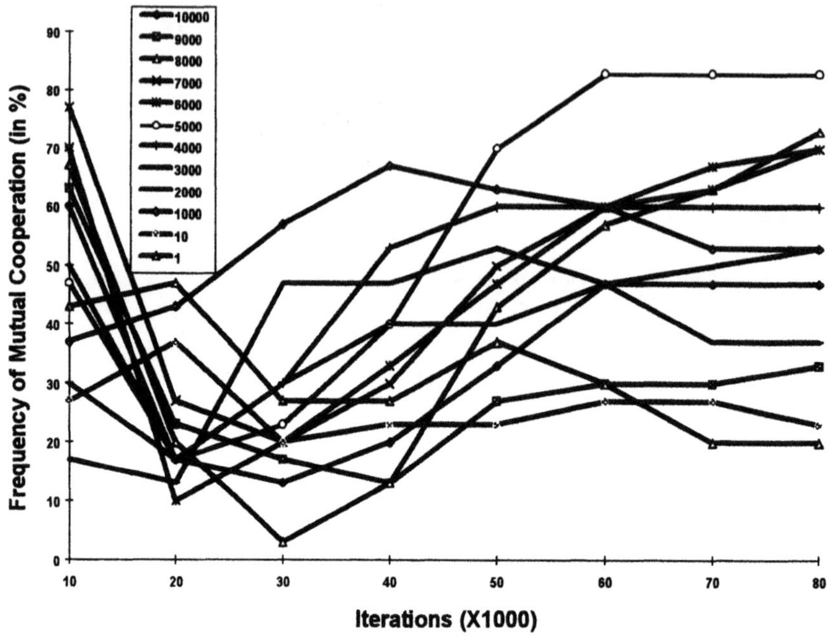

Figure 9-a: Frequencies of the Mutual Cooperation patterns (in %) under the Minimal Wining Coalition procedure for each elections condition.

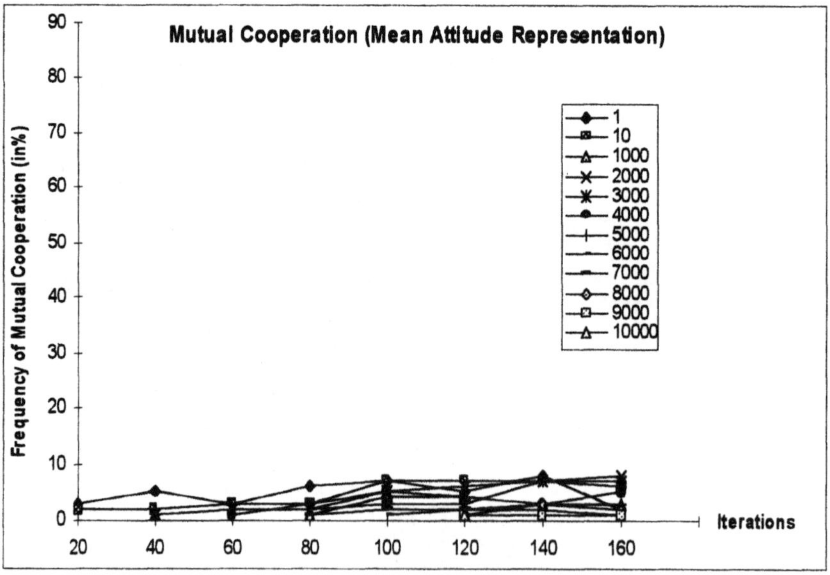

Figure 9-b: Frequencies of the Mutual Cooperation patterns (in %) under the Mean Attitude Representation procedure for each elections condition.

The patterns of mutual cooperation obtained under the Mean Representation method (Fig. 9-b) shows that the rates of cooperation obtained under all election conditions were very low. The highest rate (about 8%) was obtained under the 2000 election period.

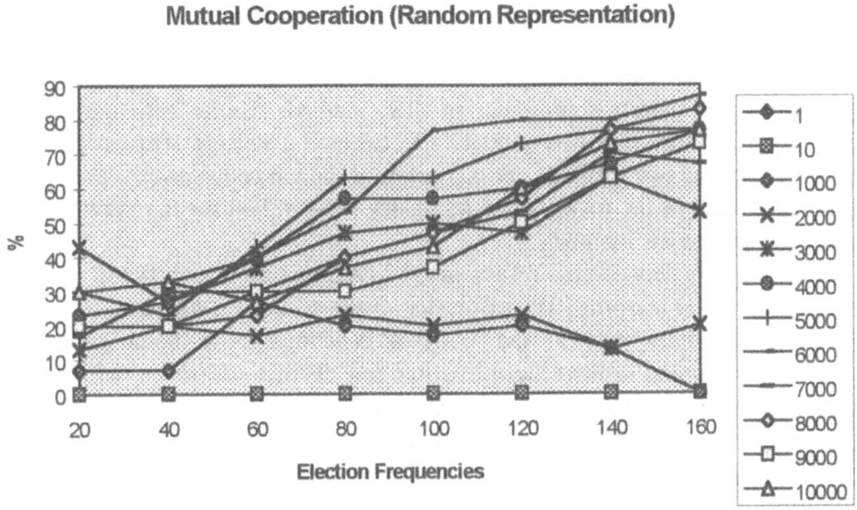

Figure 9-c: Frequencies of the Mutual Cooperation patterns (in %) under the Random Attitude Representation procedure for each elections condition.

Finally, the investigation of Figure 9-c corresponding to the Random Representation method reveals that the mutual cooperation patterns emerging under this condition are quite similar to the ones obtained under the MWC condition. Furthermore, the graphs corresponding to the various election frequencies show that: (a) for most election frequencies there exists a steady increase in the level of cooperation; and (b) in terms of their influence on the evolution of mutual cooperation, the twelve election frequencies can be classified into three distinct groups: the first includes the 1 and 10 election frequencies under which no mutual cooperation was obtained. For the second group, comprised of the 1000 and the 2000 election frequencies, the levels of mutual cooperation are relatively low, and do not exceed a 20% rate. For the last group, which includes all remaining frequencies, the obtained levels of cooperation are relatively high, ranging from about 50% for the 3000 election frequency, up to about 80% for the 6000 and 5000 election frequencies.

12.3 Discussion

The inspection of the 1080 simulations reported in this study shows that the dynamics of inter-group conflict evolve into several phases of well-defined patterns. The results show that the prospects of mutual cooperation are quite sensitive to the election frequencies in the interacting groups, and to the representation method through which the two groups choose their delegates.

An intriguing finding of this study is the similarity in the results concerning the evolution of mutual cooperation under the Minimal Winning Coalition (MWC), and Random Representation (RR) methods. Under both methods, the evolution of cooperation was strongly affected by the election frequencies. In addition, under the two methods optimal levels of mutual cooperation were obtained near the middle of the investigated frequency range (5000 for the MWC method, and 6000 for the RR method).

The similarity between the results of the MWC and the RR methods is, in part, a result of social learning. The inspection of the social learning dynamics shows that the continuous "pushes" and "pulls" of constituents (on the cooperativeness continuum) resulted in their clustering into few distinct subgroups, while leaving most of the cooperativeness continuum almost non-populated. When such configurations were reached, delegation under the MWC method ensured the election of delegates whose attitudes toward cooperation were close to the attitudes of constituents in highly populated clusters (i.e., to "winning" coalitions). The same result was reached under the RR method, although due to a different underlying process. Recall that under the RR method delegates were chosen from "populated" attitudes by a random process. Accordingly, the probability of choosing a delegate with a given attitude toward cooperation was positively related to the number of constituents holding the same attitude.

Social learning accounts for the clustering process described above, as well as for electing delegates who represent majority attitudes. Notwithstanding, it does not explain the high rates of cooperation obtained under the MWC and the RR methods. Our results suggest that the emergence of cooperativeness under these representation methods can be mainly attributed to the *institutional* mechanism of periodic elections.

Furthermore, a careful study of the Mutual Cooperation pattern under the MWC and RR methods suggests that election frequency can enhance mutual cooperation in at least two possible ways: (1) longer periods between elections allow the delegates to exercise a more systematic and directional influence on their constituents. The build-up of such influence can result in a coherent and directional shift of group members in one defined direction. In contrast, frequent changes of delegates can result in an incoherent, "random-walk" type movement caused by the frequent changes in the delegates' positions. This results in frequent, non-orderly, "pulls" and "pushes" of the constituents in alternating directions; and (2) the effect of long election periods is dramatically pronounced when the two groups drift to very low cooperation levels and their delegates engage in a lose-lose (D-D) interaction. In such situations, long election periods provide sufficient duration for "pushing" the populations of both groups towards high cooperation levels. Conversely, very long periods considerably decrease the chances for achieving a synchronized mutual cooperation, since any small difference between the levels of cooperation of the two delegates causes an amplified influence, pushing one, or both, populations away from the existing high cooperation level. These two opposing effects of the frequency of elections yield an optimal election period for which mutual cooperation occurs with highest probability. Our results indicate that under the tested conditions, this optimum is close to the 5000 elections period point. There is no simple way for correlating this optimum (or other elections periods) with the temporal dimension of real political or social systems. Validating the reported results requires data gathered over long historical periods on the distribution of attitudes toward cooperation in real groups and their political structures. Notwithstanding, we contend that the emerging patterns in this study, their evolutionary path and optimal periods, reveal some important principles concerning the nature and underlying structures of some political systems.

In contrast to the results obtained under the MWC and the RR methods, the rates of cooperation obtained under the Mean Representation (MR) method were considerably lower under all election conditions. This result can be attributed, in part, to the somewhat conservative criterion for defining Mutual Cooperation. The inspection of the intra-group dynamics under the MR method revealed that the symmetry of the initial distributions instigated a dynamics through which the two groups were split into two equal subgroups. These groups then gradually departed - in opposite directions - from the center of the cooperativeness continuum. This state of affairs resulted in fixating the groups representatives exactly at the 50% cooperation rates. Our conservative criterion for defining the C-C phases (requiring cooperation rates of 51% or more) resulted in excluding many results from being counted as Mutual Cooperation cases. The counting of 50%-50% results as C-C patterns could have increased the C-C rates obtained under the MR method. In retrospect, we also think that the number of 50%-50% cases could have been considerably less if the simulations were run using asymmetric initial distributions.

Notwithstanding the above remarks, it might well be that the MR method will remain inferior to the MWC and the RR methods for achieving inter-group cooperation. If validated, the results reported here add to previous reservations, voiced on different grounds, against the averaging of population opinions. In reference to the relationship between constituents and their representatives, Rousseau, in his 'Social Contract', discriminated between the "General Will" and the 'Will of Everybody". In Rousseau's words "it often happens that the will of everybody, because it is looking to private interest and is thus merely a sum of particular wills, is something quite different from the general will, which looks exclusively to the common interest" (Rousseau, 1762). Rousseau goes further to suggest that very extreme opinions should be disregarded when the public "General Will" is defined.

The complexity of the investigated situation renders it unamenable to mathematical tractability. In the absence of closed form solutions, computer simulations - as undertaken here - remain the best tool for studying the evolutionary dynamics of inter-group (and intra-group) processes. Notwithstanding, the results of such simulations should be taken with care due to their notorious sensitivity to the specific parameter values under which they are run. In responding to this problem, we plan to run a series of studies to test the robustness of the results obtained thus far in our simulations project. Specifically, such sensitivity analysis requires the testing of the reported results under various initial distributions, election frequencies, and social learning rules.

A major motivation behind the present study lies in the authors' interest in studying the evolution of conflict and cooperation between real groups. We must confess that the effort taken here is too limited to allow any speculations, let alone predictions, regarding the evolution of real-life inter-group relations. Notwithstanding, we believe that the simulation, and results, reported in this study may help in exploring some of the basic aspects of inter-group conflicts. We also suggest that simulation methods can serve as useful educational tools for studying some of the basic processes underlying the evolution of inter-group conflicts, and exploring different possibilities for their resolution.

References

Amacher, R. C., & Boyes, W. J. (1978): Cycles in Senatorial Voting Behavior: Implications for the Optimal Frequency of Elections. Public Choice 33 (3), 3-13.

Axelrod, R. (1984): The Evolution of Cooperation. Basic Books, New York.

Bentham, J. (1843): Collected works. William Tait, Edinburgh.

Birch, A. H. (1971): Representation. Pall Mall Press, London.

Brams, S. (1975): Game Theory and Politics. Free Press, New York.

Derman, C., Gleser, L. J. & Olkin, I. (1973): A Guide to Probability Theory and Application. Holt, Rinehart, and Winston, New York.

Fischer, I. & Suleiman, R. (1997): Mutual Cooperation in a Simulated Inter-group Conflict. Journal of Conflict Resolution 41 (4), 483-508.

Hamilton, A. J. J. & Madison, J. (1787-8): The federalist, paper 52, (new edition 1918) Everyman's Library, London.

Johnson, N. L. & Kotz, S. (1969): Distributions in Statistics. Houghton Mifflin, Boston, MA.

Latane, B. (1981): The Psychology of Social Impact. American Psychologist 36, 343-365.

Mill, J. (1828): Essay on Government (new edition 1955). Bobbs-Merrill, New York, 70.

Nowak, A., Szamrej, J. & Latane, B. (1990): From Private Attitude to Public Opinion: A Dynamic Theory of Social Impact. Psychological Review 97 (3), 362-374.

Rousseau, J. J. (1762): The Social Contract (new edition 1954) Henry Regnery, Illinois.

Schelling, T. (1960): The Strategy of Conflict. Harvard University Press, Cambridge, MA.

Suleiman, R., & Fischer, I. (1996): The Evolution of Cooperation in a Simulated Inter-group Conflict. In: Liebrand, W. B. G. & Messick, D. M. (Eds.): Frontiers in Social Research. Springer-Verlag, Berlin, 419-438.

Part IV

Sensitivity Analysis

13. Sensitivity Analysis in the Social Sciences: Problems and Prospects

Edmund Chattoe[1], Nicole J. Saam[2], Michael Möhring[3]

"Variables don't, constants aren't". (Osbourn's Law)

13.1 Introduction

In the natural sciences and engineering, sensitivity analysis is a standard method for verifying simulation models. The term "sensitivity analysis" is generally used to describe a family of methods for altering the input values of the model in various ways and such analyses are included in the validation step of almost all technical simulations (Law and Kelton 1991, pp. 310ff). Nevertheless, a short review of simulation textbooks and other studies reveals that the term is currently used as a general catch all for diverse techniques: there is no precise definition and no special methodology currently associated with this term. A cursory inspection of recent work suggests that the situation is even worse in the social sciences.[4] Most social simulators still omit any form of sensitivity analysis. (However, there are exceptions, mostly recent: Axtell *et al.* 1996, Harrison and Carroll 1991, Latané 1996, Macy 1991, Saam 1996 and Saam and Reiter, submitted.) There is also a definite lack of methodological literature on sensitivity analysis in the social sciences.

This paper gives an overview of the general problems currently experienced in the application of sensitivity analysis to social simulation. By describing and exploring the current fragmentation of methods, we hope to pave the way to a more systematic treatment of sensitivity analysis, leading in the long run to a standardisation of algorithms and tools. This should improve the prospects for future application in social simulation, as well as improving the recognition of simulation results by outsiders.

In section 2 of the paper we provide a very general definition of sensitivity analysis, following this by a discussion of its purposes (section 3) and importance (section 4). We then describe some basic methods and algorithms that are presently being used for sensitivity analysis (section 5). Section 6 illustrates some of

---

[1] Department of Sociology, University of Surrey, Guildford, Great Britain
[2] Institut für Soziologie, Universität München, München, Germany
[3] Institut für sozialwissenschaftliche Informatik, Universität Koblenz-Landau, Koblenz, Germany
[4] This observation is borne out by the analysis of Metz (1979).

the general problems of sensitivity analysis techniques using three specific examples. In section 7 we discuss the problems that appear to result in the rather infrequent application of sensitivity analysis in the social sciences. In section 8, we conclude by sketching a list of requirements for the standardisation of sensitivity analysis and developing some concepts to characterise appropriate tools.

13.2 Definition

Sensitivity analysis is a collection of tools and methods used for investigating how sensitive the output values of some model are to changes in the input values. A "good" simulation model (or a "significant" result) is believed to occur when the output values of interest remain within an interval (which has to be defined), despite "significant" changes in the input values (which also have to be defined).

Consideration of the disparate uses of the term in the current social science literature (exemplified in the next two sections) suggests that it is not fruitful to decide the "correct" use of the term by appealing to some prescriptive definition. Instead, one purpose of this paper is to try and make as much typological sense as possible of the set of techniques currently described as sensitivity analysis and to draw out parallels and contrasts using different possible ways of "filling in" the general definition, which is intended to be descriptive rather than prescriptive. The development of this typology involves a more detailed consideration of the status of "input" and "output" along with a range of possible measures of change or stability (lack of change).

13.2.1 Input Variability

Using the most general possible definition, the following kinds of deliberate input variability can all be seen as commonly used examples of sensitivity analysis (see also Bratley *et al.* 1987, p. 125):

Random Seed Variation: Testing for the effect of random elements in the model by repeating a simulation using a different sequence of computer generated random numbers for each run.

Noise Type and Noise Level Variation: Testing for the effects of variation in stochastic elements of the model by varying the distribution of noise (from normal to uniform errors for example) or its level for a particular distribution (changing the mean or variance of a normally distributed error). Varying the stochastic elements of a model in this way differs from varying the random seed because noise distributions and levels are parameters of the model while the actual set of random numbers generated comprise variables for a particular run of the simulation.

Parameter Variation: Although adjustments to noise type and level are particular cases of parameter variation, parameters are used to refer to a much wider range of fixed or quasi-fixed elements in models. Indeed, parameter variation is the nearest we have to a "paradigm case" for sensitivity analysis, if only because the term parameter is used so loosely that very few variables under the control of the simulator definitely fall outside it. Parameters can be "physical" (the time taken between conception and birth in a demographic simulation), "cognitive" (the rate of forgetting during some decision-making task) and "behavioural" (the rule used by consumers to relate current consumption to current income). Because the term parameter is so general, it will be helpful to subdivide model parameters into different classes, for example parameters describing behavioural rules, cognitive processes, physical constants, stochastic elements and so on. In particular, the classes of parameter are very important in interpreting a given sensitivity analysis. Changing a noise parameter may only imply some form of uncertainty about the quality of the data. By contrast, changing a physical parameter like the time between conception and birth implies a much more radical "possible worlds" view of the role of simulation. Approaches of this sort, studying possible worlds to learn more about the world as it is, are favoured by both Artificial Life (Langton 1989) and Artificial Societies literature (Gilbert and Doran 1994, Gilbert and Conte 1995). Instead of attempting to mirror a social process, these approaches use simulation as an exploratory or theory building tool in which apparently quite counter-intuitive hypotheses can be developed in a coherent way. (Doran, this volume.) In the case of Artificial Life, this is important because evolution through natural selection has "thrown away" alternative biological possibilities which might exist, for example on other planets, making it hard to form an effective contrast with the currently observed system. It might be argued that the same justification applies to studies of Artificial Societies. We can only make sense of the current social environment in the context of coherent alternatives. The importance of coherent interpretations in sensitivity analysis will be addressed in more detail in section 6.

Temporal Model: In order to simplify social processes for simulation, it is often desirable to make assumptions about the order of actions and whether these take place in discrete or continuous time. It has long been known (Huberman and Glance 1993, Hegselmann, this volume) that interesting results from Cellular Automata are not necessarily robust to changes from discrete to continuous time or from fixed to random updating of cells.

Aggregation: Although not involving simulation, papers by Attanasio and Weber (1993, 1994) suggest another possible form of sensitivity analysis. Econometric studies of consumption at the aggregate level are forced to make joint hypotheses about both individual rationality and aggregation. Studies making use of consumption data at the household level reveal the instability of the econometric results (output) to changes in the level of data aggregation (input). It does not appear that simulation has yet been used for this form of sensitivity analysis. In particular, there is an important role for micro-simulation techniques (Merz 1994) in exploring the effectiveness of econometric modelling at capturing important patterns in individual behaviour. Traditionally, conflicting microeconomic data cannot be used to criticise macroeconomic models directly because its aggregate effects cannot easily be explored. Simulation permits econometric estimations based on the aggregate data generated by the model to be compared directly to distributions of individual behaviour.

Econometric and Statistical Variability: The commonest "input" to models is a given data set, for example a set of values for aggregate consumption and aggregate income. Several common econometric and statistical techniques can also be seen as forms of sensitivity analysis under the most general definition. For example, recursive fitting of data (Harvey 1990, pp. 52-5) and out of sample prediction (Hendry 1985, pp. 75-6) are both attempts to demonstrate that the output (prediction) is insensitive to particularities of some subset of the input data sample.

Action Sets: Most of the types of sensitivity analysis discussed so far make sense only in the context of "traditional" equation based approaches to modelling. However, agent based approaches like Distributed Artificial Intelligence (Bond and Gasser 1988) and Evolutionary Game Theory (Friedman 1991) allow us to investigate the aggregate effects of interactions between individual agents with differing decision processes and capabilities. Well known examples are provided by the Evolutionary Game Theory literature (Weibull 1995) and that on evolutionary tournaments (Axelrod 1987, Miller *et al.* 1994). In these cases, particular agents, strategies or programs are shown either theoretically or empirically to be resistant to changes in the environment. For almost all models in this literature, the environment consists only of other agents so resistance involves survival in the face of introduction of new agent kinds. (It does not appear that any analyses of this kind have been done for situations where the environment has an existence independent of the agents in it). Findings about this kind of survival have been formalised in the concept of an Evolutionary Stable Strategy or ESS (Maynard-Smith 1982). The same sort of "robustness" has been observed and assessed in populations of agents generating equilibria that consist of populations with mixed strategies (Lomborg 1992, Latané and Nowak 1997). A related and more detailed illustration of the kind of sensitivity analysis required for action sets will be discussed in section 6.3.

If nothing else, these disparate approaches suggest that it is difficult to draw a dividing line between "proper" and "improper" forms of variability or sensitivity analysis based on definitions or current practice alone. Everyone would agree that changing parameter values or the level of noise in an equation system is "proper" sensitivity analysis, but what is it about changing the capabilities of agents, the sampling of the data set or even the overall dynamics of the system which stops these techniques being equally proper? One simple answer is a practical one. Appropriate variability is defined not in theoretical terms, but in terms of ease of applicability. For solutions to simple equation systems, it is quite easy to change particular parameters and observe the corresponding changes in the model output. (This is true of simulated systems based on computer programs. For mathematical systems being solved iteratively or analytically, there may only be a range, or even a single parameter value, for which this is true.) For more complex systems, like models of interacting agents, the task may be much more difficult. For example, certain sorts of analysis like changes to the temporal model may require considerable alterations to the simulation. In such circumstances, changes in output behaviour could just as easily be the result of programming errors as anything else. We will return to this point in section 8. Nevertheless, this remains a technical problem, to be solved by technical means and it cannot be used as a justification for avoiding sensitivity analysis or proscribing particular kinds of model. One task of this paper is to use the general definition of sensitivity analysis to draw attention to its applicability in new areas, particularly that of social simulation.

13.2.2 Output Variability

Sensitivity techniques are not restricted to demonstrating that the simulation output is (in)sensitive to a defined variation of a single input unit. In such cases, conventional measures such as the confidence interval could be used. The confidence interval indicates the region between lower and upper thresholds within which the (single) output is located with probability 1-α (where α is the confidence level). In fact, social simulators are usually interested both in the *evolution* of each output variable (rather than just its final value) and in the evolution of *several* output variables (rather than just one). In such circumstances, one would have to calculate (combinatorially) many different confidence intervals or perform extensive and potentially meaningless aggregation, losing a lot of information in the process.[5] For this reason, the confidence interval approach is inappropriate to most social simulations and we suppose this to be the main technical reason for its general neglect in the sensitivity analysis of these models.

Beyond the simple case of confidence intervals, sensitivity measurements may for instance assert that:

- A single output variable experiences an *x* percent change in its value if the input value of another variable is changed by *y* percent. (This is the elasticity coefficient suitable for a quantitative deterministic model.)
- The *mean* of a single output variable experiences an *x* percent change in its value if the input value of another variable is changed by *y* percent (This is the elasticity coefficient for a quantitative stochastic model. In addition, the standard deviation of the mean may be given.)
- Due to its connectedness with other variables, a particular variable has an active, reactive, buffering, critical or neutral influence (Vester 1990, 1991) on the general output of the model. (These are the descriptive properties of variables for a semi-quantitative model.)
- Qualitatively different patterns (modes) of long term behaviour in an output variable result if the input value of another variable is changed. (This form applies to a qualitative deterministic model.)

[5] In physical systems, there is a greater tendency for the measured parameters to be independent since this appears to be one of the defining characteristics of measurable properties in physical science.

- Qualitatively different patterns (modes) of long term behaviour in an output variable change their frequency distributions if the input value of another variable is changed. (This form applies to a qualitative stochastic model.)
- All the output variables for a particular simulation model are comparatively (in)sensitive to changes in all the input variable values. (This form applies to an aggregate model.)

This is a short and incomplete list of the classes of sensitivity outcomes for social simulations. The complexity of sensitivity results depends on the number and kind of input variables, the parameters, the output variables, the model structure and the algorithms used to calculate the sensitivity measures. As a representative sample, recent social simulations have presented the following types of sensitivity results:

- Axtell *et al.* (1996) present a sensitivity analysis of an agent-based model. They obtain different average numbers of equilibria for their simulation model.
- Harrison and Carroll (1991) present a regression analysis based on their sensitivity results. They regress the equilibrium outcome variables on the input variables and obtain regression coefficients.
- Latané and Nowak (1997) use a combination of simulation and independently collected experimental data to perform a very general sensitivity analysis of the "clustering" results obtained in their studies of social influence. They show how factors like exponential decay of influence and intrinsic differences of influence are essential to "preserve" clusters.
- Macy (1991) presents a preliminary sensitivity analysis of his non-linear dynamical model. He obtains different patterns (modes) of long term behaviour for his outcome variables.
- Saam (1995, 1996) presents two different kinds of sensitivity analysis for her stochastic non-linear multi-level dynamical system. She obtains different frequency distributions of patterns (modes) in long term behaviour for her outcome variables which have to be interpreted qualitatively. She also obtains classifications of the sensitivity of the variables based on Vester's (1990, 1991) semi-quantitative categorisation.
- Saam and Reiter (forthcoming) present a sensitivity analysis of a stochastic dynamical multi-level model. They obtain elasticity coefficients.
- Wuensche and Lesser (1992) present a sensitivity analysis of a cellular automata model. They obtain graphical representations of changing basins of attraction.

After consideration of different kinds of changes to "input" and "output" variables, along with a range of possible measures of change or stability, we now turn to the purposes, importance and methods of sensitivity analysis under the general definition.

13.3 Purposes

The three major purposes of sensitivity analysis are corroborating the central results of the simulation, revealing possible variations in the results and guiding future research by highlighting the most important processes for further investigation (Liebl 1995, pp. 214f).

Often the methods applied in sensitivity analysis do not differ from those used in model optimisation and validation. For this reason, many authors give a broader definition of sensitivity analysis (e.g. Bernard and Plaisent 1990). The analysis described in this paper may be referred to as *sensitivity analysis strictly speaking*. What we might call *sensitivity analysis in the broadest sense* also includes various sorts of model optimisation and validation techniques:

- parameter estimation by showing which combination of parameters lead to the "desired" model behaviour;
- verifying whether the parameters and starting configuration of the simulation are the "best" ones in some sense;
- verifying whether the same or similar results would be obtained if the simulation were run again with slightly different parameters;
- verifying whether the model is the simplest that yields the desired output and is therefore to be preferred over other, more complex models.

Nevertheless, we believe that sensitivity analysis strictly speaking should be kept separate from model optimisation and model validation. It is the purpose of these techniques that makes the difference. Sensitivity analysis does not necessarily indicate which solutions are "good", rather it can save the resources of the researcher from being wasted on unnecessary calculations that entail only small changes to insensitive parameters (Meyers *et al.* 1994, p. 1634). As long as model optimisation is not based on mature optimisation algorithms, sensitivity analysis will produce (as a by-product) results that serve the purposes of model optimisation. The same applies to the process of model validation. Strictly speaking, in sensitivity analysis, we assume that model optimisation and model validation have already been carried out (Liebl 1995, p. 214).

Sensitivity analysis should also be kept separate from experimentation. Experiments are run *after* model optimisation and vary selected parameters (or other model features) systematically. The selection of these varied entities is guided by particular theories, and by the research interests of the investigators. Therefore, it is not appropriate to represent experiments as a component of sensitivity analysis as some authors have done (Axtell *et al.* 1996).

Finally, sensitivity analysis should be kept separate from "what if" analysis. In "what if" analysis (see e.g. Carley 1991) each parameter (or other model feature) is systematically varied so that all possible "what if" questions are asked for each possible alternative "value". (For a traditional parameter, these "values" are simply numerical. For shifts in temporal structure, "values" may be feasible alternatives like "continuous time" or "discrete time".) Here, the results are obtained by systematic variation of the input. Again, it is the intention of the technique that makes the difference. The major purposes of sensitivity analysis are corroborating central results of the simulation, revealing possible variations in results and guiding future research by highlighting the most important processes. The purposes do not include obtaining the results themselves.

13.4 Importance

It is the general task of a sensitivity analysis for any simulation model to distinguish between input units which have a big influence on the output and those which do not. The large influence of an input unit means nothing else than a high sensitivity of the model output to variations in that input unit. It is not absolutely necessary, but typical, that different input units are of different importance. This holds even for the simplest simulation models that try to represent well understood systems and this is one reason why sensitivity analysis is important. If the objective is to model systems that are not yet so well understood, like most of those in the social sciences, then sensitivity analysis is even more important because in most cases input data error must be assumed, input data are not available or are only available in a related form, assumptions are heroic or hypotheses about cause and effect relations in the system are only vague (Liebl 1995, p. 215). Thorough sensitivity analysis effectively involves taking an "additional" share of the responsibility for scientific criticism onto oneself. By doing a sensitivity analysis, one disposes of criticisms that one has made "trivial" errors, thus restricting critics to more substantive objections. It is standard practice in econometric work to quote certain accepted and "minimal" statistical tests. There is not currently an equivalent set of "standard" sensitivity analysis for simulations, in part because there is no such thing as a standard simulation. By changing the input systematically, sensitivity analysis helps to avoid artificial results and to *explore* relationships and mechanisms that are not yet well understood. It is most surprising that in the light of this, sensitivity analysis is not yet a standard tool for verifying social simulations.

13.5 Methods and Algorithms

As has already been remarked, sensitivity analysis is a collection of methods used for evaluating how sensitive the model output is to changes in input values. In most sensitivity analyses the input is changed *systematically* (Kleijnen 1987, p. 145). One exception is sensitivity analysis based on random experimental designs (Kleijnen 1987, p. 145 and part II. B). An overview of general statistical methods for sensitivity analysis is given by Kleijnen (1992a, 1995a, 1995b) and general methodological texts on sensitivity analysis are also available: Deif (1986), Fiacco (1983, 1984), Köhler (1996) and Ríos Insua (1990). In this section we give a short presentation of some basic methods which have actually been applied in social science simulation: sensitivity analysis through graphical methods using conventional quantitative methods and simple sensitivity indices, meta-modelling and semi-quantitative sensitivity analysis. Although these algorithms cover the broad range of input variability described in section 2, they obviously cannot deal with all the difficulties encountered in the application of sensitivity analysis to the social sciences. Some of these problems will be addressed in detail in the following section.

13.5.1 Terminology and Basic Assumptions

The input and output behaviour of a simulation model as an abstraction of a real system can be described as follows:

$$(y_1,\ldots y_l,\ldots y_m) = f_{model}(x_1,\ldots x_j,\ldots x_k, r)$$

x_j : input parameters (quantitative or qualitative)

y_l : output parameters (quantitative or qualitative)

f_{model} : black box function, mapping k input parameters to m output parameters. (This mapping can be performed by a computer program instead of a mathematical function.)

r : the seed of a pseudo-random number generator in a stochastic simulation. (By using different seeds, different outputs can be generated from each input combination.)

Corresponding to the various kinds of input variability discussed in section 2, the x_j can represent a broad range of possible model components:

- parameters in a (differential) equation based model;
- non equation based model parameters (e.g. the number of servers in a queuing model);
- noise distribution parameters (e.g. uniform, normal or exponential);
- temporal parameters (e.g. discrete or continuous);
- model level parameters (e.g. aggregate or micro model);

This variability has important consequences for the design of sensitivity analyses for different simulation systems. For example, comparing the values of an output parameter based on different model levels requires modelling facilities which allow the definition of a model on different (aggregation) levels. An example of a package providing these facilities is MIMOSE (Möhring 1996). Additionally, random seed variation can be realised by using different values of r for each combination of input parameter values.

If we assume that each input combination is replicated an equal number of times, it would be possible to neglect r, because the individual observation y could be replaced by its average. For the sake of simplicity, we concentrate on a single output parameter y. If there were several outputs, the model would have to be applied to each.

Assuming that a sensitivity analysis requires the application of f_{model} for n combinations of x_j to produce y, the data base of a sensitivity analysis can be described in general terms as follows:

$$\begin{pmatrix} y_1 \\ \ldots \\ y_i \\ \ldots \\ y_n \end{pmatrix} \begin{pmatrix} x_{11},\ldots,x_{1j},\ldots,x_{1k} \\ \ldots \\ x_{i1},\ldots,x_{ij},\ldots,x_{ik} \\ \ldots \\ x_{n1},\ldots,x_{nj},\ldots,x_{nk} \end{pmatrix}$$

The terminology described above neglects the change of the output value y during the simulation period. As already mentioned in section 2.2, for social simulations, the evolution of values is often particularly interesting and the simulator is not only interested in the value of y in the last simulation step T. This means, that the vector of the y_i's should be replaced by a matrix:

$$\begin{pmatrix} y_{11},\ldots,y_{1t},\ldots,y_{1T}(=y_1) \\ \ldots \\ y_{i1},\ldots,y_{it},\ldots,y_{iT}(=y_i) \\ \ldots \\ y_{nl},\ldots,y_{nt},\ldots,y_{nT}(=y_n) \end{pmatrix}$$

The selection of suitable combinations of the input and the sample size n is not the main focus of this paper. From a more theoretical point of view, this complex task is covered by the statistical theory for the design of experiments (Kleijnen 1988, pp. 167f, Kleijnen 1995b) and the support of this task has to be considered within the design of modern simulation systems such as the experiment description facilities provided in SIMPLEX (Wittmann 1993). These points will be discussed further in section 8.

13.5.2 Conventional Quantitative Methods and Simple Sensitivity Indices

What we call "conventional" quantitative methods are the most widely used in social simulation. These are also referred to as "tuning by hand". Bratley *et al.* (1983, p. 9) distinguish between two important approaches:

- **Sensitivity testing**: Vary just one parameter while keeping all others fixed and check that the behaviour of the model is stable.
- **Stress testing**: Set parameters of the model to strange values and check if the model "blows up" in an understandable fashion.

Based on these two strategies, the main task of quantitative sensitivity analysis is to determine sensitivity indices. The simplest sensitivity index, widely used in economics, is known as the *elasticity coefficient* ε, which indicates the change in an output value for a variable due to the change for an input variable. Generally, sensitivity indices can be defined in terms of the *differential quotient*:

$$S_j = \frac{\delta y}{\delta x_j}$$

or, if differentiation is not possible (which is generally the case in complex simulation models) as the *difference quotient*:

$$S_j = \frac{\Delta y}{\Delta x_j}$$

In practice, this kind of sensitivity analysis is often done using scatter plots with the values x_{ij} of one input parameter on the x-axis and the values y_i ($= y_{it}$) of the output parameter on the y-axis. Some qualitative effects of the input parameter on the output can be derived directly from this graph, like increasing, decreasing or constant effects over the domain of the input parameter. Using only the last output value of a simulation run, the described difference quotient can be used as sensitivity index. Other single output values, like the first value or the minimum and maximum are possible as well (Schäfer 1996, pp. 15f). To take into account the evolution of the output, all values in a simulation run have to be aggregated. For example, the *mean deviation*:

$$\Delta D_y = \frac{\sum_{t=1}^{T} | y_{it} - y_{i't} |}{T}$$

or the mean square deviation:

$$\Delta D_y = \frac{\sqrt{\sum_{t=1}^{T} (y_{it} - y_{i't})^2}}{T}$$

These, together with some standardisation measure, lead to the new sensitivity indices:

$$S_j = \frac{\dfrac{\Delta D_y}{D_y}}{\dfrac{\Delta x_j}{x_j}} \text{ where } D_y = \frac{\sum_{t=1}^{T} y_{it}}{T}$$

Plotting the different combinations *i* of an input parameter x_j on the x-axis and the corresponding sensitivity indices on the y-axis may also help to identify regions of high or low sensitivity. Figure 1 shows the described types of scatter plots (adapted from Huber *et al.* 1996).

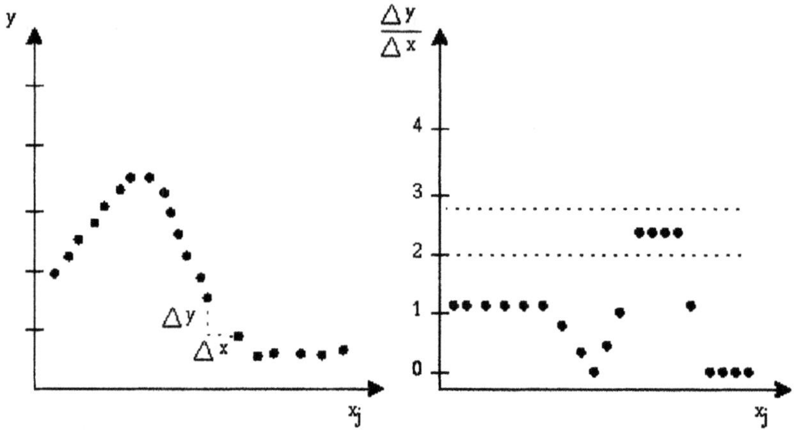

Figure 1: left - a one-dimensional example with sampled points; **right** - the corresponding values of the difference quotient Δy/Δxj

Unfortunately, sensitivity testing and stress testing suffer from the same problem. Since only the isolated effects of one or two input parameters can be observed and the number of combinations of parameter values is very limited, these models generally result in incomplete overviews of the implications of the model.

13.5.3 Meta-Modelling

Meta-Modelling Using Algebraic Functions:

A scatter plot can also be analysed by fitting a curve to the data (e.g. $y = \beta_0 + \beta_1 x_1$). This means that the simulation model is approximated by a new meta-model. Using meta-models becomes particularly useful in the case where there are several input parameters, because it allows estimation of the relative importance of input parameters as well as permitting the analysis of interaction and higher order effects. A simulation model may be approximated by a Taylor series of different degrees. The *regression function* is the most well known approximation of this type. By using a second degree approximation (Kleijnen 1995b), a regression meta-model can be described as follows:

$$Y = \beta_0 + \sum_{j=1}^{k} \beta_j x_j + \sum_{j=1}^{k} \sum_{j \neq j'}^{k} \beta_{jj'} x_j x_{j'} + E$$

x_j, $x_{j'}$: (standardised) input parameters
Y : output parameter (Because of the stochastic fitting error E, this is a stochastic parameter and therefore capitalised.)
β_0 : regression intercept
β_j : first-order effect of factor x_j
$\beta_{jj'}$: interaction effect of the factors x_j with $j \neq j'$
β_{jj} : quadratic effects of factor x_j
E : stochastic fitting error of the regression model

Linear regression, which means neglecting interaction and higher order effects, uses the ordinary least squares (OLS) criterion to estimate the parameters β_j ($j = 0,\dots,k$) by minimising the fitting error E. The relative importance or sensitivity of the input effects can be obtained by using standardised input parameters and sorting the absolute values of the β_j's.

The integration of interaction and quadratic effects requires many more runs of the simulation because the number of input parameters increases and for theoretical reasons, n must be greater than or equal to k. For example, if we want to analyse interaction effects, the number of input parameters increases as $k(k-1)/2$ ($x_j x_{j'}$, $j \neq j'$) and n must increase correspondingly. Furthermore, higher order functions are more difficult to interpret, whereas first order regression models provide a clear understanding of the effects of the input. Useful methods for checking fitted regression models include *predicting* the output for a new combination and *cross-validating* the model, which involves deleting one old combination and re-estimating the regression parameters with the remaining combinations (Kleijnen 1995b).

Other sensitivity indices than $|\beta_j|$ have been proposed, using square integrable function components (Sobol 1996) and the rank transformations of model output (Saltelli 1996).

Meta-Modelling By Learning

Just as with multivariate analysis (e.g. regression analysis, factor analysis), using algebraic functions as meta-models allows the treatment of mathematical methods in sensitivity analysis. But equally, some well known problems arise with this kind of modelling, like the selection of inadequate models (e.g. linearity) or the need to consider the underlying assumptions (e.g. integrability of functions, distribution of parameters). During the last few years, methods from Artificial Intelligence, especially from the area of *Machine Learning*, have been adapted for statistical analysis (generally known as *data mining* methods), which can also be used for meta-modelling (Pierreval 1992). The main characteristic of these methods is the absence of an explicit (mathematical) model. Machine Learning algorithms gain knowledge about the simulation model by analysing the given observations and representing them in a certain structure. For example, decision tree algorithms like ID3 (Quinlan 1986) generate rules from simulation experiments, which produce an output value depending on certain values of the input parameters. Figure 2 shows an example of a decision tree for credits granted by a bank. The input parameters are the *amount* and the *duration* of the requested credit and the *age* of the person who wants the credit. The output parameter is the decision about whether to grant the credit or not. Starting with the *duration*, each combination leads to a *credit* value. In this way, the order of the chosen input parameters reflects their individual importance for the response.

Referring to figure 1, this technique can be used as an automatic means of identifying intervals of input values with high or low sensitivity (Huber *et al.* 1994, Huber *et al.* 1996). Because the quality of Machine Learning algorithms strongly depends on the number of observations used for learning, creating meta-models by learning algorithms need efficient and powerful simulation facilities to perform a lot of simulation runs in a short time.

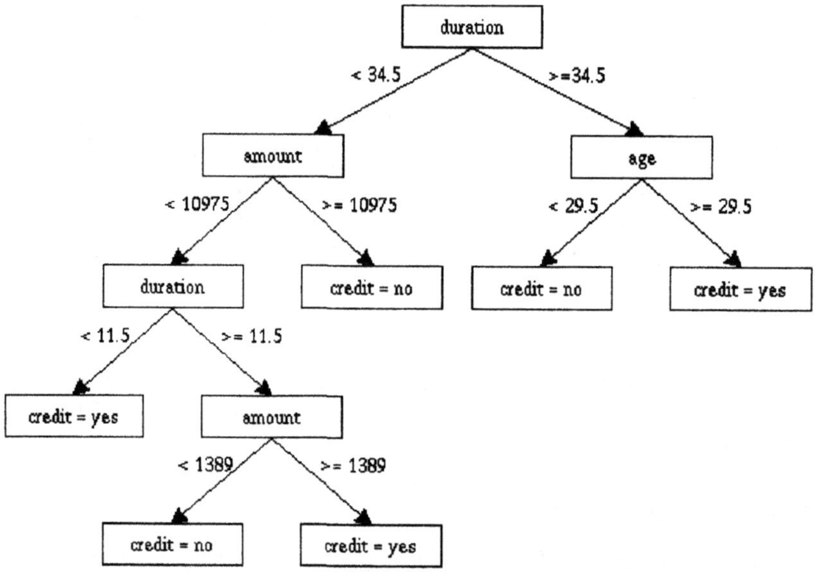

Figure 2: Decision tree for credit scoring

13.5.4 Semi-Quantitative Sensitivity Analysis

The semi-quantitative sensitivity analysis was developed by Vester (1990, 1991). To deduce statements about the sensitivity of his variables, Vester would send a questionnaire to those with expertise in the functioning of the real system and let them specify the influence of each variable on every other. All answers received were then summarised and the score for each variable divided by the number of answers. This data was used to construct an influence matrix which reflects the connectedness and intensity of connection for each variable in the system. From this matrix can be calculated the active influence (the sum of the influence that this variable has upon all those variables which are dependent on it) and passive influence (the sum of the influence for all variables which this variable depends on) for each variable. The influence values for each variable are then located in a two-dimensional system of co-ordinates. According to their locations, all variables are classified into five sensitivity categories: active, reactive, critical, buffering or neutral.[6]

In contrast to our extended definition claiming that sensitivity analysis involves systematic input changes, Vester's method does not vary the inputs at all. Therefore, this procedure is far less time consuming than conventional methods. On the other hand, the results are only approximate and subjective. The attributed roles as active, reactive, critical, buffering or neutral variables are intended to serve only in *steering* decisions. The degree of correspondence to results from conventional quantitative sensitivity tests has still to be investigated systematically.

In sum, conventional quantitative methods, simple sensitivity indices, meta-modelling and semi-quantitative methods all have severe disadvantages. Generally, the results are either incomplete or approximate. The degree of systematic procedure in all these methods is rather low. In the following section, we illustrate some of the general problems in application of these sensitivity analysis techniques by using three specific examples.

13.6 Applications

In this section, we consider a number of general problems with the application of sensitivity analysis in social science, illustrating the points by discussing a set of models of progressively increasing complexity. Although particular sensitivity analysis techniques address these problems in different ways, it is useful to con-

[6] *Active* variables are suitable for system steering, because many other variables are dependent on them but they are dependent on very few others. *Reactive* variables should not be used for system steering. They depend on a lot of other variables but have hardly any reciprocal effect. They are good system indicators. *Critical* variables are extremely highly connected. Small changes in these variables can very easily lead to uncontrollable changes in the system behaviour. Although they catalyse and accelerate changes they should be used very carefully in system steering. *Buffering* variables are not useful in system steering because their connectedness is too small to induce relevant changes in system behaviour. *Neutral* variables are too highly connected to be used safely as steering variables but they are suitable for self regulation of the system.

sider them in general terms, following the broad definition, in order to put the potential solutions into some sort of order.

13.6.1 The Simplest Case: Linear Regression

A simple linear regression describes the effect of an independent variable (the age of a school child for example) on a dependent variable (the height of that child). Using two sets of data, for the height and age of each child, the line of best fit is defined as that which minimises the sum of squared errors, assuming there is no uncertainty about the positions of the data points themselves. The slope of the line is the elasticity of height with respect to age, the proportional change in height expected for a given proportional change in age. (Trivially, linear regression implies a constant elasticity.) If we assume that there is a potential measuring error in the height of each child, but not in the corresponding age, it makes sense to investigate the effect that different assumptions about the height measurement have on the elasticity. In particular, it is important to know whether the maximum and minimum heights for the set of children, given by the measured error, produce a dramatic change in the elasticity. If this is the case, then it is important to increase the accuracy of measurement. This is an example based on a physical system and it would be straightforward to work out the actual errors of various measurement methods somewhat independently of the measuring process (calibration). This would probably not be true of any social process. (Because the regression is linear, the analysis is trivial. The steeper the slope of the graph, the greater the absolute change in the elasticity for a fixed error at each point. Nevertheless the example is sufficient for what follows. A more formal discussion for models of this kind can be found in the first part of section 5.3.).

Despite its simplicity, this case appears to be regarded as typical for the process of carrying out most sensitivity analyses. In fact, it has a number of features which are the exception rather the rule in investigating models of realistic complexity. Firstly, the interpretation of the data and the lines of best fit are obvious. The data is quantitative and the lines of best fit represent different assumptions about errors of measurement in that data. (It would make little sense to investigate completely arbitrary assumptions about these errors without additional data, though this is sometimes done anyway. For example, it is possible to show that the elasticity is highly sensitive to errors in a single outlying point, but this information is unhelpful unless we have an independent reason to be suspicious of the provenance of that point. Linear regression *assumes* that an equal weight is given to the position of each point in calculating the line of best fit so it is inconsistent to combine it with *post hoc* modifications to the data set.) There is also no problem of identification for the "inputs", "model" or "outputs" of the system since they are explicitly defined as the dependent and independent variables which are linked by the line of best fit. Although the units of elasticity are (for example) centimetres per year, the ratio of elasticities is a convenient constant without units. For this reason, comparison of the ratios is straightforward and gives a feel for whether error in the independent data is “important” or “unimportant”. Finally, it is clear that the process of constructing a line of best fit is well defined for any set of arbitrary assumptions about the measurement errors in the system. In consequence, it can be guaranteed, given a competent mathematician, that changes in the ratio of elasticities are solely due to changes in the data and assumptions about errors. This is the *ceteris paribus* requirement which addresses the relative technical difficulty of changing the condition of the system for the purpose of sensitivity analysis.

13.6.2 A More Difficult Case: A Dynamic System of Equations

We now turn to the analysis of sets of equations which are intended to represent the dynamics of some social process. Such models are almost universal in econometrics (Mayes 1981) and system dynamics (Forrester 1968, 1973). Already, the problem of identification is much more difficult. Although there may be a set of “starting conditions”, these are seldom an accurate representation of the real start of the process, usually consisting merely of the initial elements of the data set.[7] The distinction between inputs and outputs to the model is no longer straightforward, since “outputs” at the end of one “period” are “inputs” to the next. (For more complex or parallel models, even the concept of inputs and outputs within a single period may be problematic.) Even if the starting conditions are considered to be meaningful, the sensitivity analysis of dynamic systems requires comparison of numbers with trends or trends in one variable with trends in another.

As in the linear regression example, there is nothing particularly problematic about the interpretation of the variables, which are quantitative as before. Instead the problem is to “identify” behaviours both as inputs and outputs of the model. The simpler case, assuming parameter value uncertainty, is to change parameters of the model and compare the time evolution of the outputs. To use a simple consumption, saving and labour market participation example, we might observe that a change in the unemployment rate led to more frequent or more severe oscillations in the level of saving. (This can be compared with the comparative statics approach which would say that a change in the unemployment rate produces a change in the long run *level* of saving. This means that an elasticity can be defined effectively only for the comparative statics approach.) A more difficult case, assuming data uncertainty, is to make alternative assumptions about the time evolution of the input data and see how that affects the time evolution of output data. If it is difficult to characterise differences in time evolution as a

---

[7] This is a common problem. Econometric models fit data on the basis of underlying regularity of behaviour but this is incompatible with the “beginning” of any real data generation process. For example, the aggregate “investment function” cannot realistically predate the existence of modern capital using firms which almost certainly does not correspond to the time when data started to be collected about aggregate investment.

result of a parameter change it is more than twice as difficult to compare differences in time evolution.

For simple systems at least, our ability to characterise the pattern of time evolution "by eye" is rather good. Apart from differences in level, we can identify differences in variability (the gap between high and low values), stability (the number of times the series crosses a set of fixed levels) and smoothness (the second derivative of the time evolution). We can also make observations of the form "the more unstable the labour market, the lower consumption dips when a person is out of work and the longer it takes them to return to something like the original level of consumption once they find work again." These observations may be automated using techniques like pattern recognition (Schurmann 1996) and analogue complexing (Lemke 1997). Both of these techniques involve trying to characterise patterns like "bubbles", "crashes" or "shake-ups" in the time evolution of inputs and outputs. Once these "objects" have been identified, their occurrence can be correlated using traditional techniques. (Although pattern recognition techniques can be highly sophisticated, the resulting "identifications" would need to be checked, since they might not correlate with anything that a human could identify or act upon. In such circumstances, it is unlikely that they could show behaviourally significant correlations with other patterns. This approach is interesting because it may correspond to the sort of behaviour used by stock traders and other followers of time series data. If these have observed a market crash following a particular pattern of sales in one stock, this may result in following patterns of sales for other stocks.)

It does not appear that techniques of this sort have yet been used in sensitivity analysis, although there are similarities with work on meta-modelling by learning, which was discussed in section 5.3. There are several reasons for this. Firstly, at least in economic modelling, simple models (for which comparative static analysis is possible) still dominate the field. Secondly, econometric techniques usually involve jointly estimating all the parameters of the model on the basis of available data and measuring the quality of the model in terms of prediction. (In fact, some of the parameters used do not seem capable of being estimated in any independent way.) Within this framework, there is little interest in investigating the sensitivity of the results to differing assumptions about particular parameters. Although it makes sense to investigate the effect that data uncertainty may have on the jointly estimated set of parameters, it is not feasible to investigate the effects of uncertainty on parameter values determined independently. Finally, there are technical limitations on any "automated" method of recognition which may throw up correlations or patterns which are uninterpretable. While these are interesting for predictive purposes, they do not cast any light on the strengths and weaknesses of the model under consideration and thus fall outside the definition of sensitivity analysis.

13.6.3 A Very Different Case: Genetic Programming

Genetic Programming (Koza 1992, 1994) is a technique in which different solutions to a problem are represented as programs in a simplified programming language. An initial population of solutions is generated at random and the "fitness" of each solution for dealing with the problem is assessed according to some rule. Fitter solutions are chosen differentially to survive in the population and to generate new solutions by processes of "crossover" (combining parts of two "parent" solutions in a systematic way to generate a novel "offspring" solution) and "mutation" (making small random changes to a solution). The processes of fitness evaluation and removal of less satisfactory solutions is repeated until, by analogy with natural selection, the population consists of solutions that are arbitrarily "good".

Traditionally, this approach has been used to solve "engineering" problems like integration, differentiation, sequence induction and symbolic regression (Koza 1992). In such cases, the problem space is large but closed in the sense that the fitness function is directly defined for all candidate solutions. More recently, however, there have been interesting applications where fitness is measured only indirectly by performance in some simulated environment which may be effectively open through the existence of other evolving agents and environmental processes. The fitness function is well defined as it was for closed environments. The difference is that now the function measures behaviours which are a *result* of the solution program not equivalent to it. The relationship between the solution program, the actions of the simulated agent and the environment can be arbitrarily complex so that it is not trivial to make this distinction. This approach is much more closely analogous to the selection of genotypes through the performance of phenotypes in biological evolution. Simple GP models do not distinguish between genotype (program or decision process) and phenotype (behaviour) because the programmers knowledge of the problem and his or her ability to produce a "relevant" encoding for solutions makes this distinction unhelpful. For many social processes, however, there is no "intuitive" encoding.

An example is provided by Koza *et al.* (1992) in which the food foraging strategy of a lizard is simulated. In this case the operators and terminals of the GP represent particular physical or mental capabilities which the real lizard appears to have. (These may even be tested in experimental laboratory conditions.) Each complete program represents a particular possible food foraging strategy, the fitness of which is measured by the number of simulated flies it catches divided by the amount of energy expended to do so. We can take a particular environment (assumptions about the distribution and behaviour of simulated flies) and then evolve the best lizard for that environment with a fixed set of capabilities. If we then add a new capability and evolve a second lizard in the same environment, we learn two kinds of things about the quality of the first lizard. The simpler case is that the first lizard was only optimal given its capabilities, but that it can achieve better performance with the new capability. The second, and perhaps more interesting case occurs when the new capability has no effect on the optimal strategy or performance. This suggests that the strategy is optimal *for the environment*. There is clearly an analogy here with the sort of "robustness" results suggested by evolutionary game theory, where an ESS is a strategy which is not capable of being "invaded" by any other strategy from a closed set. Here, however, we have a much stronger result that is likely to occur far less often. The optimal strategy is one that cannot be invaded by any *new* capability from a potentially infinite set. (Such an optimal strategy does not quite imply a stable environment but seems extremely unlikely to occur in any open environment. This suggests an interesting new test of optimality where analytical solutions are not available.)

So far, this description has been strictly formal. It says nothing about the relationship between the real and simulated lizards or the real and simulated environments. A good correspondence of this kind is actually quite difficult to obtain, since apparently implausible behaviour in the simulated lizard can *either* reflect mistaken assumptions about the capabilities of the real lizard or about the structure of the real environment. Two related advantages of the GP approach compensate for this difficulty. Firstly, individual capabilities (move left, stick out tongue, duck) and the overall foraging behaviour (if a fly appears straight ahead then stick out tongue otherwise sleep) are both easy to interpret and can be observed independently. Secondly, the GP is such a powerful optimiser that it rapidly identifies "holes" in the specification of the agent or its environment. In work using GP to model budgeting behaviour (Chattoe and Gilbert 1996), "trivial" budgeting strategies indicated *both* what was wrong with our assumptions about the agent *and* the environment, thus providing an interesting case of "co-evolutionary" learning.

Models of this kind have very straightforward interpretations. The meanings of the operators and terminals in the simplified programming language are given explicitly by the model as are the interpretations of the "output" behaviour. Identification is also made simple by the fact that simulation enables us to look "inside" the lizard as well as observing its behaviour independently. This behaviour can also be characterised in a straightforward manner because of its common-sense interpretation. (For example, how much time does the lizard spend standing still, turning its head and moving its body in two different strategies? Does it always follow a head turn by sticking out its tongue?) It is clear that the addition of new capabilities within the simulation is somewhat modular, at least if the program is properly designed, so that the addition of one capability need not cause errors in the implementations of those which already exist. These strengths go some way to offsetting the main difficulty of comparability between inputs and outputs to the candidate programs. It is not at all clear *a priori* whether the addition of a new capability is likely to produce a "significant" change in output behaviour or even what it means to say that one capability is much more "important" than another in terms of its effect on behaviour. (This is almost certainly why behaviours and sets of capabilities are compared but links between capabilities and behaviours are rather ignored.) Similarly, two different programs may produce rather similar behaviour in terms of simple measures like proportions of body and head movements. (This can arise because of programs which are almost functionally equivalent but structured differently or because there are two functionally different programs which produce rather similar output in a given environment.)

It is clear that many of these observations will also apply to other forms of agent based simulation, where attempts are made to compare the behaviours of agents with different cognitive capabilities.

The purpose of the examples in this section was not to "resolve" the difficulties of sensitivity analysis or propose fully standardised techniques which could never exist. (Some of these issues will be discussed further in the remainder of the paper.) Instead, the object was to draw attention to different kinds of models for which different techniques will be required. It is now possible to describe the problems faced by sensitivity analysis.

13.7 Problems

In this section we discuss those problems that appear to have led to the infrequent application of sensitivity analysis in social simulation. They range from fundamental problems to merely technical ones. It is possible to describe the technical problems faced by sensitivity analysis under four headings.

13.7.1 Identification

Identification is the task of deciding which patterns constitute "inputs" to the model, "outputs" to the model or "interesting" model behaviour. These identifications are pre-conditions of any form of sensitivity analysis under the original definition. For simple linear or static systems the input, model and output are all clearly defined (linear regression) but this is by no means typical. For complex, parallel or interacting systems, both the notions of obvious "input" or "output" (dynamic models) and the "model" itself (dynamic models, GP) may be much less straightforward. The technical problem of identification should not, however, cause us to make theoretical simplifications, as it does in choosing only mathematical models which tend to steady states.

13.7.2 Comparability

All sensitivity analysis relies on the comparison between the inputs and outputs in at least two different conditions, once the task of identification has been carried out. Comparability usually refers to "before" and "after" states of specification (old model versus new model) and of behaviour. Cases where the comparison is numerical (linear regression) or based on a formal language (GP) are easier than those where trends or patterns are to be compared (dynamic models), although the tools for pattern comparison have not yet been thoroughly explored. A subsidiary issue is the nature of the comparison. For linear regression, the ratio of two elasticities is a constant without units. For GP it is typically necessary to measure changes in behaviour away from some potentially arbitrary "base case" (like random behavior) since neither the capability set nor the emerging behaviours can be quantified in any intuitive way. (That is not to say that neither feature can be quantified. One could, for example, plot the number of capabilities available against the proportions of time the agents spent in different actions. Such an approach would probably not be useful however, since it is the "meanings" or effects of the capabilities and behaviours that determine fitness.)

13.7.3 Interpretation

Interpretation refers to the link between aspects of the simulation and aspects of the "real world" which that simulation is supposed to represent. For simple cases (linear regression), the interpretation is either obvious or given by definition. (The line of best fit just *is* the line that minimises the squared errors. It isn't supposed to correspond to anything in the world and it is traditionally bad statistics to suppose it demonstrates an underlying causal regularity.) For more complex cases, the consistent interpretation of the model is more problematic. For dynamic models, for example, the values of parameters are jointly estimated on the basis of a particular data set and in some cases, there is no available technique for measuring them independently. In such cases, it makes no sense to attempt sensitivity analysis of parameter values for anything other than data errors. Estimation alone provides no guidance for other parameter values which one could try and constraining one parameter to some arbitrary alternative value simply distorts the estimated values of the other parameters. GP models suffer from a different problem. Although the interpretations of individual operators, terminals and emerging behaviours are very clear and they can be observed and measured independently, the complexity of the relationship between the program, the behaviour and the environment makes overall interpretation technically difficult, though not theoretically problematic. Although tools are available to facilitate the task (for example, it is possible to "parse" programs in various ways to remove irrelevant parts), it is very hard to look at a particular program and describe how it produces the observed behaviour under different environmental conditions.

13.7.4 Modularization

We have already discussed in passing the "technical" requirement that certain sorts of sensitivity analysis should not induce inadvertent changes elsewhere in the system. For most simulations it should be easy to make repeated runs with a different random number seed but much harder to alter the model from discrete to continuous time or move from "mindless" agents to satisfying ones. It has been argued elsewhere (Chattoe 1995) that one of the advantages of computer programs (simulations) as representations of social processes is that they facilitate this kind of modular design in a way that mathematics and everyday language do not. Given the very general definition of simulation discussed in this paper, it is hard to draw a line between adequate sensitivity analysis on a particular model and effectively testing a whole family of related models. One way to draw the line is to expect sensitivity tests of every declared "parameter" but not only is this impossibly time consuming, it also assumes an "equations and parameters" approach to modelling which cannot usefully be applied to other kinds of simulation.[8] It does not seem that this issue can be resolved in a technical manner, but we raise it nonetheless, in the hope of promoting discussion about the creation of an appropriate norm.

13.7.5 The Robustness Requirement

"We are usually looking for *robust* solutions, not just the best one."
(Vignaux, 1995)

Conventionally, in order to have the validity of a simulation model accepted, we are asked to demonstrate the *robustness* of the simulation results. This means that in sensitivity testing small deviations on the input side do not yield large deviations on the output side. In other words, the system behaviour is characterised by some degree of stability. This request for robustness is not typically supported by the knowledge we have about social reality at present:

- The non-linearity paradigm strongly suggests that many social systems have to be regarded as stochastic non-linear dynamical systems (Albert 1995, Kiel and Elliott 1996, Stewart 1989). We know that small deviations of the input values for these systems typically lead to large deviations in output values. The request for robustness may therefore reflect an outdated view of the social world.
- Assuming that we obtain a frequency distribution for different kinds of output, even if there is only a very small probability that some output (event) may occur in the social world, this event may happen despite, in fact exactly because of, this small probability.

For example, Saam (1996) simulates coup sequences in Thailand in the period between 1932 and 1992. She uses a stochastic non-linear dynamical system of equations. Due to self-organization effects within the system, she obtains different types of coup sequences as simulation results. For each combination of input values there exists a frequency distribution for types of coup sequences. For example, type X may have a probability of 15%, assuming that the inputs have

[8] One interesting possibility for testbeds is "back calibration". A group of simulators working on, for example, Iterated Prisoner's Dilemma games could build a test bed and get shared results. Inevitably they would want to extend the testbed, but the check on the quality of the extension and its results would be whether they could still replicate the *previous* results arbitrarily closely. If not, it would be a sign that the extended testbed had not been properly developed. Ideally, it should be possible to reproduce the old results *exactly*. (This is the econometric concept of encompassing.)

certain values. If only one of these values is changed, the probability of type X may increase to 20%. In the process of trying to reproduce the empirical coup dynamics of Thailand, one obtains a type, call it type T, which comes closest to the empirically observed coup sequence. To give explanations about why this coup sequence may have happened in Thailand, it is now necessary to find those combinations of input values that lead to positive probability values for type T. Unfortunately, from the point of view of explanation, there is not a single combination of input values which leads to type T, but several. Under certain input conditions the probability of type T reaches 100%, under others it is less. In the social world we often observe only one representative of a special kind of social system. That our model fits "best" may not be assumed from the best simulated reproduction of the empirical output at the highest probability. To return to the coup example, the model that produces 100% probability of type T uses some input values that have definitively not been observed in Thailand. Instead, under conditions of empirically observed input values, the probability of type T was about one in eight. Thus, although the 100% probability result would be the "best" from the technical point of view, because it would always produce the best fit with the empirical data, it is not the best model from the empirical point of view. Reconsidering and reinterpreting the model using the empirical situation in Thailand, this implies the following: In Thailand, there was a certain (but rather small) probability that the coup sequence which actually occurred would happen. There were other probabilities, even larger ones, for other types of coup sequence. However, it was not the type with highest probability that was realised, but another type with lower probability. Therefore it is not appropriate to investigate only high probability scenarios. The best model may in fact be one with a low probability.

In sum, in the social sciences, the "best" solution need not necessarily be robust. Obviously, our argument holds for simulations that try to reproduce individual empirical phenomena which may be described as non-linear and stochastic. There is only one kind of simulation application in the social sciences where robust simulation results will be required and that is in models used for social planning. Here, one is very much interested in the stability of results. Until very recently, nobody even considered building models of chaotic systems for social planning because it was believed that they could not be used for social steering.

13.7.6 Equivalence Of Results

In section 2 we gave an outline of the types of model input and output along with a range of possible measures of change or stability, discussed further in section 5. These disparate approaches give rise to several questions. If applied to one and the same model, what is the relation between the results of different methods? Do they confirm each other? Which would be the most appropriate technique? Should several techniques be applied consecutively?

We are still in need of research into the formal or empirical equivalence of the results from the methods described. As Saam (1995, 1996) has pointed out, the results of different sensitivity methods may not necessarily confirm each other. Rather, the results may be complementary. For example, semi-quantitative sensitivity analysis produces results that relate only to the sensitivity of variables, while regression analysis may treat parameters as well. Whereas a semi-quantitative analysis of all variables for even a complex simulation model may be performed in an acceptable time, it is practically impossible to run a complete regression analysis of a complex simulation model. One may therefore perform a semi-quantitative analysis in order to ensure that each variable is at least covered. The results produced will be comparatively rough. In addition to this one may run regression analyses on selected parameters or variables to sharpen the results of critical input units.

It appears that we currently lack standardisation (theoretical unification and classification) of sensitivity measures in order to make sensitivity results more communicable and comparable and this poses serious problems in general application of sensitivity analysis. In the last section, we provide some tentative discussion that may facilitate the development of such a standardisation.

13.8 Conclusions

The aim of this paper has been to give an overview of current problems in the application of sensitivity analysis to social simulation. In attempting a more systematic treatment of sensitivity analysis, we wanted to support the standardisation of algorithms and tools, which can be seen as an important step towards improving the prospects for future application of these techniques as well as facilitating the acknowledgement of simulation results by outsiders.

The overarching problem for sensitivity analysis is its many different interpretations and applications. These differences exist both between the physical and social sciences and within particular disciplines. They prevent communication and cross-fertilisation among disciplines as well as deterring widespread use of a method which appears to be an essential prerequisite for producing valid and reliable simulation results. The first step towards a more precise and standardised view of sensitivity analysis is the description of its main characteristics, which also leads to concrete requirements for the design of modelling and simulation software systems. In our view sensitivity analysis can be characterised by:

1. Algorithms for calculating sensitivity indices using different approaches (simple indices, regression analysis, decision trees algorithms, non quantitative approaches) and the standardisation (theoretical unification and classification) of these algorithms in order to make sensitivity results more communicable and comparable. This also provides a good starting point for developing new algorithms.
2. Strategies for collection and comparison of data suitable for sensitivity analysis. These include:

- The user-oriented definition, organisation, and execution of multiple simulation runs.
- Complexity reducing algorithms which decrease the number of runs necessary to carry out a sensitivity analysis systematically and which bunch results. (The results should indicate all variables and constraints which are critical, the limits at which the current behaviour changes, the degree of change before these limits are reached and the magnitude of change to be expected beyond the limit.)
- Experimental facilities for comparing different sensitivity indices and selecting regions of high sensitivity by using statistical algorithms and graphics.

3. Its dependence on the level of model complexity and the chosen modelling formalism. Models in social simulation lead to particular validation problems, which can only be partially solved by the application of the present sensitivity analysis algorithms and strategies like input variability (section 5.1). In addition there are specific problems of application (section 6) and general issues about the robustness and complexity of results in chaotic or non-linear models (section 7).

As regards the architecture of modelling and simulation tools it should be remarked that most of the methods described in the previous sections are (more or less) implemented in modern simulation systems (systematic parameter variation, variation of random seeds, probability density functions) or in statistical analysis systems (regression analysis, decision trees algorithms, graphics). Usually, however, these facilities are not integrated and explicitly assigned to sensitivity analysis, because, as discussed in section 3, they are also used for other simulation tasks (model validation, model optimisation and "what if" analysis) as well.

From a more practical point of view, the realisation of sensitivity analysis as a separate part of the experimental framework for simulation would also be an essential contribution to the standardisation of sensitivity tools. It is the example provided by applied statistics which offers standardised measures and standardised software (most importantly SPSS) which we have in mind here. Only by standardisation could a norm be established in which it is required to report certain coefficients for any "respectable" simulation. Standardisation proved to be an important component in the spread of applied statistical techniques as well as to the level of research and the acknowledgement of its results by outsiders. Mature methods for sensitivity analysis (as well as methods for model validation, which we have not been able to cover here) are one prerequisite to scientific take-off in the field of social simulation.

Partly for reasons of length, this paper has done little more than *begin* to systematise the problems and techniques in sensitivity analysis for social simulation. In particular, it raises more questions than it answers. We hope that, nevertheless, the search for some systematic framework may facilitate further discussions and lead to an increase in the application of sensitivity analysis techniques to social simulation.

References

Albert, A. (ed.) (1995): Chaos and Society. Amsterdam, Presses de l'Université du Québec.

Attanasio, O. P. & Weber, G. (1993): Consumption Growth. the Interest Rate and Aggregation. Review of Economic Studies 60, 631-649.

Attanasio, O. P. & Weber, G. (1994): The UK Consumption Boom of the Late 1980's: Aggregate Implications of Microeconomic Evidence. The Economic Journal 104, 1269-1302.

Axelrod, R. M. (1987): The Evolution of Strategies in the Iterated Prisoner's Dilemma. In: Davis, L. D. (ed.): Genetic Algorithms and Simulated Annealing. Pitman, London, 32-41.

Axtell, R. R., Axelrod, J., Epstein M., & Cohen, M. D. (1996): Aligning Simulation Models: A Case Study and Results. Computational and Mathematical Organization Theory 1, 123-141.

Bernard, P. M. & Plaisent, M. (1990): Validation Processes in Simulation. In: Belardo, S. & Weinroth, J. (eds.): Simulation in Business and Management Volume 21. SCS Publications, Erlangen, 84-88.

Bond, A. H. & Gasser, L. G. (eds.) (1988): Readings in Distributed Artificial Intelligence. Morgan Kaufmann, San Mateo, CA.

Bratley, P., Fox, B. L. & Schrage, L. E. (1987): A Guide to Simulation, second edition. Springer, Berlin.

Carley, K. (1991): Designing Organizational Structures to Cope with Communications Breakdowns: A Simulation Model. Industrial Crisis Quarterly 5, 19-57.

Chattoe, E. (1996): Why are we simulating anyway? Some answers from economics. In: Troitszch, K. G., Müller, U., Gilbert, N. & Doran, J. (eds.): Social Science Microsimulation. Springer, Berlin, 78-104.

Chattoe, E. & Gilbert, N. (1996): The Simulation of Budgetary Decision Making and Mechanisms of Social Evolution. Paper presented at the conference ISA Essex '96: Fourth International Social Science Methodology Conference, Essex, July.

Deif, A. S. (1986): Sensitivity Analysis in Linear Systems. Springer, Berlin.

Doran, J. (this volume).

Fiacco, A. V. (1983): Introduction to Sensitivity and Stability Analysis in Non-linear Programming. Academic Press, Paris.

Fiacco, A. V. (ed.) (1984): Sensitivity, Stability and Parametric Analysis. North-Holland, Amsterdam.

Forrester, J. W. (1968): Principles of Systems. MIT Press, Cambridge, MA.

Forrester, J. W. (1973): World Dynamics. MIT Press, Cambridge, MA.

Friedman, D. (1991): Evolutionary Games in Economics. Econometrica 59, 637-666.

Gilbert, N. & Doran, J. (eds.) (1994): Simulating Societies: The Computer Simulation of Social Phenomena. UCL Press, London.

Gilbert, N. & Conte, R. (eds.) (1995): Artificial Societies: The Computer Simulation of Social Life. UCL Press, London.

Hanssmann, F. (1987): Einführung in die Systemforschung: Methodik der modellgestützten Entscheidungsvorbereitung, third edition. München.

Harrison, R. & Carrol, G. (1991): Keeping the Faith: A Model of Cultural Transmission in Formal Organizations. Administrative Science Quarterly 36, 552-582.

Harvey, A. C. (1990): The Econometric Analysis of Time Series, second edition. Phillip Allan, London.

Hegselmann, R. (this volume).

Hendry, D. F. (1985): Monetary Economic Myth and Econometric Reality, Oxford Review of Economic Policy 1, 72-84.

Hillier, F. & Lieberman, G. J. (1986): Introduction to Operations Research, fourth edition. Oldenbourg, München.

Huber, P. & Szczerbicka, H. (1994): Sensitivity Analysis of Simulation Models with Decision Tree Algorithms. In: Kaylan, A. R., Lehmann, A. & Ören, T. I. (eds.): Proceedings of the European Simulation Symposium (ESS'94) Volume 1. SCS Publications, Erlangen, 43-47.

Huber, P., Szczerbicka, H. & Barton, R. (1996): Locating Sensitive Regions by Learning from Simulation Data. In: Saltelli, A. & von Maravic, H. (eds.): SAMO95: Theory and Applications of Sensitivity Analysis of Model Output in Computer Simulation, 25th-27th September, 1995, Belgirane, Italy. European Commission, Brussels.

Huberman, B. A. & Glance, N. (1993): Evolutionary Games and Computer Simulations. Proceedings of the National Academy of Sciences of the United States of America 90, 7716-7718.

Kiel, D. L. & Elliott, E. (1996): Chaos Theory in the Social Sciences: Foundations and Applications. The University of Michigan Press, Ann Arbor.

Köhler, J. (1996): Sensitivity Analysis of Integer Linear Programming. Discussion Paper, Fachbereich Mathematik und Informatik, Universität Halle-Wittenberg.

Kleijnen, J. P. C. (1987): Statistical Tools for Simulation Practitioners. Dekker, New York.

Kleijnen, J. P. C. (1988): Simulation: A Statistical Perspective. John Wiley & Sons, Chichester.

Kleijnen, J. P. C. (1992a): Sensitivity Analysis of Simulation Experiments: Regression Analysis and Statistical Design. Mathematics and Computers in Simulation 34, 297-315.

Kleijnen, J. P. C. (1992b): Techniques for Sensitivity Analysis of Simulation Models: A Case Study of the CO_2 Greenhouse Effect. Simulation 58, 410-417.

Kleijnen, J. P. C. (1995a): Sensitivity Analysis and Optimization of System Dynamics Models: Regression Analysis and Statistical Design of Experiments. System Dynamics Review 11, 275-288.

Kleijnen, J. P. C. (1995b): Sensitivity Analysis and Related Analyses: A Survey of Some Statistical Techniques, Journal of Statistical Computation and Simulation 57, 111-142.

Koza, J. R. (1992): Genetic Programming: On the Programming of Computers by Means of Natural Selection and Genetics. MIT Press/A Bradford Book, Cambridge, MA.

Koza, J. R. (1994): Genetic Programming II: Automatic Discovery of Reusable Programs. MIT Press/A Bradford Book, Cambridge, MA.

Koza, J. R., Rice, J. P. & Roughgarden, J. (1992): Evolution of Food Foraging Strategies for the Caribbean Anolis Lizard Using Genetic Programming. Working Paper 92-06-028, Santa Fe Institute, New Mexico.

Langton, C. G. (ed.) (1989): Artificial Life: The Proceedings of an Interdisciplinary Workshop on the Synthesis and Simulation of Living Systems, Los Alamos, New Mexico, September 1987. SFI Studies in the Sciences of Complexity, Volume VI. Addison-Wesley, Redwood City, CA.

Latané, B. (1996): Dynamic Social Impact. Robust Predictions from Simple Theory. In: Hegselmann, R., Müller, U. & Troitzsch, K. G. (eds.): Modelling and Simulation in the Social Sciences from the Philosophy of Science Point of View. Kluwer, Dordrecht, 287-310.

Latané, B. & Nowak, A. (1997): Self-Organizing Social Systems: Necessary and Sufficient Conditions for the Emergence of Clustering, Consolidation and Continuing Diversity. In: Barnett, G. A. & Boster, F. J. (eds.): Progress in Communication Sciences, Volume 13: Persuasion. Ablex, Norwood, NJ, 43-74.

Law, A. & Kelton, W. D. (1991): Simulation Modeling and Analysis, second edition. McGraw-Hill, New York.

Lemke, F. (1997): Knowledge Extraction from Data Using Self-Organizing Modeling Technologies. <http://www.scriptsoftware.com/km/paper2.html>

Liebl, F. (1995): Simulation: Problemorientierte Einführung, second enlarged edition. Oldenbourg, München.

Lomborg, B. (1996): Nucleus & Shield: The Evolution of Social Structure in the Prisoner's Dilemma. American Sociological Review 61, 278-308.

Macy, M. (1991): Chains of Cooperation: Threshold Effects in Collective Action. American Sociological Review 56, 730-747.

Mayes, D. G. (1981): Applications of Econometrics. Prentice-Hall, Englewood Cliffs, NJ.

Maynard-Smith, J. (1982): Evolution and the Theory of Games. Cambridge University Press, Cambridge.

Merz, J. (1994): Microsimulation - A Survey of Methods and Applications for Analyzing Economic and Social Policy. FFB Discussion Paper Number 9, Universität Luneburg, June.

Metz, W. (1979): Die Validierung sozioökonomischer Computersimulationsmodelle: Empirische Befunde und Gestaltungskonzepte. Wison, Köln.

Meyers, S. D., Stephens J. C., Legler, D. M., Miles, K. F. & O'Brian, J. J. (1994): The Sensitivity to Parametric Variation in Direct Minimization Techniques. Monthly Weather Review 122, 1632-1639.

Miller, J. H., Rust, J. & Palmer, R. (1994): Characterising Effective Trading Strategies: Insights from a Computerised Double Auction Tournament. Journal of Economic Dynamics and Control 18, 61-96.

Möhring, M. (1996): Social Science Multilevel Simulation with MIMOSE. In: Troitzsch, K. G., Müller, U., Gilbert, N. & Doran, J. (eds.): Social Science Microsimulation. Springer, Berlin, 123-137.

Pierreval, H. (1992): Rule-based Simulation Meta-models. European Journal of Operations Research 61, 6-17.

Quinlan, J. R. (1986): Induction of Decision Trees. Machine Learning 1, 81-106.

Ríos Insua D. (1990): Sensitivity Analysis in Multi Objective Decision Making. Springer, Berlin.

Saam, N. J. (1995): Computergestützte Theoriekonstruktion in den Sozialwissenschaften. Konzeptbasierte Simulation eines theoretischen Modells am Beispiel militärischer Staatsstreiche in Thailand. Unter Anwendung des Mehrebenen-Ansatzes der Synergetik. SCS Publications, Erlangen.

Saam, N. J. (1996): Multi-level Modeling with MIMOSE: Experience from a Social Science Application. In: Troitzsch, K. G., Müller, U., Gilbert, N. & Doran, J. (eds.): Social Science Microsimulation. Springer, Berlin, 138-154.

Saam, N. J. & Reiter, L. (submitted): Lotka's Law Reconsidered: The Evolution of Publication and Citation Distributions in Scientific Fields. Scientometrics.

Saltelli, A. (1996): On the Use of Rank Transformation in Sensitivity Analysis. In: Saltelli, A. & Von Maravic, H. (eds.): SAMO95: Theory and Applications of Sensitivity Analysis of Model Output in Computer Simulation, 25th-27th September, 1995, Belgirane, Italy. European Commission, Brussels.

Schäfer, A. (1996): Sensitivität und Simulation. Diplomarbeit, Universität Passau.

Schurmann, J. (1996): Pattern Classification: A Unified View of Statistical and Neural Approaches. John Wiley & Sons, New York.

Sobol, I. M. (1996): Sensitivity Analysis of Non-linear Models Using Sensitivity Analysis. In: Saltelli, A. & Von Maravic, H. (eds.): SAMO95: Theory and Applications of Sensitivity Analysis of Model Output in Computer Simulation, 25th-27th September, 1995, Belgirane, Italy. European Commission, Brussels.

Stewart, I. (1989): Does God Play Dice? The Mathematics of Chaos. Basil Blackwell, Oxford.

Vester, F. (1990): Ausfahrt Zukunft. Strategien für den Verkehr von morgen. Eine Systemuntersuchung, third edition. Heyne, München.

Vester, F. (1991): Ausfahrt Zukunft: Supplement; Material zur Systemuntersuchung. Studiengruppe für Biologie und Umwelt GmbH, München.

Vignaux, G. A. (1995): Sensitivity Analysis.

<http://www.isor.vuw.ac.nz/~vignaux/teaching/sensitivity/sensit/sensit.html>
Weibull, J. W. (1995): Evolutionary Game Theory. MIT Press, Cambridge, MA.
Wittmann, J. (1993): Eine Benutzerschnittstelle für die Durchführung von Simulationsexperimenten. SCS Publications, San Diego, CA.
Wuensche, A. & Lesser, M. J. (1992): The Global Dynamics of Cellular Automata: An Atlas of Basin of Attraction Fields of One-Dimensional Cellular Automata. Addison-Wesley, Reading, MA.

14. Tests of Dynamic Social Models with Time Related Surveys: an Experimental Approach

Christof Schatz[1]

'God does not usually tell us from what distribution the data come.'
(Bratley et al. 1987, p. 135)

Abstract

In Social Science Modeling Theory the integration of empirical aspects seems to be poor. One reason may be the paralysing effect of highly complex and unsolved methodological problems. In this contribution a first step in developing tools for testing models with empirical data is taken. A non-linear formal model explaining job-keeping expectations of employed people is used as example. Three problems estimating parameters are discussed: the Complexity Problem, the Micro-Macro-Link and the Ambiguousness Problem. Despite these problems the test arrives at a definite positive falsification result. Hence this falsification provides an example for tests, which lead to systematic ways developing empirically founded formal models and theories.

14.1 Introduction: Validation of Dynamic Social Micro Models (DYSOMMs)

The title of the International Seminar "Social Science Microsimulation: Tools for Modeling, Parameter Optimisation, and Sensitivity Analysis" 1997 reflects the importance of quality tests applied on social models. The most common, mostly used and most elaborated quality test of models in natural and social sciences are validations with empirical data.

The reader of this book (and of former seminar reports) can recognise from the contributions the kind of models addressed in this context: Micro models based on game structures, cellular automata, artificial intelligence, stochastic dynamics. Subsequently I will pool these types of models under the term 'DYSOMMs' (DYnamic SOcial Micro Models): The models of this type are dynamic (autodependent in time) in contrast to static state models, (which are used in econometrics, empirical sociology and psychology). And they are *Micro* Models in contrast to the macro models of economic theories and the models of the System Dynamics approach. In spite of the validation issue mentioned above and in spite of the large validation efforts undertaken by researchers using macro models and/or

[1]Chair of Sociology, TU München. Address: Lothstr. 17, 80335 München, Tel: ++49/89/289-24219, Fax: ++49/89/289-24302, Email: schatz@lrz.tu-muenchen.de.

static state models, only few of the DYSOMMS presented during the seminar addressed the subject of validation with empirical data or even applied such a validation.

This fact only mirrors the situation in social micro modeling sciences: Except a particular class of econometric micro modeling conceptually weakly connected to the DYSOMMs (presented e.g. by Orcutt et al. 1986) there are only very few publications about empirical validation of DYSOMMs on the one hand. Only a few authors asked for empirical tests or validations of the derived conclusions, (i.e. Ziegler 1972, Troitzsch 1990, Saam 1995, Vries 1998). On the other hand the need for such validations was expressed several times during the seminar and is reflected in its title.

What is the reason for this gap between wisdom and reality? The author supposes this reason to be the same as for other gaps of this type: A lack of methodology. In this introduction some reasons for this (hard) assumption shall be given.

In the following, some presumingly undiscussed facts are collected:

1. There is a large amount of statistical methods for analysing empirical data of pretty much all kinds.
2. In Physics the validation of even complicated dynamic micro models such as kinetic gas, plasma, star or elementary particle systems provides minor methodological problems.
3. In social or economic empirical sciences the meaning of 'model' is mostly restricted to models described by static linear function equations, of which simulations are almost trivial.

From this facts the lack of DYSOMM validation is hard to understand. A first consideration may be: Why not to improve point 3 by applying DYSOMMs instead of these simple linear equations? The methods we can take from point 1. And point 2 shows that this way has a fairly good chance to come to a successful end.

The main thesis of this article is: This consideration may not be true. The error is claimed to be given by the difference between the type of mathematical structures underlying DYSOMMs on the one side and the types of equations underlying the available statistical methods on the other. The "prototype" of equations describing interesting dynamics in a DYSOMM are equation defining the relations between the attributes of one actor and the situation around this actor including in general all the other actors with all their attributes. Examples for this kind of self-evolving systems are the voter models (e.g. in Troitzsch 1990), where one voter orientates himself on the political attitudes of all others. Or game situations where the own decision depends on the momentary game situation which emerges from the decision history of all other participants (many examples, the most famous being Axelrod 1984). A shared frame for all DYSOMMs is a system of coupled differential equations, each equation describing the behaviour conditions of one actor. In the case of ordinary deterministic differential equations these micro models represent roughly, what the 'system dynamics' approach represents in the world of macro simulation systems. But the mathematical requirements extend these of macro differential equations:

Characteristics of a Deterministic Ordinary Differential Equation System as a Regression Model:
1. Function equation
2. Multivariate (i.e.: multiple independent AND multiple dependent variables)
3. Time dependent
4. Non-linear operators for scalar or vector variables.
5. Dynamical
6. Linear differential operators.
7. Coupling of micro variables.
8. Micro parameters related to actors
9. Ev. distributed micro parameters.

From mathematical analysis and computational simulation view these requirements are modest. Most evolutionary game simulations, cellular automata or stochastic dynamic models require more. In this sense we start with a simple approach. Anyway, from a statistical point of view this set presents something not trivial. The standard tool to analyse the correspondence between the theoretical model and empirical data are regressions. Are we able to do regressions of DYSOMMs with standard tools? To answer this question these tools have to absorb the *accumulated* properties 1 through 8 (ev. 9). The following paragraph shows a short screening of statistical methodology with regard on this question.

We start with the most simple form of the first both points: Linear equations. Regressions methods for multiple variant systems are standard. In the last two decades the methodology for non-recursive multivariate regression systems, especially structural equation systems (EQS, LISREL, AMOS, et al., see e.g. Hayduk 1987, Bentler 1989, Joreskog 1989, Jaccard et al. 1996) have been developed. There is a large methodology for regressions with event histories (Dieckmann 1984, Blossfeld et. al. 1986), time series (Greene 1997, pp823-870) or panel data (Fahrmeier 1994), if they have multiple variable structure. If they are multivariate and non-recursive, there is, as far as I know still no tool and no methodology. Point 3 clarified. Non-linear regressions in the form of transformed linear equations (e.g. log linear) or probit/logit fits are fairly common in econometrics or social sciences (Bruederl 1996, pp 208-228, Greene 1997, pp 871-947), but only common as far as they can be reduced to linear equations. Nonetheless there are possibilities to do further non-linear fits on base of function equations (Greene, pp 450-495). Even multivariate, e.g. in LISREL 8. (Bates 1988, Gifi 1990, Joereskoog 1994). Can it be done on the same run time-dependent? The author does not know a publication proposing such a methodology. Point 4. Now we come to a crucial point: A time dependent equation, where the dependent variable is a function of itself at previous times, (autoregressive models.) There are methods to do linear multivariate recursive time dependent dynamic regressions ('dynamic system regression') within some limits, which are given by the requirement for convergence. (Greene 1997, pp 764-770) Still, these methods do not allow non-linearity. Point 5. Differential operators require in addition to regress *changes* of the dependent variable against a function containing this same variable at a previous point in time. In 1976 Doreian and Hummon published an article proposing the estimation of such systems, if the right side of the equation is linear. (See also Doreian and Hummon 1979). If it is non-linear we are in the range of the "classic" systems dynamics approach. With the growth of this field also statistical methodology grew up. See e.g. (Qureshi1982), (Novovicova 1983), (Kolobov 1990). These methods cannot be discussed here, but it should be mentioned that they don't offer the opportunity for gaining best unbiased full efficient LM estimators, but serve only as check of the statistical quality regarding improvements on base of different numerical algorithms. They require and depend strongly upon an initial guess. (Almost all algorithms use variants of the 'Newton-Rephson' algorithm[2] for to solve the arising normal equations. If these equations are non-linear, the root finding problem is very similar to the global optimisation problem explained below.)

Regarding the problem of point 7 firstly we seem to come to an end: The author did not find anybody suggesting a method to estimate parameters of a regression equation $y_i(t_k) = F(y_j(t_k), a_{ij})$, $i \neq j, a_{ij} =$ regression parameters, y_i, y_j not being different variables of the same actor, but the same variables of different actors. However, this is the wrong perspective. There is an excellent and wide spread methodology, estimating parameters of stochastic differential equations, which deal directly with the probability density functions rather than interpreting deterministic variables as random variables (See mainly Hamerle, Singer, Nagl 1993; Cherkasov 1995). But the Fokker-Planck approach (Haken 1990, pp.172), used as macro formulation of the micro relations, (in it's discrete form called 'Master equation', applied in several multi-level simulations), has a decisive disadvantage for modeling micro dynamics: It presumes the actors to be exchangeable. The equation does not distinguish between actors having at the moment the same income but completely different histories. That means it is impossible to relate a certain parameter to a certain actor. It is only possible to relate to a cer-

---

[2] See: Flannery, P. Prian; Press, William H.; Teukolsky, Saul A.; Vettering, William T.: Numerical Recipes in C. The Art of Scientific Computing. 2nd Edition. University of Cambridge 1992: Press Syndicate, pp. 279

tain value, because distinct actors don't exist within this model. A similar problem is quite well-known in evolutionary game theory as the "changing fitness landscape" (Stanley 1994) and it is the reason, why interesting evolutionary game simulations, e.g. many "Axelrod systems" cannot be formalised as stochastic differential equations. This point of "individualism" of agents concerning their parameters and their network relations, the fact that there are different type of actors with different attitudes and characters arranged in different spatial patterns is often the core of DYSOMMs (see e.g. Hegselmann 1996). Of course one could have the idea to model each actor as his own subsystem, each described by one (partial differential) equation. But that would become more complex than the calculations in the next chapter and the author does not know anybody who has done this before. A way 'of compromise' could be the usage of sub-populations, containing actors of identical characteristics. But even in such a system with only a few coupled multiple equations one will meet similar problems as they will be described below. However this is an alternative way, requiring a lot of stochastic analysis, but worthwhile to follow in another publication.

Altogether one can recognise that standard regression methods with deterministic equation models are far away from fitting the mentioned accumulated requirements. The best prospective seems to be the stochastic differential equation methodology, far away from being an approved tool, but still closest to the goal. The crucial point seems to be point eight together with point nine, being both sensible arguments within the most Micro Models.

If in the following very complicated things are done, one may ask: "Why this complexity?". Both sides of this question, which is only *one*, have to be regarded: The first side with a simple differential equation for micro behaviour, which is much too simple to absorb all theoretical experiences of social psychology. The other side with thousands of unknown variables and a complex fit topology in a huge dimensionally space. This section attempted to show that these sides are inseparable. And to give thereby a guessed reason for the weak validation activity in spite of the strong wish.

A further question may be: Have we to do regressions? Are there no simpler methods for validation with empirical data? Of course, there are. (See the contribution of Georg Mueller in this volume.) But they are no substitutions for regressions just as Chi-Square tests are no substitution for multivariate regressions in standard econometrics. Typical dynamical behaviours such as oscillations, collapses, sudden explosions or network patterns can hardly be tested or estimated using only one scalar test variable.

14.2 A Methodological Experiment

In section 1. a wide gap between purely theoretical models and tests with empirical data by regression methods was shown. There are two purposes of the experiment in the following chapters. But before these purposes will be explained, it should be mentioned what this experiment is *not* for:

- It is not to give an answer on the mentioned question for statistical methodology.
- It is not to do sociological research on social confidence network in socio-economic systems. Before using sophisticated dynamic models, the application of standard static models would, of course be far more appropriate.
- It is not to analyse a certain data set, starting "at the scratch". The data set SOEP, used here, is the most analysed data set in Europe (more than thousand publications about).

It may be hard to accept, but here the socio-psychological issue serves only as "playing material" for the methodological question. (In principle one could do the same experiment with meta-simulated data, but some typical problems with "real data" would be missed this way.)

This article reports about a methodological experiment. The methods applied are not results of recent statistical or numerical developments, but just a trial. The following research question is used to do, what the author was able to do about comparing (relatively) simple micro models with empirical data by numerical fits. On base of the discussion above this can not be a statistical comparison. It is a purely numerical procedure without any knowledge about statistical significance. Its purposes shall be

- to add a small piece to the collection of validation ideas, which cover the ground of the methodological gap. This collection may grow step by step together towards a consistent validation methodology.
- to show up more concretely the problems of validation by regression. To measure the limits, the size of the task within the limits and the chances to manage this task sometime in the future.

14.3 The Sociological Subject: Job Confidence

Although it is intended to 'misuse' the socio-psychological issue and the empirical data set, it shall be shown, that the applied quantitative model is not an 'artificial' one, but derived from a larger research context.

In social systems with rising levels of unemployment not only researchers ask: How do people react to this rising level? The psychological and sociological research on this subject "...has generated literature consisting of several hundred articles and numerous books (...). This material has three broad characteristics: It is descriptive, it is diffuse and fragmentary, and it is almost entirely concerned with the unemployed man." (Kelvin 1985, p. 34) Although the material is diffuse, one result of this research is distinct: The unemployed themselves develop neither class structures nor social collective activities (Aiken 1968, Aschenbach 1994). Therefore it is not expected in first order that the origin of social movements and political changes will be found in unemployed people.

But not only the unemployed are affected by unemployment. Psychological studies show that the employed are also involved in redundancies and bankruptcies (Armstrong-Stassen 1994, Stevens 1991). Firstly by changed social working conditions such as increased working demands, secondly by direct contacts with the unemployed (family members, friends, colleagues) and thirdly by information flows: Employed people have expectations for their future, mainly for the future of their jobs. This latter expectation is here called 'job confidence'. People build up job confidence by interpreting current information. They take current events as indicators for estimating their job security or their chances for new jobs. Job confidence based on unemployment events in the social environment is therefore a specific reaction. The question is now: How do people get their job confidence? Which are the constituting factors?

The Rational Choice formalism is an appropriate starting point to formalise this question, whereas the theory to be constructed will not be a decision theory. (It's a part of it!) The central variable of RC theories is the expectation utility $E(U) = \sum_k p_k U_k$, where p_k is the probability of an event with utility $_k$. In our case, the possible events of the 'experiment job future' are 'to have a job' or 'not to have a job'. Hence we have to identify U with 'job utility' and p with 'job probability' or job confidence. And the question is: In which way does the actor get his p? This question could be partially answered by evolutionary game theory, where p within a RC solution concept is a strategic parameter, influenced by learning processes. However for such processes repeated games are required. But most of the employed people in modern industrial societies were never unemployed and have job confidences anyway.

Trust theories give a better approach. Due to J.S. Coleman (Coleman 1990) trust is based on a deal with a time lag: Actor A gives something to B and doesn't know for sure if he will get something back. Instead he assumes a certain probability and this is called 'trust'. The question is the same: How does A estimate the probability? Especially if he hasn't had any experience with B? Coleman mentions two types of information sources: Other actors who deal with B and other actors who do not. In any case information is important and it comes from 'outside', from the social environment. To understand the job confidence dynamics one needs to specify the information network around the confiding actors.

14.4 The Model

The focus here is on employed people in an industrial society and their job confidence, defined as job-keeping expectations. These expectations are probability values, called c. (c stands for 'confidence in keeping the job for the next 2 years'). Which factors determine c? Two basic assumptions are made:

1. The main influence factors are factors of the social environment. I call them c_E. These may be information from media or from job colleagues or the job expectations of family members or friends.

2. The sensitivity to this social environment is variable and changes with the anticipated well-being of the actor. If somebody expects to feel well in future, he will not be very concerned about his social environment. On the other hand, if he feels bad or expects to feel bad (anxious), he will be very sensitive.

Both basic assumptions are used to build the following quantitative model:

14.4.1 The Kernel Model

Future Value: $E(U) = cU$

Differential Property: $\frac{dc}{dt} = -\delta(c(t) - c_0(t))$

Coupling to the environment: $c_0 = c_B + \gamma\,(c_E(t) - c_B)$

Sensitivity: $\gamma = a_\gamma\,(E(U) - U)^2$

Normalisation:

$c < 0 \Rightarrow c := 0$

$c > 1 \Rightarrow c :=$

U = Contribution of job to the general well being of the actor. ==> 0 U 1.
$E(U)$ = Expectation Value of U.
δ = Flexibility of c.
c_B = Bias confidence in job stability.
a_γ = Sensitivity constant

All variables and constants have well defined units. All equations combined, the Kernel Model is a normal differential equation in time with 3 independent parameters (δ , c_B, $a_\gamma U^2$) and the initial condition $c(t_0)$ for each actor.

Fig. 1 shows $c(t)$ for an oscillating c_E. The dynamics are of the standard coupled oscillation type. Fig. 2 shows $c(t)$ for different constant c_E. The curves split into a singular and an asymptotic type. These two types appear not only in this particular 1 actor model, but in all endogenous models.

For an empirical test, a 1-actor-model is not sufficient. The dynamic effects result from information flows between different actors. In a 2-actor model the job confidence of one actor is the environmental job confidence c_E of the other. In a model with more than 2 actors a model for c_E is needed: The 'Network' Model. (The name refers to the existence of an information network, not to a standard network analysis.)

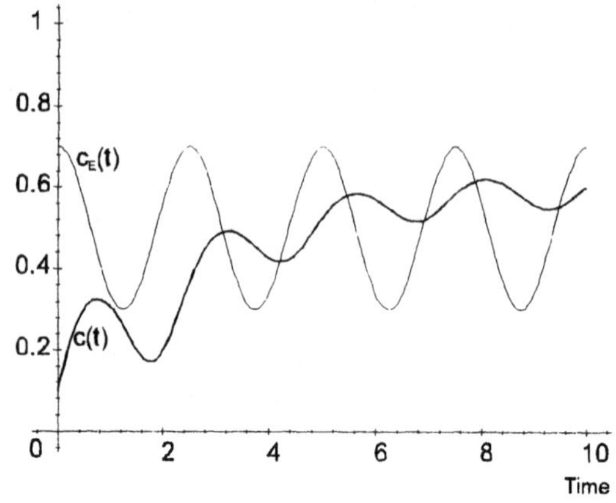

Figure 1: c(t) due to the Kernel Model at given oscillating c_E.. (c_E is a cosine function).

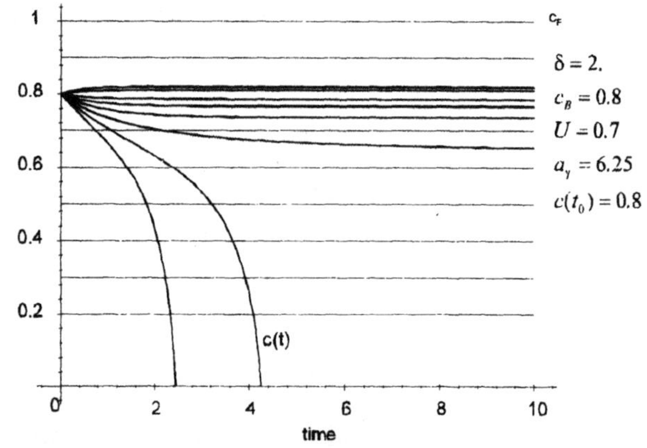

Figure 2: c(t) for given constant c_E on different levels.

Two types of dynamics: Singular (2 lower curves) and asymptotic (5 upper curves).

14.4.2 The 'Network' Model

For description of multiple actor dynamics we need to specify the relations between the actors due to c_E. For the empirical test, I chose the following model:

$$c_{Ei}(t) = \alpha_m \left(\frac{1}{N-1} \sum_{i \neq j} c_j(t) \right) + \alpha_p \frac{dp_{NS}}{dt} + \alpha_q q_{UE}$$

p_{NS} = Net Social Product of the global system.
q_{UE} = Level of Unemployment of the global system
$\alpha_m, \alpha_p, \alpha_q$ = Network constants.
N = Number of actors

The model assumes that c_E of actor i is proportional to the mean job confidence of all others, proportional to the rate of Net Social Product and proportional to the Unemployment Rate.

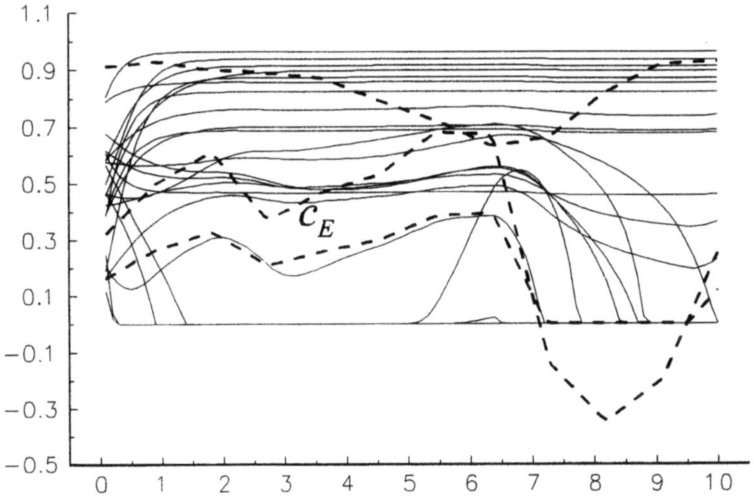

Figure 3: Example of c(t) for 20 actors, coupled with the Network Model (1).
The dashed lines show the unemployment level of West Germany between 1984 and 1994 (upper dashed curve), c_E (dashed curve in the middle) and the NSP rate in West Germany (lower dashed curve).

An example of a multiple actor system based on the Kernel and Network Model is shown by Figure 3. Its dynamics are characterised by the following points:

In many cases the confidence is saturated either at c=1 or at c=0.

In most not saturated cases, c follows c_E with a (positive) time delay due to the differential property.

There are strong non-linear effects. These can be inflation as well as deflation.
The dynamics depend more on the combination of given parameters in a population than on the one dimensional distributions of these parameters. Therefore there are no simple relations between the dynamics and the parameter distributions such as 'the higher c_B the higher the risk for a deflation'.

The main aspect for theory construction as well as for applications in political sciences: A self-evolving 'collapse' effect, which could be intuitively associated with sudden eruptions of system scepticism and thereby caused right-wing extremism very known in the newer histories of Germany, France and other countries. The claim would be: There is an explanation for these eruptions even if the system conditions only change slightly. Can we relate this explanation in some way to empirical data? Can we find an indicator, whether the data fit to this assumption or not? The following section goes into this question.

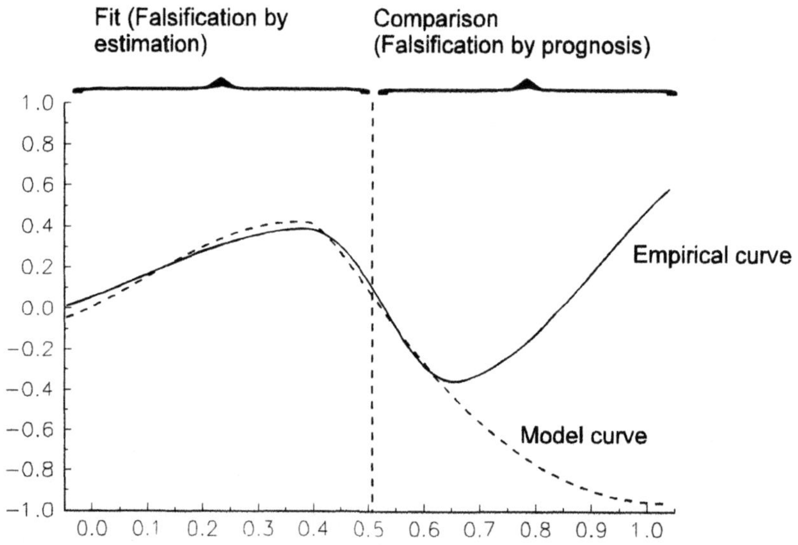

Figure 4: The straight line indicates the data, the dashed line the model curve. In the first part in front of the perpendicular dashed line the calibration is done. The scheme shows a successful calibration. After the perpendicular line the model curve is extrapolated. The scheme shows a great deviation of the model prognosis from the data and hence a positive falsification.

14.5 The Empirical Test

14.5.1 Definitions and Problems

First, it should be clarified, what is meant by 'test'. According to Harbordt[3] a complete test procedure includes the splitting of data into two parts, one to be used as a calibration base, the other to test the model prognosis, (see Figure 4). The first part already has falsification character: In case the model cannot be sufficiently fitted to the data, the model is falsified. In the following I apply only this first type of falsification test, falsification by calibration. But it is important to emphasise that this type of test is not sufficient. There are always (purely mathematically defined) models which fit any data set almost perfectly, (the so called 'orthogonal function systems' such as Taylor-, Fourier- or Legendre-systems). Therefore, a calibration test shows whether or not a model is able to reproduce the data, (i.e. whether it is able to 'explain' the data). But it is not an indicator for the correct theoretical structure of the model (i.e. it doesn't show, whether or not the 'explanation' is correct.)

Further the stochastic aspect has to be envisaged. A perfect reproduction of the data by the model cannot be expected. So how can we deal with errors? In standard parametric statistics one compares explanations from the test model with explanations from a 'zero model'. If the probability is sufficiently small that the 'zero model' generates the measured data (i.e. the related stochastic parameters such as mean values or variances) the test model is accepted. However, in physics or chemistry this method is rarely applied. In complex multidimensional systems the zero model is too vague a criterion. Instead, the physicist usually estimates the measurement errors of his instruments and all other disturbing factors and calculates from them the data errors. Further he estimates the input parameter errors of the model and calculates the model errors. If the test model can be fitted within these errors, the model is accepted. In our case, the only possibility to measure the input parameters of the model is the calibration itself. That would mean the acceptance of any fitness: The calibration test is no longer a test.

In the following sections I don't apply stochastic criteria: Firstly by reasons already mentioned in section 1, secondly in order not to complicate things which are complicated enough, thirdly in order to maintain the test character of the calibration, even if the limit of fitness is not fixed by any theory. We will see that *falsifications* are very well possible anyway.

The Complexity Problem

Whereas the Kernel Model has a relatively simple structure (no stochastics, only a few parameters), in a multiple actor system the complexity of the model increases greatly with the number of actors. Therefore two antagonistic require-

---

[3] Harbordt, Steffen 1974: Computersimulationen in den Sozialwissenschaften, Bd. 1, S. 183, Reinbek bei Hamburg: Rowohlt.

ments have to be dealt with: On the one hand reduction of complexity is needed to carry out the fit procedure and to enable a clear perspective on the results. On the other hand, increased complexity makes the model more realistic. One cannot even hope to get reasonable results if the model is reduced to only two actors. Further the Kernel Model structure generates converging job expectations if homogeneous parameters are chosen. (I.e. for each actor the same). Therefore one has to fit a model with at least 10 actors, which means 70 parameters to fit. A simultaneous fit with such a large number of open parameters is a huge optimisation problem.

The Micro-Macro-Link

Chances of solving the optimisation problem depend heavily on the quality of data. Most time-related micro data in the social sciences are available from repeated surveys. If the interviews are repeated with the same population ('panel survey') then time series for each actor can be extracted. One use of these individual time series could be the effort to linearise the fit problem.

For the present model I used data with 6 data points and panel structure. Each of 1,400 interviewed persons was identified with a modeled actor. The total amount of open parameters is 9,800. The actors coupling in the model is the Mean-Confidence-Term in the 'Network' Model. I tried to approximate this term by an iteration using as a starting point the interpolated data. From them the overall mean confidence for each time point was calculated. I fitted the model person by person (each fit function has 7 parameters and a fixed mean confidence). Finally the new Mean-Confidence-Terms were calculated from the resulting confidences. This procedure was repeated five times. (How the person by person fits were done, is explained below.)

Table 1: l_{OLS} indicates the averaged ordinary least square of the fit.

Step	l_{OLS}
1	0.343
2	0.372
3	0.115
4	0.361
5	0.110

The convergence properties of this algorithm are poor. (See Tab. 1). In general the problem of global optimisation and global root finding in non-linear systems is the rub in fitting and regression efforts[4]. See the following paragraph. The algorithm applied here at least generates some improvement in l_{OLS} (sum of the ordinary least square) compared with step 1.

The Ambiguousness Problem

In general complex global optimisation problems have non-convex optimisation functions and fit topologies with a large number of relative minima. The chance of finding an absolute minimum is almost zero. The problem is illustrated by Figures 5a and 5b. The task in both cases is to find the minimum. In figure 5a with only one minimum, there exist canonical numerical methods even if the function is not analytically known. In the case of Fig. 5b, there exist methods to find from a given point the next relative minimum. This search may be quite difficult if the valley with this minimum has a 'nasty' irregular shape with almost constricted parts. But even if the minimum has been found it is not clear whether it is a relative or absolute one. The only possibility is to look for and to compare other relative minima. In fact, 'virtually nothing is known about finding global extrema in general.' (Flannery 1992, p. 309) Genetic algorithms attempt to learn something about regularities in the location and depth of the minima and to increase the search efficiency. However if there is nothing to learn there is nothing to increase.

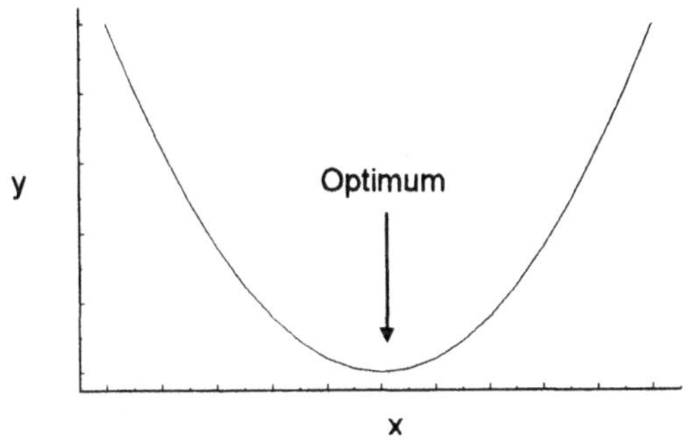

Figure 5a: The ideal (one dimensional) function for global optimisation: A function with only minimum and without flat regions.

[4] How nasty the search for roots can be already in simple systems you can try yourself: Find from the equation system $y_i = x_i / (\sum_i x_i)^2$ and given (y_i) the vector (x_i). This in, let's say, more than 10 dimensions.

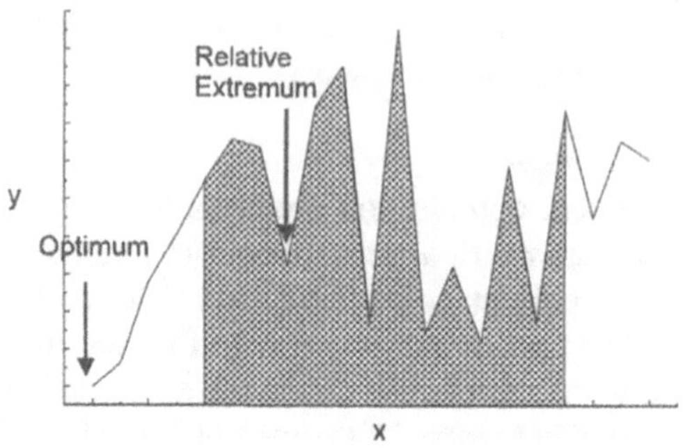

Figure 5b: Typical shape of a fit problem appearing in applications: Several narrow valleys with relative minima. If the search is restricted to the grey segment, the algorithm cannot find the absolute minimum to the left.

Thus the results are restricted to a set of more or less low relative minima. This is the first reason why it is so hard to undertake statistical estimations: If your algorithm has not found the best estimator, and if nothing is known about the probability for efficiency, you do not know anything about errors[5].

Before bothering about statistical problems the following should be addressed: How should one deal with this 'crowd' of relative minima? Which minimum represents the 'true' parameters? Is there only one 'true' set of parameters? The decision may be to take the lowest minimum as the location for the 'true' parameters. But the assumption that the nearer the relative minima are to the absolute minimum the deeper they are, is an arbitrary one. The probability of each point (not only the minimum points!) being in the vicinity of the absolute minimum is a priori exactly the same. In addition there may be slight distortions of the data or slight errors in the model structure, which are sufficient to increase the 'true' minimum above another minimum which is located far away in parameter space. Thus it is better not to concentrate only on one minimum. But what to do with multiple minima? My approach to do a cluster analysis is explained below.

14.5.2 The Data

SOEP [6] data were used for the empirical test. SOEP is a survey based on a representative sample of around 10,000 people in Germany, repeated every year since

[5] Which does not mean, that such estimations are not possible. They are just generate data, including random errors, from your model and try to find the applied parameters from the data with your fitting algorithm. The problem is just computing power to cover the huge parameter space with enough test points.

[6] Socio Economic Panel of Deutsches Institut für Wirtschaftsforschung Berlin.

1984 with the same population. Of the several hundred questions I consider only three:

Q1: Are you working at present? (I'm employed full-time, part-time employed, in military service, (...), not employed)

Q2: Are you unemployed? (Yes/No)

Q3: Do you expect to lose your job in the next 2 years? (Definitely / probable / quite improbable / definitely not)

Only in every second wave is Q3 included. Therefore there are six waves available. Two selections were made: First those persons were selected who had participated in each of these six waves. Secondly for technical reasons those persons were selected who were fully employed in all six waves. The second selection is expected to be a strong distortion of representation. On the other hand it has to be considered that our test model includes no unemployment experiences, even in the social environment.

Table 2: Col. 2 shows the mean job confidences of all respondents who were present in all six waves. Col. 3 shows the means of only those, who were fully employed in all six waves. Col. 4 gives the differences, Col. 5 the relative differences and Col. 6 the differences relative to the variance of the first mean (all persons).

Wave	Mean *e* 1.S.	Mean *e* 2.S.	Δe	$\lvert 2\Delta e / (e_1 + e_2) \rvert$	$\Delta e / \sigma_{e1}$
2	3.472	3.499	-0.027	0.8%	-36.7%
4	3.477	3.516	-0.039	1.1%	-53.0%
6	3.519	3.564	-0.045	1.3%	-61.2%
8	3.511	3.564	-0.053	1.5%	-72.1%
10	3.405	3.457	-0.052	1.5%	-70.7%
11	3.306	3.330	-0.024	0.7%	-32.6%

Table 3: Distribution of employment states over all six waves.

Category	Mean	Standard Deviation
Full Employed	46.8%	1.2%
(Short-time work)*	0.6%	0%
Partially employed	7.3%	0.7%
(Partially employed and short-time work)*	0.1%	0%
Education	3.5%	0.6%
Non-regular work	2.1%	0.2%
Motherhood	1.6%	0.25%
Military Service	0.4%	0.09%
Not employed	38.7%	1.6%

* Only wave 11.

The difference in the mean values of job confidence, measured by question Q3, between all persons and those who are fully employed the whole time, is shown in Table 2. Whereas the absolute difference in job confidence seems to be small this difference is large relative to the overall variance which is small as well. Some univariate statistics of the data are given in tab. 3 and figure 6.

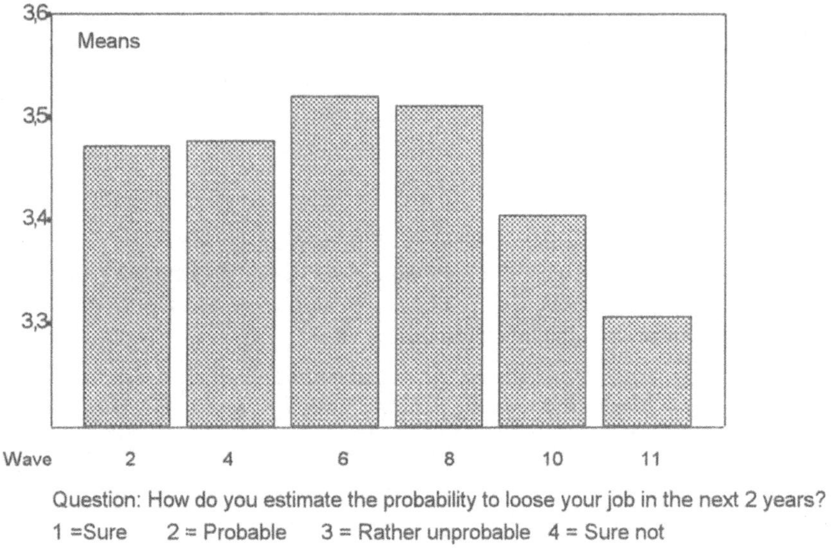

Figure 6: Expectations for stable employment, all six waves. (The wave number corresponds to the year: 2=1985, 4=1987 and so on.)

14.5.3 The Fit Procedure

The fit procedure proceeds in three nested loops: local optimisation in the inner loop, global optimisation in the intermediate loop and iteration in the outer loop.

Local optimisation

From a selected point, the next relative minimum in l_{OLS} has to be found.

$$l_{OLS}(c_i) = \sqrt{\frac{1}{6}\sum_j (c_i(t_j) - x_i(t_j))^2}$$

$x_i(t_j)$ = Data point of actor i in time point j

$c_i(t_j)$ = Modeled job confidence of actor *i* in time point *j*, depending on 7 parameters $a_\gamma U^2$, c_B, ...

I applied a non-standard multidimensional minimisation procedure, the Downhill-Simplex-Method (Flannery 1992, pp. 326). This method is in general quite slow, but very robust. In any case this method arrived at a minimum. Typically around 130 calculations of l_{OLS} were needed for one search.

Global optimisation
Here I applied a very simple 'learning' algorithm, called 'quadmin'. First I selected a hyper box out of the parameter space. Then, in each dimension the limiting line segment was split in half and $2^7 = 128$ 'sub-boxes' were generated in this way. From each sub-box a random sample of 10 points was taken and the local optimisation was started there. The results were entered in a list. The 'sub-box' with the lowest mean minimum value in the list was split again into 'subboxes' and the whole procedure was repeated. With 3 such 'quadmin' iterations, there were at the end 384 sub-boxes and 3840 local minima. The advantage of this method is a good ratio between minimisation of calculation needs and governance of the parameter space in combination with a (simple) learning structure.

Cluster analysis
What can be done with almost 4000 minima? The hundred best minima were selected and a cluster analysis was made. I applied the 'Single Linkage Method': From each pair of minima their distance[7] was calculated. The best pair was combined into a 'cluster' and the gravity point of this cluster was taken as the new co-ordinate. Afterwards the list contained 99 points: 98 single point and one cluster. This procedure can be repeated until all 100 minima are in one cluster. In each turn the inner variance of the clusters is compared with the overall variance of all points. In the optimal case the difference between both variances, the cluster contrast, remains large until it decreases suddenly to zero. At this step the optimal clustering point is found. In our case, this point cannot be found by hand after each global optimisation. (There are 1400 actors.) A pattern recognition algorithm was applied, which bases on the 'second derivation' of the cluster contrast as a function of the clustering step. These algorithms were calibrated with 'test clusters' which overlapped by about 4/5 of their scattering segments until the algorithm was able to separate all points correctly.

During the cluster procedure the PR algorithm checked the cluster contrast and marked the first three cluster steps, in which a corresponding drop-off occurred. Mostly directly after the first mark, further drop-offs followed. The first three

[7] As distance I chose the usual 2-norm: $d(\underline{x},\underline{y}) := \|\underline{x} - \underline{y}\|_2$.

marks were recorded. Object of the further analysis was the cluster structure, which emerged from the first marked step. From this the biggest cluster was considered as the result of the total global optimisation.

Iteration

As mentioned above, for a 'person by person' fit an iteration of the mean confidence term in the 'Network Model' is needed. The following method was chosen:

1. Interpolate the $c_i(t)$ data points for each actor
2. Calculate the mean confidence $\bar{c}(t)$
3. Global optimisation with one 'quadmin' iteration.
4. With the best fit from 3. go back to step 2.

After repeating steps 2., 3. and 4. four times, do the 'final iteration':

5. Global optimisation with three 'quadmin' iterations
6. Select the 100 best minima of each actor
7. Do the cluster analysis for each actor.

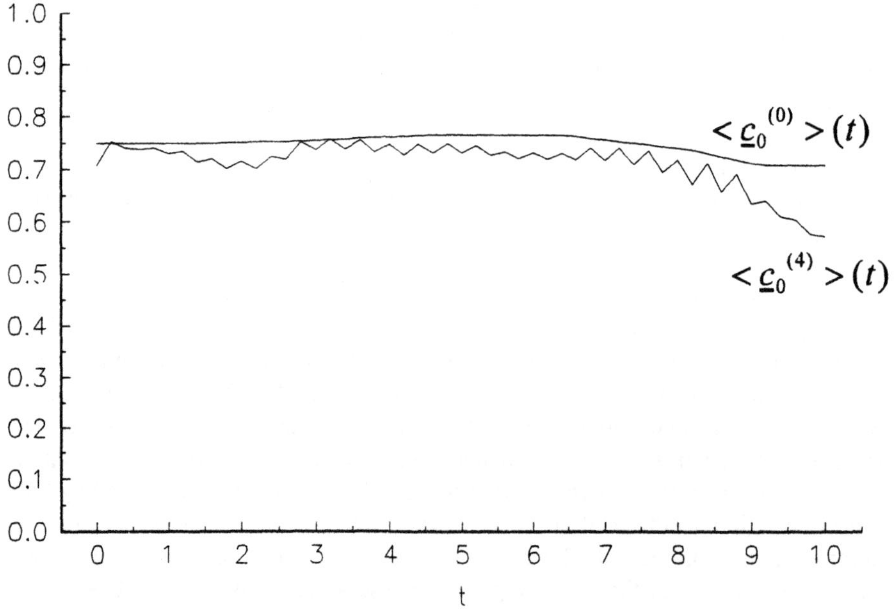

Figure 7: The Proximum $< \underline{c}_0^{(4)} > (t)$ as result after the first four global minimum searches. $< \underline{c}_0^{(0)} > (t)$ gives the history of empirical mean confidences, interpolated between the waves. It is to respect: $< \underline{c}_0^{(4)} > (t)$ is the mean fit of 1400 c(t) curves, not an approximation of $< \underline{c}_0^{(0)} > (t)$

The first four iterations are done with a simplified method, because here the only important data are the confidences of the actors and it doesn't matter by which parameters they are generated. In the final iteration the parameters have to be approximated as accurately as possible.

14.5.4 The Results

Figure 7 shows the evolution of the best fitted model after the 4th iteration step in comparison with the interpolated mean values of the surveyed confidences. One recognises that the overlaid micro fits tend to overreact on the tail changes in mean confidence. The distribution of best fits after the 5th iteration (figure 8) gives a mean error of l_{OLS}=0.08. (Without trivial fits taken into account. The difference to the number in tab. 1: There the goodness of the best 100 fits are averaged.) Compared with the overall variance of around 0.1, this error is quite large. This does not reflect the insufficiency of the fit algorithm, (which is given by the unknown location of the absolute minimum in parameter space, not necessarily as well by the insufficient fit quality of the found relative minima!), but more the limits of the model to absorb the dynamics of the empirical data. This means: Although the data seem to have relatively low movement in time, if one regards only the macro level, even 7 parameters per person don't fit this movement, if these parameters belong to a coupled micro model! That is a clear difference to static regression models and a chance to falsify dynamic models even without stochastics.

Tables 4a and 4b give information about the best hundred fits for those actors whose confidence changes at least one time during the 6 waves. Table 4a shows data, which characterise the distribution of means: Table 4b refers to the distribution of standard deviations within the set of the hundred best fits. The means of Table 4a have the following meanings: It takes about 1.8 years (on the average) until an actor has adapted his job confidence to new conditions and shrunk his original confidence down to 1/e (δ =1.5)[8]. The actors have a base confidence of around 0.5. That means they expect with roughly 50% probability (independently of the environment) to keep their jobs the next two years. $a_\gamma U^2$ 30 can have different meanings. It may mean that the job for each actor governs 100% of his life satisfaction (U=1) and a deviation of the expected job value by 10% from the optimum generates γ=30% (a_γ=30). Or U is lower and a_γ is higher, (the effect is the same). γ=30% says that a positive difference of 10% between the job confidence of the social environment and the base confidence increases the reference confidence (C_0) by 3%. The initial job confidence in 1984 was around 75%. A change of 10% in the mean confidence of others changes the actor's own confi-

[8] It is not 1.5 years because t_{max}=12 years is recalibrated to t_{max}= 10 model time units.

dence by 3.8%. Both the other couplings are to be understood in the same way. l_{OLS} 8% shows that the confidence reproduction by the model has an average error of 8% (for each actor).

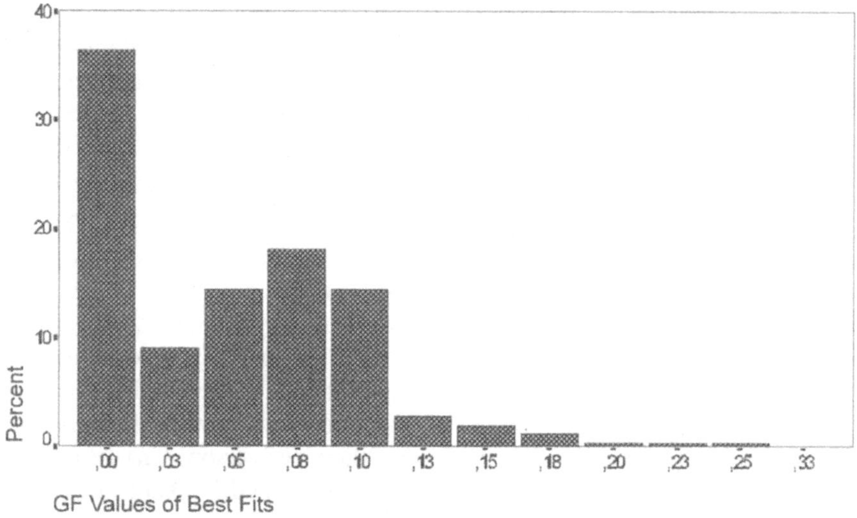

Figure 8: Goodness of Fit after the 5th iteration of global minimum search. Displayed are only best fits. Around one third of the fits are just perfect, because in this cases the confidence didn't change. The l_{OLS} -average of the rest is 0.08. It is to respect: This is the average from 1400 different c(t)-curves, not from the mean curve.

Table 4a: The hundred best fits: Distributions of means for all parameters, (last iteration)

Parameter	Mean	Stddev	Minimum Value	Maximum Value
δ	1.5553	1.7789	.0	8.8372
c_B	.4617	.3144	.0	.9837
$a_\gamma U^2$	31.5508	26.3118	.0	91.4092
$c(t_0)$	.7654	.1812	.0	.9989
α_m	.3790	.4807	-.999	.9276
α_p	-.1745	.4557	-.999	.9772
α_q	-.498	.4532	-1.000	1.022
l_{OLS}	.0759	.0681	.00	.4053

Table 4b: The hundred best fits: Distributions of standard deviations for all parameters, (last iteration).

Parameter	Mean	Stddev	Minimum Value	Maximum Value
δ	1,04	,77	,0	3,7
c_B	,13	,08	,0	,4
$a_\gamma U^2$	13,28	6,05	,0	36,0
$c(t_0)$	,08	,07	,0	,3
α_m	,29	,14	,0	,7
α_p	,30	,12	,0	,8
α_q	,33	,11	,0	,8
l_{OLS}	,02	,04	,0	,3

If one compares Table 4a and 4b, it can be recognised that the average mean and the average standard deviation of most parameters have similar size. That indicates that the minima are not located in one narrow cluster. On the other hand, if one compares the deviation of the average means with the means of the best fits the relative differences of 5 parameters are 10 percent or less. So the distribution of the minima seems not to be totally irregular.

More insight into the distribution of minima is given by the cluster analysis. The average size of the biggest cluster at the first mark is 55 3. (The latter number is the mean error). The first mark itself is on the average at step 86 2. That shows a clear cluster structure of the minima with less than 20 clusters in most cases. Around one half of the minima are represented by the biggest cluster.

Fig. 6 shows the distributions of all seven parameters. Each actor is represented by the means of biggest cluster. c_B is characterised by a bimodal distribution with modi at $c_B \approx 0.15$ and $c_B \approx 0.8$. The high ratio of people with high c_B in combination with high momentary confidences is a pre-condition for high deflation probabilities.

For all network parameter, a large number of actors have 'inverted' couplings: A rising mean confidence or a rising NSP rate decreases their job confidence and a rising unemployment quote increases it. In case of the NSP rate even more than half of all actors show this relation, which cannot be understood by basic socio-psychological considerations. Something seems to be wrong. This result, that 'something is wrong', is the (pre) success of the applied method: We arrived at a positive falsification and we have now something to learn.

14.5.5 Prospects

What may be the reasons for these inverted couplings? In general there are four sources of errors:

The approximation procedure
The Network Model
The Kernel Model
The simple socio-psychological assumptions.

Also there are various possibilities for making crosschecks:

Testing the approximation procedure by a set of variations: Do the parameter distributions remain the same?

Testing the approximation procedure by a Monte Carlo Simulation: Is it possible to reproduce a given parameter distribution?

A sensitive point may be the small variance of data and the low resolution of measurement, (the answers to Q3 have only 4 categories.) How does the approximation efficiency depend upon this factor? How can the model be modified to give reliable information on the basis of data with such small variances and resolutions? One partial answer to this question was given already in 5.4..

How does the model deal with incomplete information ? In standard statistics unbiased estimators enable the scientist to eliminate distortions from incomplete information by increasing the sample size. At a certain point all distortions 'disappear' in the variance. We cannot a priori assume the same property for the non-linear model. Tests have to be made.

Reduction of the nearly 10,000 fit parameters by stochastification of the model or grouping actors in some way.

Testing the models by comparison with results of 'classical' time series analysis.

Modification of the Network Model by micro factors. The SOEP data contain data on household level in such a way that the job confidence of other household members is available and identifiable.

Modification of the Kernel Model formulas.

Introduction of new external factors in the Kernel Model. The job confidence may depend not only upon the job value (U), but upon the general well being of the actor, which is influenced by other factors.

14.4 Conclusion

Perhaps it is an inconvenient and irritating way, if the author avoids the illusion to have solved every problem. But there are problems, which are big enough not to allow a quick and handsome solution (and the corresponding dextrous article. In fact scientists are often confronted with the reproach to select their problems according to their own ability to solve them.). But problems are not solved as well by ignoring the fact that they exist. That is what happened until now, a few exceptions mentioned in section 1. This article did not give a compact answer. But it attempted to pose a good question, to show up the essential points of the problem and to give some suggestions to start the search for solutions.

These suggestions are just the very beginning of this work. The complexity of the crosscheck procedures require a whole team of researchers instead of only one, a large amount of computing power and - most important - a great deal of systematic work on calibration and test methods for micro-macro-models.

What could be shown is: Although there are some problems to overcome, model theory in social sciences is not restricted to the use of empirical data on a descriptive level. Tests can be efficiently done as in other sciences such as physics, chemistry or 'classical' sociology. What is needed in the future is the systematic research on problems of

optimisation methodology
analysis of optimisation results
crosschecks and of checking robustness.
prognosis tests.

Figure 9: Mean value distributions of all parameters (based of biggest clusters).

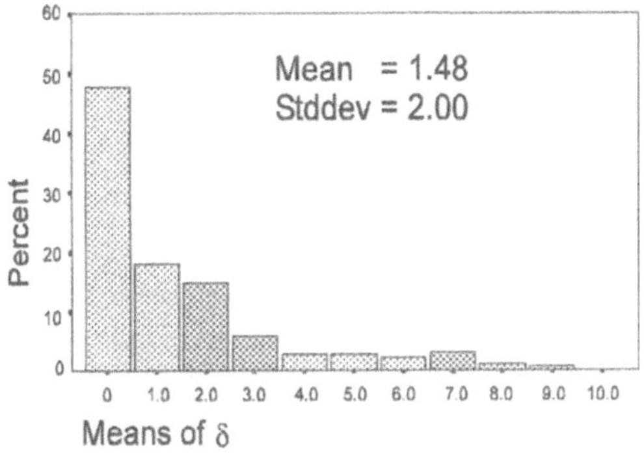

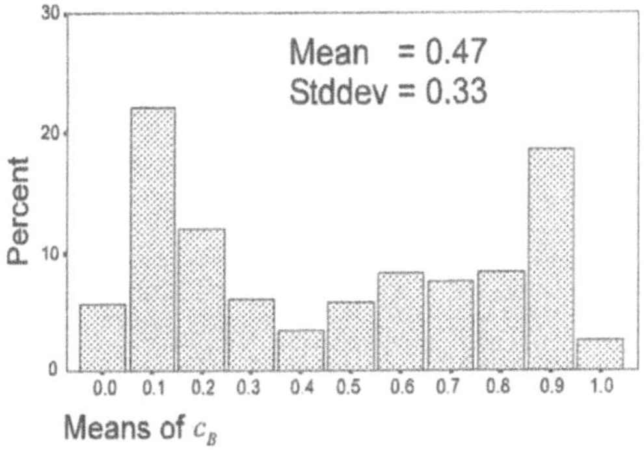

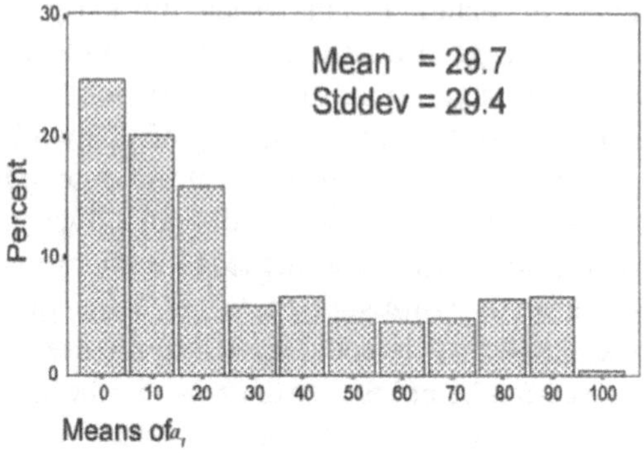
Mean = 29.7
Stddev = 29.4
Percent
30
20
10
0
0 10 20 30 40 50 60 70 80 90 100
Means of a_i

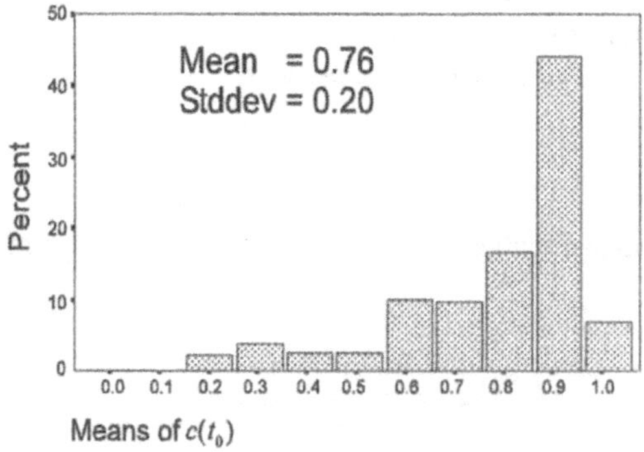
Mean = 0.76
Stddev = 0.20
Percent
50
40
30
20
10
0
0.0 0.1 0.2 0.3 0.4 0.5 0.6 0.7 0.8 0.9 1.0
Means of $c(t_0)$

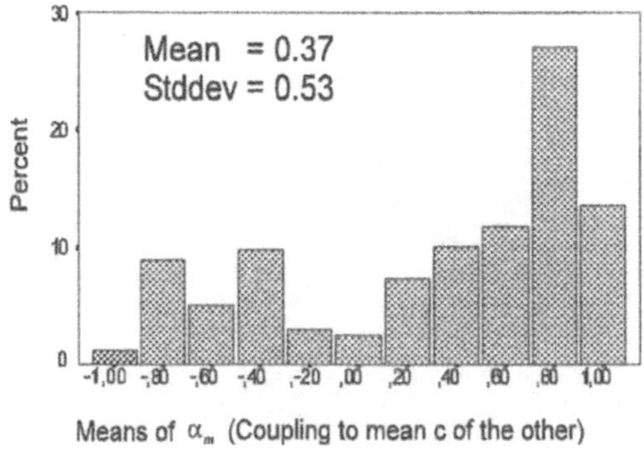
Mean = 0.37
Stddev = 0.53
Percent
30
20
10
0
-1,00 -,80 -,60 -,40 -,20 ,00 ,20 ,40 ,60 ,80 1,00
Means of α_m (Coupling to mean c of the other)

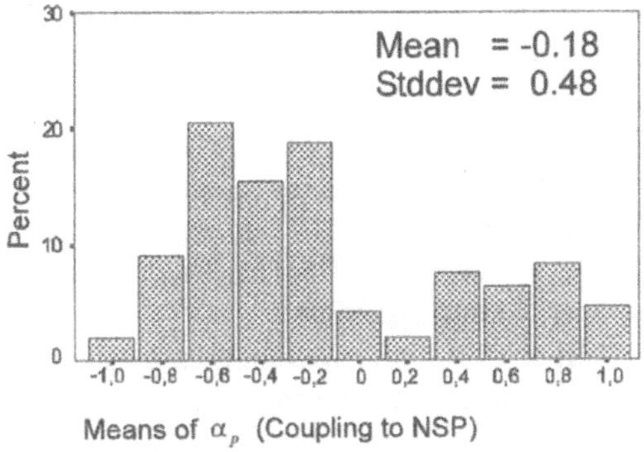

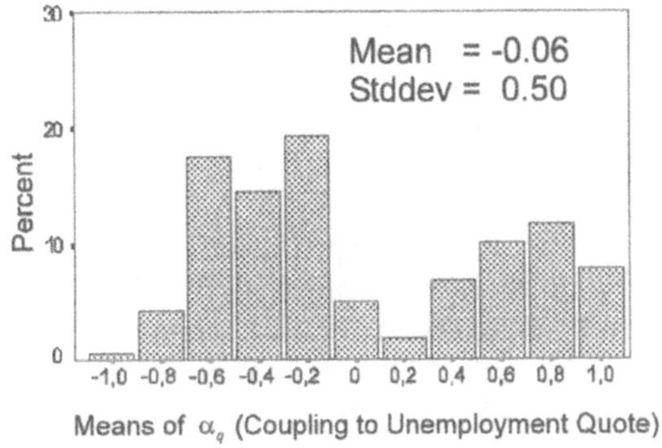

References

Aiken, M., Ferman, L. A. & Sheppard, H. L. (1968): Economic Failure, Alienation and Extremism. Ann Arbor, The University of Michigan Press.

Armstrong-Stassen, M. (1994): Coping with Transition: A Study of Layoff Survivors. Journal of Organizational Behavior 15, 597-621.

Aschenbach, G. & Frey, D. (1994): Arbeitslosigkeit. Aus: Frey, Dieter; Greif, Siegfried (eds): Sozialpsychologie. Ein Handbuch in Schlüsselbegriffen, Beltz, Basel, Weinheim, 529-542.

Asher, H. B. (1983): Causal Modeling. Series Quantitative Applications in the Social Sciences. Sage, London.

Bates, D. M. & Watts, D. G. (1988): Nonlinear Regression Analysis and Its Applications. Wiley, New York.

Bentler, P. M. (1989): EQS Structural Equations Program Manual. BMDP Statistical Software, Los Angeles.

Blossfeld, H. P. & Rohwer, G. (1995): Techniques of Event History Modeling: New Approaches to Causal Analysis. Erlbaum, Mahwah, NJ.

Bruederl, J. (1996): Multivariate statistische Verfahren in der Bevoelkerungswissenschaft. In: Dieckmann A.; Müller, U.; Nauck, B. (Hg.): Handbuch der Bevoelkerungswissenschaft. De Gruyter, Berlin.

Bratley, P., Fox, B. & Schrage, L. E. (1987): A Guide to Simulation. New York.

Cherkasov, I. D. (1995): Estimation of the Coefficients of a Stochastic Differential Equation. Mathematical Sciences 75 no. 1, 1453-1460.

Coleman, J. S. (1990): Foundations of Social Theory. The Belknap Press of Havard University Press, Cambridge/Mass.

Dieckmann, A. & Mitter, P. (1984): Methoden zur Analyse von Zeitverlaeufen. Teubner, Stuttgart. (English Version availible).

Doreian, P. & Hummon, N. (1979): Estimating Differential Equation Models on Time Series: Some Simulation Evidence. Sociological Methods and Research 8, 3-33.

Doreian, P. & Hummon, N. (1976): Modeling Social Processes. Elsevier, New York.

Fahrmeir, L. & Tutz, G. (1994): Multivariate Statistical Modelling Based on Generalized Linear Models. Springer, Berlin.

Flannery, B. P., Press, W. H., Teukolsky, S. A. & Vetterling, W. T. (1992): Numerical Recipes in Pascal. Cambridge University Press, Cambridge.

Gifi, A. (1990): Nonlinear Multivariate Analysis. Wiley, New York.

Greene, W. H. (1997): Econometric Analysis. Prentice-Hall, London.

Haken, H. (1990): Synergetik, Berlin, New York: Springer

Hamerle, A., Singer, H. & Nagl, W. (1993): Identification and Estimation of Continuous Time Dynamic Systems with Exogenous Variables Using Panel Data. Econometric-Theory 9 no. 2, 283-295.

Hayduk, L. A. (1987): Structural Equation Modelling With LISREL: Essentials and Advances. The Johns Hopkins University Press, Baltimore.

Hegselmann, R. (1996): Solidaritaet unter Ungleichen. In: Hegselmann, R. & Peitgen, H. O. (eds.): Modelle sozialer Dynamiken. Ordnung Chaos und Komplexitaet, Hoelder-Pichler-Tempsky, Wien, 105-128.

Jaccard, J. & Wan, C. K. (1996): Lisrel Approaches to Interaction Effects in Multiple Regressions. Series Quantitative Applications in the Social Sciences. Sage, London.

Joereskog, K. G.(1994): Lisrel 8. Presentation on the Congress "Statistical Instruments for Social Sciences" Amsterdam.
See also: http://www.GAMMA.rug.nl/sibweb/lisrel2.html

Joereskog, K. G &; Soerbom, D. (1989): Lisrel 7. A Guide to the Program and Applications. SPSS Inc., Chicago, Illinois.

Kelvin, P. & Jarrett, J. E. (1985): Unemployment, Its Sociopsychological Effects. Cambridge University Press, Cambridge.

Orcutt, G. H., Merz, J. & Quincke, H. (eds.) (1986): Microanalytic Simulation Models to Support Social and Financial Policy. North Holland, Amsterdam.

Otter, P. W. (1988): Structural, Dynamic Modelling in Unobservable Spaces of Covariance-Stationary Stochastic Processes. Time Series Analysis 9, no. 1, 59-72.

Vries, P. H. (1998): On the Understanding of the Self-Organization of Cognitive Systems: A Strategy For Computer Simulation and Empirical Research. In: Liebrand, Nowak, Hegselmann (eds.): Computer Modeling of Social Process. Sage, London, 169-194.

Saam, N. J. (1996): Multilevel Modeling with MIMOSE. Experience from a Social Science Application. In: Troitzsch, K. G., Müller, U., Gilbert, G. N., & Doran, J. E. (eds.): Social Science Microsimulation. A Challenge to Computer Science. Springer, Berlin, 138-154.

Stevens, Rebecca (1991): Locus-of-Control, Attribution, Social Stress and Self-Perception. An Analysis of the White Collar Unemployed, Reemployed and Spouses. The Humanities and Social Sciences 6.

Troitzsch, K. G. (1990): Modellbildung und Simulation in den Sozialwissenschaften. Westdeutscher Verlag, Wiesbaden.

Ziegler, R. (1972): Theorie und Modell. Der Beitrag der Formalisierung zur soziologischen Theoriebildung. Oldenbourg, München.

15. What Is the Use of Gossip? A Sensitivity Analysis of the Spreading of Respectful Reputation

Mario Paolucci[1], Mario Marsero[1], Rosaria Conte[2]

15.1 Introduction

The evolution of social norms is explained by many authors as a solution to problems of either coordination (cf. Lewis, 1969) or cooperation (Binmore, 1994) or both (Ulman-Margalit, 1977). In the social dilemmas literature (cf. Kerr, 1995), social norms include commitment, reciprocity, equity, and roles. While ethological studies (Eibl-Eibesfeldt, 1967/1978; de Waal, 1982) show that norms evolve also as means to control aggression in both human and primate sociaties (see, for example, the finder-keeper precept), this functionality of norms has been essentially ignored by the social scientific literature.

Of late, the role of social norms in reducing collisions and aggressions has been investigated by means of computer simulation in a sub-field of AI, namely the area of multi-agent systems (Shoham & Tennenholtz, 1992a and b; Conte & Castelfranchi, 1995; Walker & Wooldridge, 1995; Castelfranchi, Conte & Paolucci, 1998). While some of these authors (Shoham & Tennenholtz, 1992b; 1994; Walker & Wooldridge, 1995) have concentrated upon the spread of norms over the system, others (Conte & Castelfranchi, 1995; Castelfranchi et al., 1998) focused on the specific functionality of social norms.

In (Conte & Castelfranchi, 1995), the role of a norm of precedence was observed in a population of agents competing for scarce resources in a spatially defined world. The norm under study prevented agents from attacking other agents eating food units flagged as their own. A normative strategy (not directly advantageous for the executors) was confronted with a purely utilitarian strategy (always advantageous for the executors) to understand the consequences of belonging to one or the other community. Findings pointed out that both strategies reduce aggressions; but while the utilitarian strategy performs better than the normative on a measure of efficiency (higher average strength of the population), the normative strategy is not surprisingly more equitous than the

[1] Vigoni Program "The evolution of elementary morals" (Germany, University of Bayreuth/Italy, Institute of Psychology, Cnr) - email: paolucci@istat.it

[2] Division of AI, Cognitive and Interaction Modelling - PSS (Project on Social Simulation), IP/Cnr, email: rosaria@pscs2.irmkant.rm.cnr.it

utilitarian (lower variance in the distribution of such strength among individuals). In Castelfranchi et al. (1998), the difference in efficiency between the two strategies was found to be neutralised by the spreading of the cheaters' reputation, which we call here "gossip" since agents accept this type of knowledge from one another whitout checking its validity. Checking would anyway actually be useless in this case, because the transmitted knowledge is trivially true.

More precisely, normative agents were allowed to record the behaviours of their fellows and spread the news to other normative agents. In such a way, cheaters (agents not complying with the norm) were identified by compliant agents even before encounters. Therefore, compliant agents were spared the costs of direct learning. Normative reputation acted as preventive social knowledge.

The simulation model contained a number of different objects: a finite world, a finite number of agents and of resources (food units). A number of parameters have been chosen: (a) structural (including the value of agents' initial strength, the nutritional value of the food units, the costs of attacking and being attacked); (b) social, that is, the density of the agents in the simulated world seen under two different aspects: global density (varying the number of agents while keeping constant the world size), and relative density (that is, varying the ratio between the number of agents exhibiting the different strategies, while keeping constant the global size of the population); (c) agent-architectural, that is, the built-in strategy executed by one in-group with regard to an out-group: how does a normative agent behave with regard to a utilitarian?

Given the complexity of the parameter space, the results obtained initially ought to be confirmed through successive experiments. Also, the hypothesis concerning the role of reputation in the efficiency of normative aggression control needed to be confirmed through a more complete exploration of the parameter space. In this paper, we will present the results of further exploration of the model described above. Since we are not interested in checking how realistic our model is, we will leave aside the structural parameters (namely, agents' initial strength and the food value) and will concentrate on the effects of the (essentially density, both global and relative) and cognitive-architectural parameters (strategies and sub-strategies of the agents). This choice is inspired by two research questions, each responsible for one given set of parametrs.

First, which role do the social structural conditions play with regard to the efficiency of norms? In the cohesion-compliance theory (Homans, 1974), social conditions were found to affect social order in a rather positive way. According to that theory, in fact, the higher the cohesion (and therefore the social density), the higher the compliance with the intra-group obligations. In the present study, we wanted to test a different or more specific hypothesis, namely that social density affects norm efficiency because it favours the spread of reputation. To test this hypothesis, we will observe the effect of density on norms efficiency with and without communication about reputation.

Secondly, how does reputation work exactly? More precisely, which type of reaction allowed by this knowledge proves most efficient in redistributing the costs of compliance? In the previous study, the mechanisms through which the spread of reputation improved the efficiency of norms were not explicitly investigated. While findings showed that the spread of reputation allows for the costs of normative compliance to be redistributed, "retaliation" behaviors which normative agents can possibly exibit were not distinguished and compared. In that context, only one retaliation behavior was applied, namely smart retaliation (attack only weaker agents independent of food possession). But what about more elementary and plausible forms of retaliation, where the punishing agent does not take into account the target's strength? Finally, we were bound to check the model for stability. A model that is examined in a region of instability will give unacceptable answers; starting from two sets of paramenters, apparently equal, it could bring to completely different regions, and this means that even the small approximation of computer number representation should be taken into account. Therefore, we decided to realize and present here an extended exploration of some of the parameters of our model. This exploration, indeed, has both confirmed our assumption about the stability of the model, and produced new interesting results.

15.2 Details of Computer Simulation

The game is composed of a variable number of discrete moves; all agents choose one action and perform it simultaneously. The world is a square, torical grid, of variable size, where agents cannot move diagonally (i.e. the metric is such that two squares joined by an edge are at distance 2). Each agent has a strength value (which grows with the agent eating food, and decreases with its moving and attacking), stable strategy (whether complying with the norm - *respectful* or *normative*- or not - *cheater* -) and reputation (corresponding to the idea that each normative agent has of any other agent). The number of agents is chosen at the beginning of the game. Food units are 1/4 of the number of the cells, with fixed nutritional value. Their number is constant during the game: food eaten reappears randomly on the grid. Each food unit may have up to 5 owners, decided on the basis of proximity at the creation time.

At every turn, any surviving agent (that is, any agent whose strength is greater than zero) chooses an action among the following, in order of preference. The first choice is always of eating a food unit in its own cell. If no food is available in one's cell, the agent will move toward a "free" food unit (in a cell that is not occupied by another agent). To choose the direction of motion, first, the region of the grid one cell away from one's own, then the region two cells away from one's own, will be checked. If no food unit is available, the agent will try to snatch a food unit from the nearest eating agent, if any. Strategies differ only in attacking: while the utilitarian (cheaters) agents attack only the weaker, the normative (respectful) agents will respect a norm of private property, and will not attack agents which are eating their own food. If no action is feasible, the agent will pass its turn. All actions have a cost in strength.

The following list contains all the main numerical parameters with default values (bracketed). A high nutritional value of the food was maintained to prevent a high mortality of the agents in the first steps, which would have caused difficulty in comparing the experiments. In general, with these settings, death is rare and average strength grows more or less in a linear way with time. The default parameters are set up as follows:

- Initial strength (40)
- Cost for moving (1)
- Cost for attacking and receiving attacks (4)
- Food value (20)
- Number of iterations of the run (2000 steps)
- Agents density (50%)
- Food density (25%)
- World's dimension (10*10 grid)

Results are compared along efficiency (which popolation is stronger?) and fairness (what is the distribution of strength among the population?).

The results for populations composed of agents sharing the same behavior are shown in Table 1 (stand-alone experiment), where we report average strength from two sets of 50 experiments. In the first set the population is composed of 50 Respectful agents; in the second, of 50 Cheaters. The σ (sigma) is the standard deviation over the repeating experiments; "Dev. Std" is the standard deviation over the agents population.

Table 1: Stand-alone, two experiments

Kind	Strength	σ	Dev. Std	σ
Respectful	5578.16	27.83	609.94	29.87
Cheater	4687.98	136.22	1784.28	58.86

In the stand-alone experiment, the normative agents perform better in both strength and fairness.

If agents with different strategies are allowed to co-exist in the same experiment, how will they interact? Utilitarians, using an individualistic algorithm, find no utility in identifying the strategy of the other agents. On the contrary, for normative agents, which sustain all the costs of aggression control to the benefit of the whole population, it is necessary to resort to some retaliation strategies (cfr. Castelfranchi et al., 1998). We have compared four such strategies:

- absolute: keep to the norm with any agent. It is a losing strategy.
- quasi-rational: blind retaliation; attack cheaters, independent of their strength. It is a punitive strategy.
- rational: smart retaliation; attack cheaters only if weaker, that is, behave as a cheater with cheaters.
- super-rational: like quasi-rational but independent of recognition of enemies. Knowledge is acquired at the beginning of the game. It is a control strategy.

If agents with different strategies meet in the same world, we expect normative agents to perform less well, since they bear alone the costs of social control, and act as a cushion population with regard to utilitarians.

In this study, the importance of knowledge transmission, implemented as a sort of gossip, is also tested (but see again Castelfranchi et al., 1998). Respectful agents recognize non-compliant agents (cheaters) when these violate the norm, and record them as such. Therefore, when meeting with one another, normative agents will be allowed to either exchange this information (Xinfo) or not (No Xinfo). One should realize that a normative agent identifying a cheater will always pay a cost: it will certainly lose the match that allows for recognition, since cheaters attack only when they are in advantage.

Results from four sets of 50 experiments, both with a mixed population of half Respectful-half Cheaters, are shown in Figure 1. Quasi-rational (*qsrat*) and rational (*rat*) behaviours are confronted, both without (*No Xinfo*) and with (*Xinfo*) exchange of knowledge.

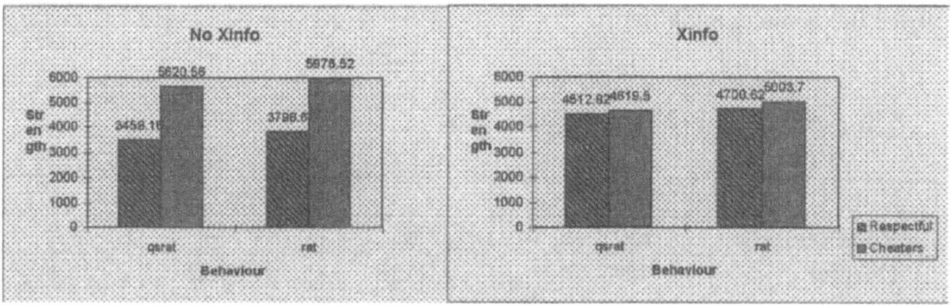

Figure 1: Mixed population, with and without Xinfo.

It is clear that the retaliation strategy of respectful and the spread of knowledge proved crucial in balancing the efficiency of the different behaviors (we will concentrate on the comparison between quasi-rational, or blind retaliation, and rational, or smart retaliation; the other two behaviors are essentially test behaviors). Early recognition of cheaters and the consequent reaction are necessary for respectful agents to perform acceptably. When applying a retaliation strategy, the respectful retain a much better distribution of strength when they exchange information about the reputation of cheaters.

In the following paragraph, we will choose of the parameters for exploration.

15.3 The Parameter Space

Our model, as seen above, is characterized by a quite large parameter space. The parameters we have decided to vary here are:

- Number of iterations
- Global Agents density
- Relative density (respectful/cheaters ratio)

In each of the cases above one of the parameters will vary while keeping the others constant, and the results from the quasi-rational and rational conditions, with and without information exchange, will be compared. When left unspecified, all parameters are intended as default values. We will present here only a small subset of all the explorations, which constitute, in their most detailed form, of a block of 172 tables. The results will be shown as graphs.

15.3.1 Iterations

The ending time, or number of iterations, is checked to look for confirmation of stability in time. For each fixed number of iteration, a different set of experiments is run, in order to obtain statistical indipendence. A change in the results would mean that the results previously published (Castelfranchi et al., 1998; Conte & Castelfranchi, 1995) had been obtained before stability. Even worse, it could reveal intrinsic instability of the model. Luckily, no such indication appears. Results for the quasi-rational behaviour are shown in Figure 2 and Figure 3, respectively without (*no Xinfo*) and with (*Xinfo*) exchange of knowledge. Each figure shows the averages of 70 experiments; each experiment is run with default settings, and with a balanced mix of strategies (50%/50% of Respectful/Cheater). From the difference between the figures we can appreciate the importance of the communication exchange in the case of quasi-rational behavior. Evidently, the normative algorithm works only if knowledge is spread soon enough (at the end, knowledge about cheaters is pratically complete). The model shows stability: stable relative superiority of cheaters, stable linear growth of strength.

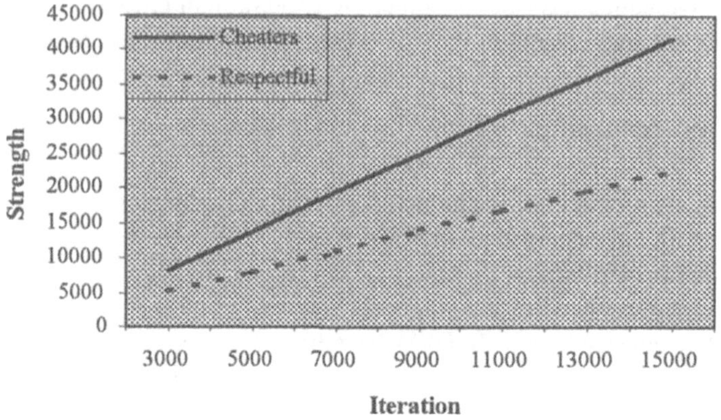

Figure 2: No Xinfo, quasi-rational.

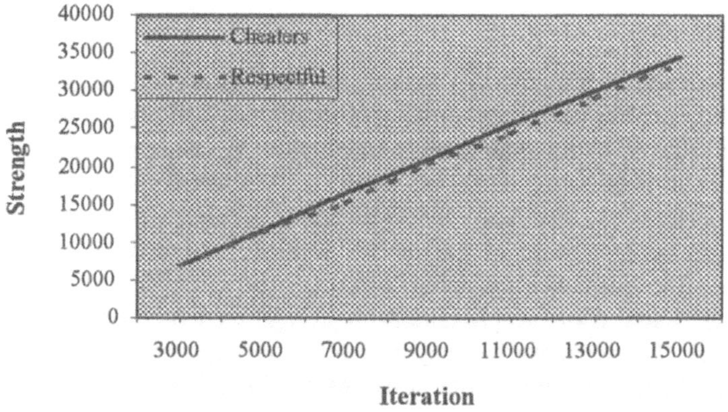

Figure 3: Xinfo, quasi-rational.

With the rational behavior (only the condition with information exchange is shown, the figure is the average of 280 runs) the situation in Figure 4 is very similar to the above: the information spread plays a crucial role, and the respectful obtain scores very close to those obtained by the cheaters, although always lower.

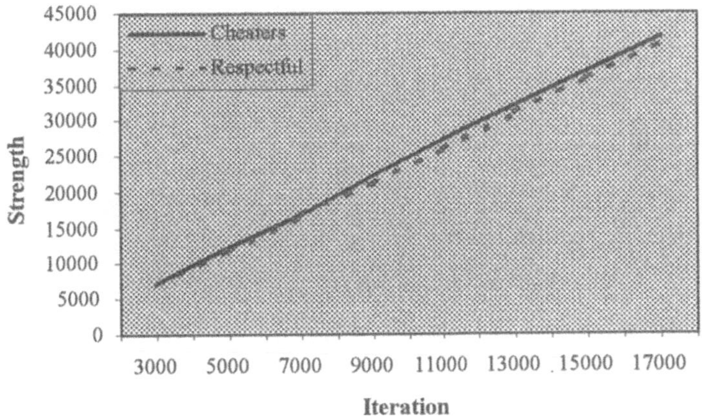

Figure 4: Xinfo, rational

15.3.2 Agent Density

Exploring the global density is even more interesting. This will show how the model reacts to extreme overpopulation and underpopulation situations. Again, in this case, without information exchange there is a stable inferiority of respectful agents. But the respectful step over the cheaters in the extreme high-density situation, as can be seen in the following longer runs (6000 steps experiment, see Figure 5; Each figure plots the results from 70 experiments).

Therefore, the quasi-rational behavior (which is a punitive, blind-retaliation strategy) seems to work better in troublesome conditions (high density means lesser per-agent food).

In the case of the rational strategy, as can be seen from Figure 6, the step-over does not occur, even if the two strategies behave similarly at about 60% global density. While being more effective at our default density of 50%, the rational strategy is not as effective in critical conditions. The increased aggressiveness of quasi-rational agents proves more powerful in densely crowded environments.

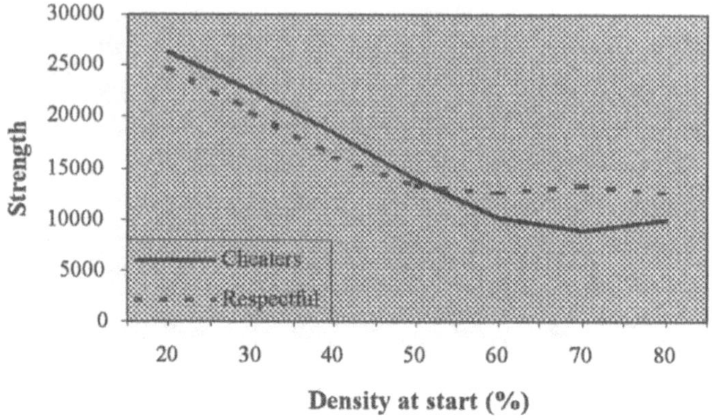

Figure 5: Xinfo, quasi-rational

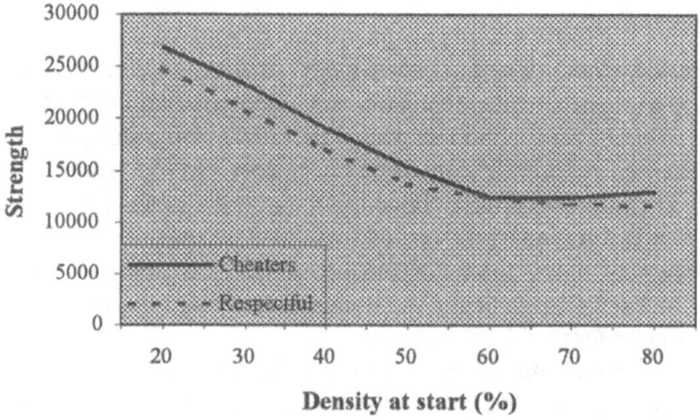

Figure 6: Xinfo, rational

15.3.3 Relative Population Ratio

So far, density is varied while keeping the relative ratio of the two populations constant at 50% (the number of respectful and cheaters was the same). We check now what happens when their relative ratio is varied, while keeping the usual global density of 50%; the total number of agents is fixed at 50. Without information exchange (see Figure 7, from 220 experiments), the distance between cheaters and respectful is wide and stable. However, a population with a lower number of cheaters performs globally better than the opposite. With information exchange (Figure 8; averages from 620 experiments), a pattern of crossing occurs: the strategy in relative majority takes the lead, with the take-over occurring right before reaching the point of equal density.

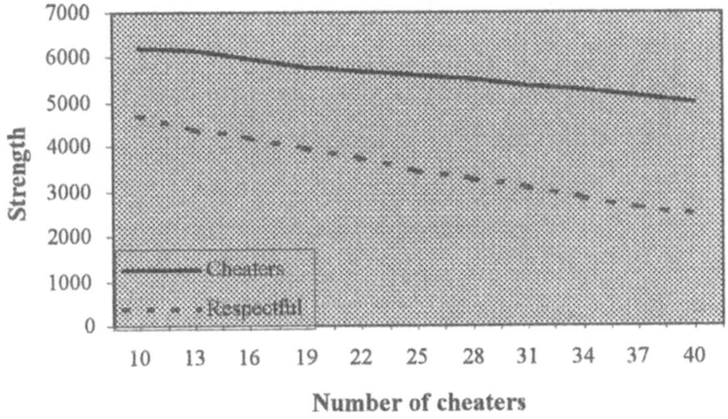

Figure 7: No Xinfo, quasi-rational

For the rational behaviour the results are similar, but the takeover is less marked and takes place only with a much stronger majority (Figure 9, 220 experiments). In our frame, an aggressive behavior pays: the punitive quasi-rational strategy takes more advantage from numerical superiority. How to explain such a result? A possible explanation of this finding lies in the difference between the costs sustained by the agents which receive attacks and those sustained by the attackers. Indeed, in the puinitive strategy, aggressors are bound to inflict useless attacks, since the only difference between this strategy and the rational one lies in resorting to aggression even when the victim is stronger. Why, then, does it perform better than the rational strategy? The reason is probably because quasi-rational inflict to cheaters higher costs than those that cheaters are inflicted by the rational agents. And these are not compensated by the higher costs sustained by quasi-rational with regard to those sustained by the rational. Inflicting aggression, indeed, is always less costly than receiving aggression, even when the aggressor does not obtain the food. This is so because the food eater has sustained costs for obtaining its food unit which may include aggression. On average, those costs plus the ones caused by the current aggression exceed the costs sustained by the aggressor, which will only amount to the costs of the aggression itself plus, possibly, those of moving from one cell to the other.

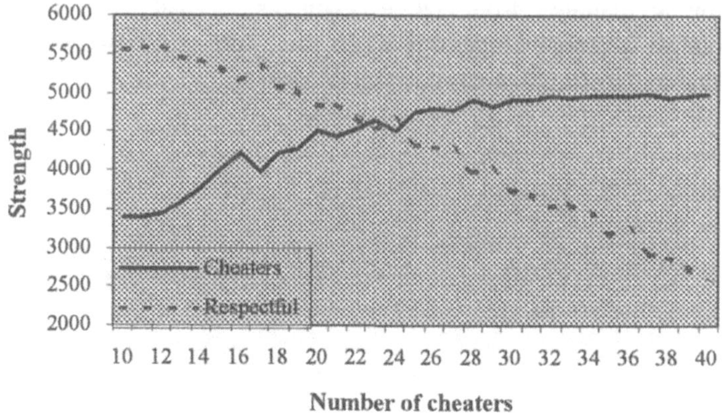

Figure 8: Xinfo, quasi-rational

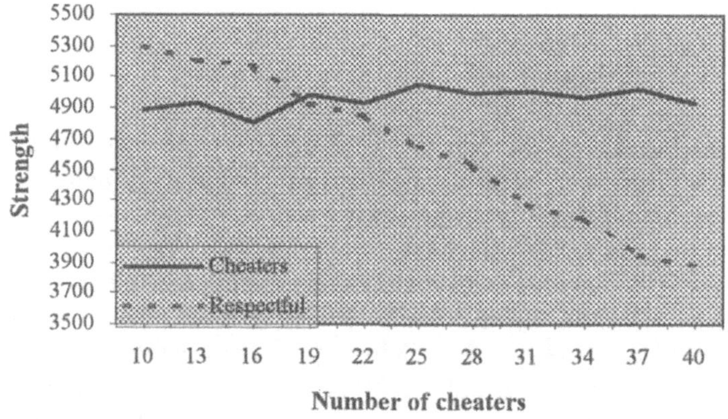

Figure 9: Xinfo, rational

15.4 Conclusion

We have shown here a partial, but relevant exploration of the parameter space of a previously used model (Castelfranchi et al., 1998; Conte & Castelfranchi, 1995), checking for variation of numerical parameters (density, end time) under different algorithms (the agents' behaviors toward a given norm and their retaliation strategies). The results confirm those previously estabilished; moreover, some interesting new features have emerged.

First, normative agents perform better in critical structural conditions (high global density) than in less troublesome ones. This reminds us not only of the Hobbesian grounds for normative regulation - norms are called for in a really tough world, in which personal utility does no longer guarantee the growth of global outcomes - but allows for a specification of the cohesion-compliance theory: normative agents perform better when density is high because gossip spreads more easily and faster.

Secondly, the rapid spread of knowledge is crucial for the functioning of normative systems, when compared with non-normative opponents. In all the experiments whitout information exchange the performance of respectful agents has never been compatible with that of cheaters, not even in dramatic density conditions. In other words, gossip, viewed as an artifice to hasten the spread of knowledge (prejudice) may have an important, overlooked role in rendering the normative behaviour competitive with the non-normative one.

Finally, the rational strategy of retaliation, which proved to be the most efficient in the equal population (50%/50%) condition, has been outperformed by the quasi-rational in any uneven situation. Somewhat surprisingly, our findings show that a punitive strategy, although blind, may be more effective than a more rational one in redistributing the costs of compliance.

References

Binmore, K. (1994): Game-theory and social contract. Vol. 1: Fair playing. MIT Press, Cambridge.

Castelfranchi, C., Conte R. & Paolucci M.(1998): Normative reputation and the costs of compliance. Journal of Artificial Societies and Social Simulation, 1(3). <http://www.soc.surrey.ac.uk/JASSS/1/3/3.html>.

Conte, R. & Castelfranchi, C. (1995): Understanding Norm Functionalities by means of simulation. In: Gilbert, N. & Conte, R. (eds.): Artificial societies: the computer simulation of social life. UCL Press, London.

Eibl-Eibesfeldt, I. (1967/1978): Grundri der vergleichenden Verhaltensforschung: Ethologie. Piper, Munchen.

Homans, G. C. (1974): Social Behaviour. It's elementary forms. Harcourt, N.Y.

Kerr, N. L. (1995). Norms in social dilemmas. In: Schroeder, D.A. (ed.): Social dilemmas. Perspectives on individuals and groups, Praeger, London, 31-49.

Lewis, D. (1969): Convention. Harvard University Press, Cambridge, MA.

Shoham, Y & Tenneholtz, M. (1992a): On the synthesis of useful social laws for artificial agent societies (preliminary report). Proceedings of the AAAI Conference, 276-281.

Shoham, Y. & Tenneholtz, M. (1992b): Emergent conventions in multi agent systems: Initial experimental results and observations. Proceedings of the 3rd International Conference on KR&R. Cambridge, MA, 225-232.

Shoham, Y. & Tennenholtz, M., (1994): Co-Learning and the Evolution of Social Activity. Stanford University.

Trivers, R. L. (1971): The evolution of reciprocal altruism. The Quarterly Review of Biology 46, 35-57.

Ullman-Margalit E. (1977): The emergence of norms. Oxford University Press, Oxford.

Waal, F. de (1982): Chimpanzee politics. Jonathan Cape, London

Walker, A. & Wooldridge, M. (1995): Understanding the emergence of conventions in multi-agent systems. Proc. of the 1st ICMAS Conference, Kaufman, San Mateo, CA.

Part V

Applications

16. Modelling Greening Investors:

Economic Development, Opinion Formation, and Technological Change in a Multilevel Simulation Model

Kai–H. Brassel, Ottmar Edenhofer,[1] Michael Möhring, Klaus G. Troitzsch[2]

The processes of invention, innovation and diffusion of technologies is of enormous importance for assessing environmental policy. The models presented in this paper identify crucial factors of energy consumption, energy productivity, unemployment and labor productivity. Therefore, the models could be used for exploring policy measures like energy taxes, permits and subsidies. The paper also addresses some methodological questions. It is demonstrated how one can build a fairly complex simulation model by a graduated process of refinement and extension. Starting with a mathematically well understood system of differential equations, in the first step a selection process between two technologies is simulated. In the second step, continuos inventions are introduced. In the last step, a multilevel model endogenizes innovations by introducing a population of firms that adjust their decisions about investments through opinion formation. This last model indicates an explicit interaction between the macroeconomic environment and the firms at the microeconomic level.

16.1 Introduction[3]

Both economists and social scientists are in search of an appropriate model for understanding the iterative process of invention, innovation and diffusion of technologies. We begin from the general definition of invention as the creation of things previously non-existing or the exploration of procedures using knowledge in a novel way. Innovation is the first commercial use of the inventions, leading to diffusion and adoption by a growing number of firms in the market. The processes

---

[1] Technical University of Darmstadt, FRG, e-mail: {brassel | ed}@ifs.tu-darmstadt.de.

[2] University of Koblenz•Landau, FRG, e-mail: {moeh | kgt}@informatik.uni-koblenz.de.

[3] For helpful discussions, we are grateful to Carlo Jaeger. We thank Irene Peters and Justin Hopson for outstanding assistance and detailed criticism. We also thank Rolf Gruetzner and an anonymous referee. The usual disclaimers apply.

of invention, innovation and diffusion in economy are not only a thorny issue in social science, but also of enormous importance in environmental policy. Some economists argue that carbon emission reduction policies will become cheaper over time as technological progress takes place. Therefore, ecological innovation could overcome the trade-off between economic growth and sustainable development maintained in many neoclassical models (Nordhaus 1994). The models presented in this paper cannot be used for environmental policy assessment; rather, they identify crucial factors such as energy consumption, energy productivity, unemployment and labor productivity in developed economies where firms, unions and energy suppliers are embedded in a conflictual social network.

In the first step (section 2) we set up a pure macro model to simulate a selection process between two existing technologies. The diffusion of technologies can be appropriately studied in this framework. In section 3 we introduce the invention of new technologies, assuming that each technology will be adopted for production the moment it is invented (thereby identifying invention with innovation). The economy is held in a permanent state of disequilibrium by continuous innovation. It is this state of disequilibrium that ensures the growth of wages and energy prices, and that prevents the profit rate from falling. In section 4 we further refine our model by introducing a population of firms whose investment decisions are in part determined by an opinion formation process. In this model, firms choose between technologies already invented for the purpose of innovation. We explicitly consider the interaction between the macroeconomic environment and the firms at the microeconomic level.

With this sequence of models, we not only explore economic and sociological questions, but also address some methodological issues. In particular, we demonstrate how one can build a fairly complex — but still understandable — simulation model by a graduated process of refinement and extension. Starting with a mathematically well understood system of differential equations, we end up with a multilevel model containing dozens of typed objects interacting on three levels of aggregation. With regard to their implementation, these models, of course, need to satisfy a great variety of requirements. In terms of computing time, it would be most efficient to implement each model within a specifically suited simulation environment. However, we decided to implement all of them within one modelling framework. This approach should make it easier for the reader to compare the models and to understand their design.

The modelling and simulation system used for this purpose is the MIMOSE framework. MIMOSE (*MI*cro- and multilevel *MO*delling *S*oftwar*E*) was developed within a project funded by the Deutsche Forschungsgemeinschaft (DFG). It consists of a model description language and an experimental framework for the simulation of models (Möhring 1996). The main purpose of the MIMOSE project was the development of a modelling language for tackling the special demands of modelling in social science. The language concept of MIMOSE specifically supports the modelling of real life entities (as objects) at different levels of aggregation, integrated into one model, and the stepwise refinement from macro to multilevel models. Special attention was given to the description of *nonlinear, quantitative and qualitative relations, stochastic influences, birth and death processes*, as well as *micro and multilevel models*. The language concept is intended to support the development of structured, homogeneous simulation models, improving the transparency of the "model programming process", facilitating the understanding of model descriptions and the corresponding simulation results.

MIMOSE is also used effectively to teach in modelling and simulation tasks. The approach has been applied to different simulation models concerning a broad range of social science research topics including migration, chaotic behavior, cooperation, gender desegregation and interaction in small groups. Broader applications to environmental and epidemiological models in biology are also possible.

16.2 Social Conflict and Technological Change

16.2.1 Problem

The diffusion of technologies is well studied in economics and social science (Metcalfe 1995; Hall 1994, 266-298). Normally, in diffusion models it is assumed that the adoption process accelerates up to an inflection point at which a critical mass of firms have adopted the innovation. After this point, the adoption process begins to decelerate until it reaches a limiting value. This empirical regularity can be expressed by a more or less S-shape curve or a logistic differential equation. The analysis of diffusion phenomena has been driven by a desire to understand such empirical regularities as these (Karshenas/Stoneman 1995, 269). The general question that requires an answer is: Why does adoption of new technology take time? There are several theoretical concepts that approach this question (for an overview see Karshenas/Stoneman 1995). Some of these explain the diffusion process within the rational actor paradigm. We cannot discuss the problems of this approach here; however, it should not be too difficult to accept that approaches which studied the diffusion process in a more evolutionary framework could have some merits. The evolutionary approach that we propose rejects the assumption of full information and classical rationality and, instead, postulates limited information and bounded rationality.

In this paper we will extend the evolutionary approach of Goodwin (1989; 1991). In our model, one technology is potentially better than the other, but that potential will not be achieved unless effort is put into improving prevailing practice (Silverberg 1984; Silverberg/Lehnert 1993; see also Allen 1975; Allen 1976). As a first approximation, we treat diffusion as a confrontation between one old and one new technology. The most profitable technology generates a surplus that is invested for its further use. Over time, the best practiced technologies will become dominant as the process of selection maps out a diffusion path. In sum, the variation of technologies is modelled completely exogenously and only the selection process is modelled endogenously. In section 3 and 4, we will extend this simple evolutionary approach.

What determines the profitability of a technology? One answer is the movement of factor prices. Unlike in neoclassical models, factor price relations in our model are driven by the social conflict between unions, energy suppliers and firms in a nonlinear manner. In our model social conflict induces technological change. There are already some remarkable proposals to represent *induced technological change* in a conflict model in the tradition of Goodwin. Shah and Desai (1981) offer a model which contains induced technical change, but unfortunately their notion of technological change is disembodied.

The concept of *disembodied technological change* means that technical change happens through an exogenous change in the quality and characteristics of inputs. Some economists call this kind of technical progress "manna from the heavens". However, there is some empirical evidence that technical progress is correlated with investment in capital goods. Therefore, it seems more appropriate to introduce dated capital goods into the model. Each investment adds capital of the current year's vintage to the existing capital stock. The concept of *embodied* technical change is related to the assumed degree of *substitutability* between input factors. Once capital has been invested, the production process is frozen in a specific capital good. We assume that each technology is characterized by fixed productivities. This implies that ex-post substitution is not possible. However, some substitution between capital, labor and energy is possible ex-ante, even if not to such an unrealistic extent as in pure neoclassical models, because substitution requires new investments.

In sum, the ability of firms to improve their capital stock depends on the amount of profits they make, since technical progress needs investment which must be financed through profits. Hence, we model technological change as *embodied* in capital goods. The ability of energy suppliers and unions to capture the profits is a function of their relative bargaining power. The assumption that wages and energy prices are only determined by power relations is open to debate. This does not imply that the scarcity of resources, labor and energy is irrelevant. On the contrary, the increasing scarcity of a resource strengthens the power of macro actors (unions and energy suppliers) who supply it. This latter aspect refers to the other important aspect of our approach: *induced* technological change. In the model presented in the next section, technological change is induced by the persistent danger of increasing factor prices. Therefore, entrepreneurs try to gain more autonomy by improving energy and labor productivity.

16.2.2 Model

Surprisingly, these relatively complex mechanisms can basically be described with a macroeconomic model consisting of only four coupled differential equations. The system's state is determined by the following variables:

- w: labor price (wages),
- p: energy price,
- k1: employment induced by old technology,
- k2: employment induced by new technology.

To keep things this simple, it is assumed that workers consume all of their wages and firms invest all of their profits. Only the production firms invest, and thus accumulate capital; energy suppliers do not. There is also a "reserve army" of unemployed persons who both provide services that "reproduce" labor (e.g., homemakers and child rearers), and act as a source of cheap labor that firms use to hold down wages if possible. Goods are produced when capital is combined with wage labor by entrepreneurs.

It is important to note that throughout economics today the dominant form of modelling is based upon the rational actor paradigm. Here we choose a plausible alternative to this paradigm: a form of *reactive behavior*. The macro actors react to situations of disequilibrium, but they do not have the ability to change the structure of the conflictual network of which they are a part. Firms can improve labor and energy productivity in order to avoid a reduction of their profits. With high unemployment and a low demand for energy, the chance for firms to shift their costs onto workers or onto energy suppliers is improved. In more general terms, if one actor has the power to "externalize" its costs (via wages, energy prices and product prices), then the other actor must bear these costs. For more details on the macroeconomic approach that serves as a starting point for our modelling effort, see Edenhofer/Jaeger 1998.

16.2.3 Implementation

The basic elements of a MIMOSE model are typed objects, which formally represent real entities like individuals, groups, or organizations. An object (type) is structured by a set of attributes, such as the formal representation of real properties, e.g. age or attitude. The values of all object attributes at a given time represent the state of an object. The behavior of a simulation model in MIMOSE is generally defined as the state changes of all objects over time, whereby each object attribute takes a state transition function, which evaluates the attribute value in each simulation step.

To implement the model described above within this framework, one must first decide, which types of objects shall occur. Adopting the macroscopic view, we let the model consist of only one single object, say of type "society". State and behavior of this object are defined by the expressions depicted in figure 1.

Note that in this macro model there exists no explicit representation of the actors (firms, unions, energy suppliers) as objects. Instead, their behavior is implicitly coded into the state transition functions of the overall society-object. Again, refer to Edenhofer/Jaeger 1998 for details.

Even though MIMOSE supports state transition functions defined by means of differential equations, we decided to use an equivalent set of difference equations. MIMOSE supports the formulation of difference equations by two mechanisms.

Lagging the value of an attribute by one time step can easily be done by adding "_1" to the attribute's name, and the length of a time step is accessible via DT (see figure 1). Note that in this paper, reserved words of the MIMOSE model description language are underlined, model parameters are in italic font, and all remaining identifiers denote attributes or object types. (A comprehensive explanation of the MIMOSE language and its predefined functions is given in Möhring et. al. 1994.)

```
society :=
{
/* production and diffusion */
w  : real := w_1 + DT * (-h*w_1 +    f*w_1*k);   /*wages*/
p  : real := p_1 + DT * (-g*p_1 + phi*w_1*k);   /*energy price*/

k1 : real := k1_1 + DT * k1_1/c1 * (1 - w_1/a1 - p_1/b1);
k2 : real := k2_1 + DT * k2_1/c2 * (1 - w_1/a2 - p_1/b2);

k  : real := k1 + k2;
}
```

Figure 1: MIMOSE model of the diffusion of a new technology.

Given the model description in figure 1 and a list of parameters and initial values, a reader with some programming experience should be able to reproduce the results presented in the following section. In doing so, he or she must keep in mind that MIMOSE as a model description language is a functional, not a programming language. The symbol ":=" does *not* assign a value to a location in the computer's memory, but — like a mathematical definition — simply binds the identifier on the left to the expression on the right. Thus, the evaluation sequence of statements (above: k1, k2, k, w, p) is automatically determined by the system, *not* by the succession of statements in the model description. While it might be a little confusing at first for those accustomed to working with ordinary (procedural) programming languages, this feature, which follows the descriptive-language-approach, makes it easier for people without programming skills to read and to write simulation models.

As mentioned above, the behavior of our macro actors is implicitly modelled within the state transition functions of the society-object. More precisely, the parameters *f*, *h*, *phi*, and *g* represent the actors' relative power with respect to each other. The meaning and values of these parameters are as follows:

- *f* (1.0) describes the adjustment of wages to variations in employment,
- *h* (0.8) indicates the relative power of unions that influences the stationary level of employment (increasing values of *h* mean that unions loose power),

- *phi* (0.3) describes the rate of adjustment of energy prices to variations of energy demand *and* wages (we assume that energy suppliers observe the wage policy of unions),
- and *g* (1.0) refers to the power of energy suppliers (an increasing value of *g* means that energy suppliers loose power).

These values will be used in all the models and simulation runs presented in this paper. Another set of parameters describes the characteristics of technologies:

- *a1*, *a2*: labor productivity of old technology 1 and new technology 2,
- *b1*, *b2*: energy productivities,
- *c1*, *c2*: capital productivities.

16.2.4 Results

The mathematical treatment of the model described above does not yield a closed analytical solution. Nevertheless, it allows the analysis of the system's fixed points and its behavior near these fixed points. The Routh-Hurwitz criterion would offer the sufficient and necessary conditions for stability near these fixed points. This criterion also indicates what kind of parameter constellations would violate this conditions (for mathematical details see Edenhofer/Jaeger 1998).

Figure 2 shows some results gained from a simulation run of the diffusion model using the parameters *a1* = 1.0, *b1* = 2.0, *a2* = 2.5, *b2* = 4.0 and initial values w = 0.6, p = 0.6, k1 = 0.95, k2 = 0.01. All simulations reported in this paper were computed with a capital productivity of 4.0 and with two steps per time unit (DT = 0.5).

As figure 2 indicates, at the end of this diffusion process the growth rates of wages w and energy price p become zero after the transition period. This occurs because the new technology *completely* substitutes the old one. During this transition process, the growth rate of wages and the price of energy will lag behind for more than half the period of substitution. Corresponding to the dynamics of wages, profits take most of the benefits of higher productivity in the earlier parts of the transition, but are doubly squeezed when productivity no longer grows. The level of employment follows the business cycle whereby k_i measures the portion of employment in the sector which uses technology i. It is noteworthy that social conflict influences the success of technologies. A technology with somewhat higher energy efficiency cannot penetrate the economy if strong conflicts persist on the labor market. On the other hand, if prices can be increased by energy suppliers, then technologies with higher energy efficiency have a greater chance of prevailing (this assertion is proved in Edenhofer/Jaeger 1998).

In the present model, such diffusion is shaped by social conflict and changing productivities. If firms are endangered by increasing labor costs they prefer to launch labor saving technologies. Therefore, societies which reduce conflicts in labor markets may improve the conditions for launching more energy productive technologies because the danger of increasing labor costs is relatively low. In the

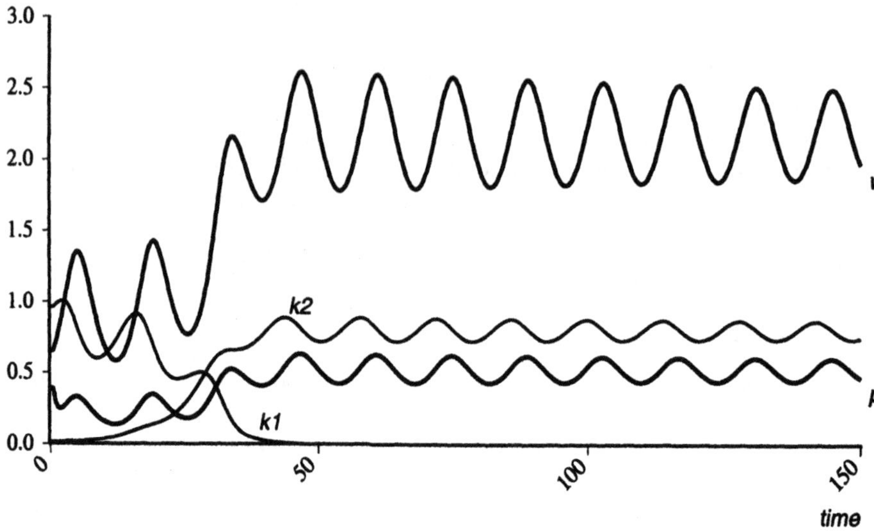

Figure 2: Wages (w), energy price (p), and the diffusion of a new technology (k2).

extremes, social conflicts can be eliminated if unions are destroyed or if the labor market is managed by social consensus. Social conflict emerges because workers increase their wages if demand for labor is high and the reserve army is small. As a result, the profits of the firms are squeezed and their investments and even capital capacity will be reduced. The reduction of investment also causes a lower demand for energy and a fall in energy prices. Because of reduced investment the demand for labor also falls and wages decrease. Now profits increase and after a time lag employment and wages increase, too.

Without further innovation, the rate of profits and the growth rate of labor and energy productivity become zero. This leads quite naturally to a possible explanation of long waves in economic growth. Suppose that an innovation generates new opportunities for profits. After several business cycles, these opportunities fade away. Increasingly, risk averse entrepreneurs are reluctant to finance additional investment because their expected profits shrink. Risk seeking entrepreneurs, however, engage in a search for innovations. As a result, a new innovation wave is generated. However, continuous innovation induces a permanent disequilibrium. Therefore, the analysis of the stability of fixed points is no longer the appropriate method for exploring the behavior of the system; computer simulation now becomes a method in its own right (Troitzsch 1997, 44-46). This is the subject of the next section.

16.3 The Invention of New Technologies

16.3.1 Problem

Inventions and innovations are mainly driven by the supply side of the economy. The demand side, the consumers, play only a passive role. The legacy of this assumption is widely debated in the literature (Matcalfe 1995, 432; Hall 1994). We do not intend to review this debate, but wish to clarify our view of the innovation process. We justify the significance of the supply-side factors by arguing that the growth of the capital stock drives the growth of the various other flows in the system. Under these conditions, capacity adjustments are borne by the labor market. Within firms, all factors of production are fully utilized, and idle capacity shows as unemployment. This assumption captures an important feature of the contemporary economy, but it does so in a simplistic way. During the recession period, firms are assumed to disinvest their capital stock so that full utilization is maintained. Weakening this assumption would lead to increasing returns to scale wherever the capital stock is underutilized. This would be a highly desirable feature of the model on the grounds of realism. However, finding a sensible representation for it constitutes a major research task because increasing returns to scale require careful reflection on the dynamics of oligopolistic markets and because it is far from trivial to define a sensible investment function for such circumstances (for this discussion see Edenhofer/Jaeger 1998).

Obviously, it would be preferable to have a propensity to invest which also depends on effective demand, rates of interest and the like. However, for our first exploration we assume that the demand side is passive, so that according to Goodwin, the so called Say's Law is taken for granted in the long-run (Flaschel 1993; Solow 1990). Dropping Say's Law without further amendments would lead to an insupportable mountain of unsold goods or a collapse from lack of supplies (Goodwin 1989, 22).

In the previous section we omitted the processes of invention and innovation. We modelled the confrontation of two technologies only and observed which technology survived. In reality the diffusion of technologies is overlapped by the invention and innovation of new technologies, so that usually more than two technologies are in use at the same time. At a first pass we assume that invention and innovation can be identified and that both happen exogenously. Therefore, the model is enriched with a stochastic invention process which generates new technologies usually displaying improved productivities. As in the former section, firms are macro actors, whose production activity is described by multiple technologies.

16.3.2 Model

Figure 3 depicts the functional dependencies between the attributes that make up the society-object of the diffusion model of section 2. This "base model" of greening investors only deals with two technologies, a new one and an old one.

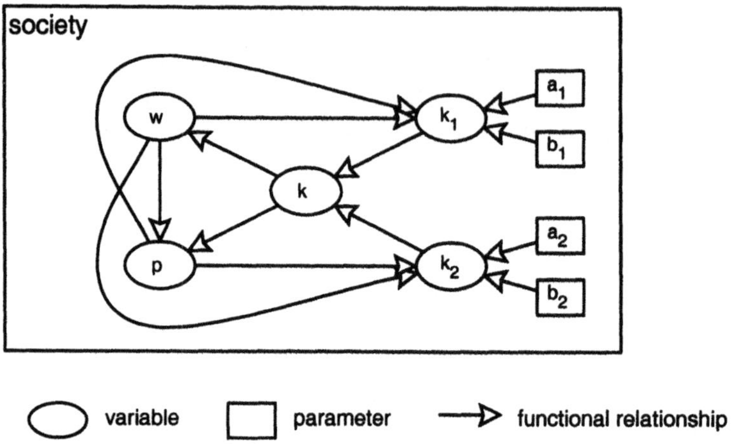

Figure 3: Scheme of the diffusion model (macro-model).

In addition to the diffusion of technologies, the model presented in this section simulates a stochastic process of continuous invention of new technologies. The number of technologies that are simultaneously productive is not restricted. At this stage of development, the model is not intended to distinguish between the process of invention and the process of innovation. As a matter of fact, each technology will enter the state of production just at the moment of invention.

Figure 4 schematically shows the invention model. It contains *two* types of objects, a single society on the one hand and multiple technologies on the other. Thus, the process of production and diffusion is distributed over the two types of objects. Every technology computes *its* capital stock and employment (k) based on its specific labor and energy productivity (a, b) and on current wages (w) and price of energy (p). These global attributes, in turn, are computed in the macro-object "society" based on the aggregated k's of all productive technologies.

The process of invention occurs in parallel. Each time a new technology is invented at the macrolevel, the new labor and energy productivity (newa, newb) is determined stochastically and then registered by the corresponding technology-object which also changes its state from "idle" to "invented". The next section will show how this two-level-model is realized in detail.

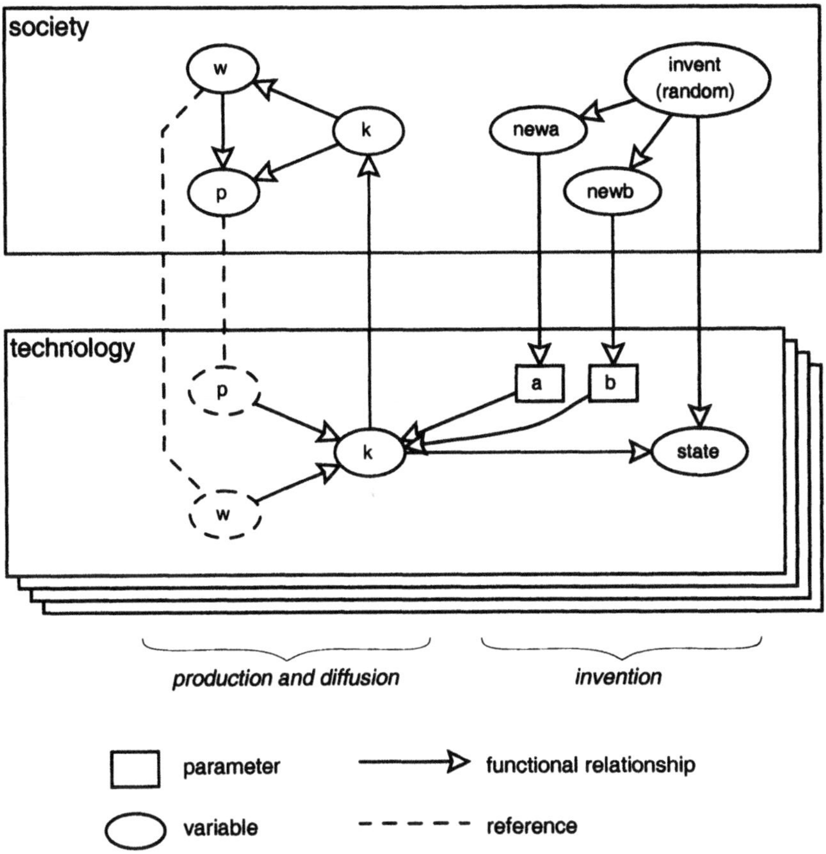

Figure 4: Scheme of the invention model (two-level-model).

16.3.3 Implementation

Let us start with the description of our new object type "technology" presented in figure 5. The first expression in line 4 establishes a reference to the society-object named `soc`. Thus, every technology is able to access the attributes defined on the macrolevel via the common point notation.

For reasons of data output and analysis, each technology-object is existent during the whole simulation run. To distinguish between technologies that are not yet invented, active, and already gone, the attribute `state` keeps track of the technologies' life cycle. A somewhat complicated case-expression is required to handle the events which cause a change of state at specific moments of time. Note that in MIMOSE, a state transition has to be defined for *each* time step; i.e. in the absence of an event, the new state must be explicitly defined to be the old one. In this context the constant attribute `key` is used to uniquely identify technologies.

```
technology :=
{
/* structure */
soc : list of society;

/* state */
state : int
      := case state_1 of
             0: 1 if soc.nInvented=key else 0; /*idle*/
             1: 2;                              /*invented*/
             2: 2 if k_1 >= lowerlim else 3;    /*active*/
             default: 3;                        /*gone*/
         end

/* invention (=innovation) */
a : real := soc.newa if state = 1 else a_1; /*labor prdct.*/
b : real := soc.newb if state = 1 else b_1; /*energy prdct.*/

/* production and diffusion */
k : real
   := case state of
         1: 0.95 if key = 1 else startk;
         2: k_1 + DT * k_1/c * (1 - soc.w_1/a - soc.p_1/b);
         default: 0;
      end;
e : real := k*a/b;
};
```

Figure 5: Invention model: object type "technology".

Each technology now executes its own production function, which is essentially the same as in the diffusion model, if the technology is in its productive state. If it is either idle or gone, the production output is zero. If the technology was just invented, its production output is defined as 0.95 for the first technology or `startk` for the others (the model has to be initialized with one active technology). Labor and energy productivities are local to individual technologies and not global parameters. Productivity values `a` and `b` are defined just at the moment of invention of the technology.

Figure 6 shows the expressions that define the behavior of the society. To get a diffusion process from the start, a second technology is always invented at the first time step. Thereafter, an invention occurs with probability *`progprob`* per time unit. If an invention occurs, `invent` assumes the value 'true' and new productivities for the technology just coming into existence are computed within a bandwidth controlled by parameter *`var`*. (Note that in MIMOSE Boolean expressions result in integer values.)

```
society :=
{
/* structure */
techs : list of technology;

/* invention (=innovation) */
invent    : int := (prob(1,progprob*DT) || (count = 1))
                   && (nInvented_1 < length(techs));
nInvented : int := (nInvented_1+1)
                   if invent else nInvented_1;
level : real := (level_1*progress) if invent else level_1;
newa  : real := uniform(1,level*(1-var),level*(1+var))
                if invent else newa_1;
newb  : real := uniform(2,level*(1-var),level*(1+var))
                if invent else newb_1;

/* production and diffusion */
k  : real := pluslist(techs.k);                    /*labor use*/
w  : real := w_1 + DT * (-h*w_1 +    f*w_1*k);/*wages*/
e  : real := pluslist(techs.e);                    /*energy use*/
p  : real := p_1 + DT * (-g*p_1 + phi*w_1*e);/*energy price*/
};
```

Figure 6: Invention model: object type "society".

Unlike the diffusion model, the invention model requires that capital stock k and energy use e have to be aggregated before they can serve as input to compute wages and price of energy. In order to do this, pluslist simply sums up the corresponding values of all technologies.

16.3.4 Results: How Disequilibrium Makes the Economy Go Around

Figure 7 shows a trajectory of inventions generated by a simulation run with parameters *progprob* = 0.05, *progress* = 1.1, *var* = 0.5 and a starting value of level = 1.0. The reader will not be able to reproduce exactly the same simulation results, because all models contain probability functions. However, using the same parameters and initial values should lead to similar results. The results given in figures 7 to 10 were produced with parameters *startk* = 0.025 and *lowerlim* = 0.02, and with initial values w = 0.6 and p = 0.6. Because the number of technologies was limited to 15 (and because the last invented technology had considerably worse characteristics than its predecessor), the diffusion of new technologies stopped roughly around time 200.

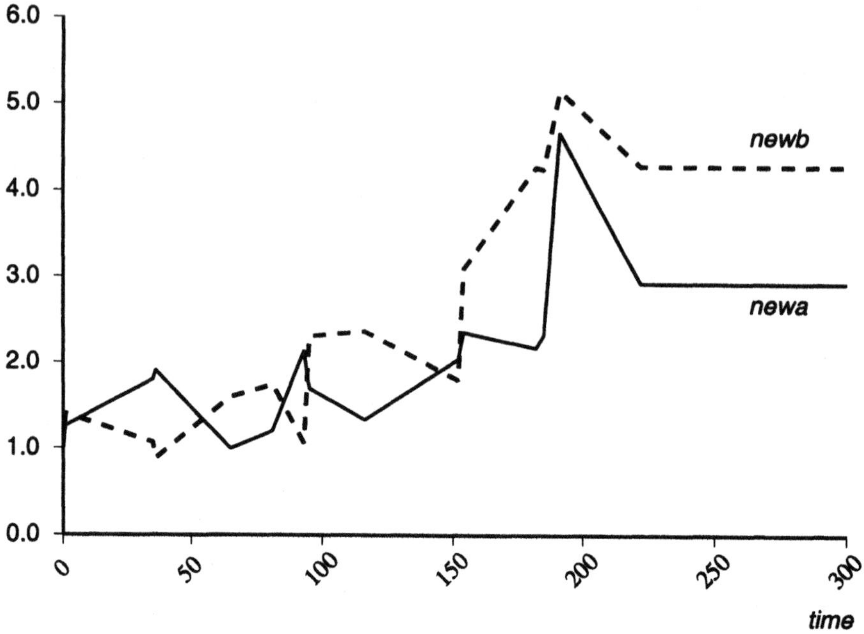

Figure 7: Stochastic invention of technologies.

The invention model offers some first insights into continuous disequilibrium processes. The stochastically driven events of invention and innovation defer the fixed points of the systems. However, the fixed points are never achieved because of innovations and the underlying social conflict in the model. As figure 8 shows, innovation driven growth has no resemblance to a balanced growth process. According to Schumpeter (1934, 207-239) this nonlinear growth process is linked to a structural transformation as indicated in our model. It is this process of disequilibrium which makes the economy go around: inventions and innovations induce a positive trend, especially in wages and price of energy, because of the improvements in labor and energy productivity (figure 9). There is some empirical evidence that there must be long waves in the growth process of capitalist societies (see for interesting empirical details of long waves Nakicenovic 1987, 76-81). There are forces toward a new equilibrium; but that cannot be achieved because of unevenly distributed innovations and their diffusion driven by social conflict.

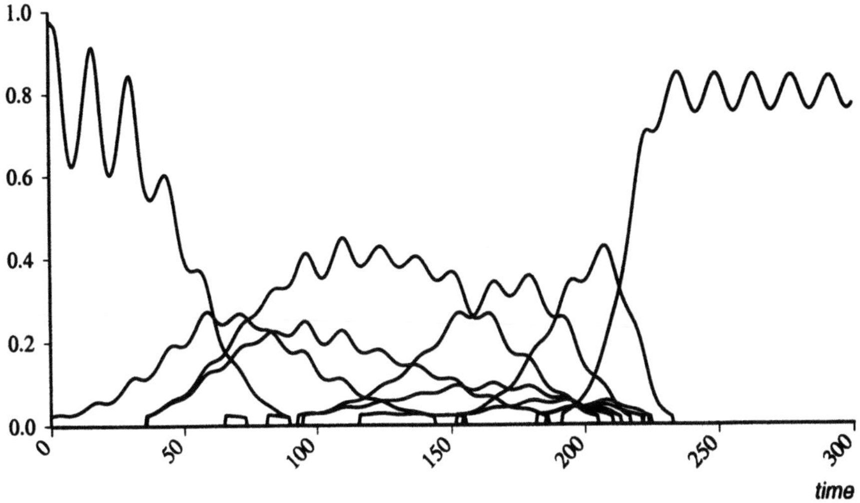

Figure 8: Invention and diffusion of multiple technologies measured by employment (k).

Robert Solow reproduces six stylized facts that, according to Kaldor, characterize the process of economic growth (for references and interpretations see Simon 1986, 24). Our model is able to reproduce the following stylized facts. We will not show all the respective simulation results because the reader should able to reproduce them by implementing the model given above. The seventh fact is supplemented by us and can be observed in our model.

1. Real output per person (or per person-hour) grows at roughly the same rate over fairly long periods of time (which is caused by a positive growth rate of labor and energy productivity; see figure 9 and 10).
2. The stock of real capital grows at roughly the same rate exceeding the rate of growth of labor input (labor input k in our model is more or less constant, see figure 10; the growth of capital can be reproduced by trivial additions to the model listed in figure 5).
3. The rates of growth of real output and the stock of capital goods tend to be about the same, so that the ratio of capital to output shows no systematic trend (capital productivity is also constant in our model).
4. The rate of profit on capital has a horizontal trend (in our model this is an implication of the movement of wages and energy prices; see also figure 10).
5. The rate of growth of output per worker can vary quite considerably from one country to another because of different rates of technical progress (which is modelled here endogenously).

6. Economies with a high share of profits in their income tend to have a high ratio of investment to output (in our model this is an implication of our assumed investment function, see figure 5).
7. In contrast to the other six statements, there is no consensus in the economic community about the role of energy in the growth process. Most of the neoclassical models neglect energy as a factor of production. However, technical progress has made it possible to use ever larger energy potentials and to transform these potentials into commercially useful forms of energy. Therefore, more energy-intensive production processes require an increasing energy flow. The model in this chapter does not necessarily predict an increase of energy use. However, it does indicate that a substantial reduction in energy use e would require a growth rate of energy productivity which exceeds the growth rate of labor productivity (see figure 10).

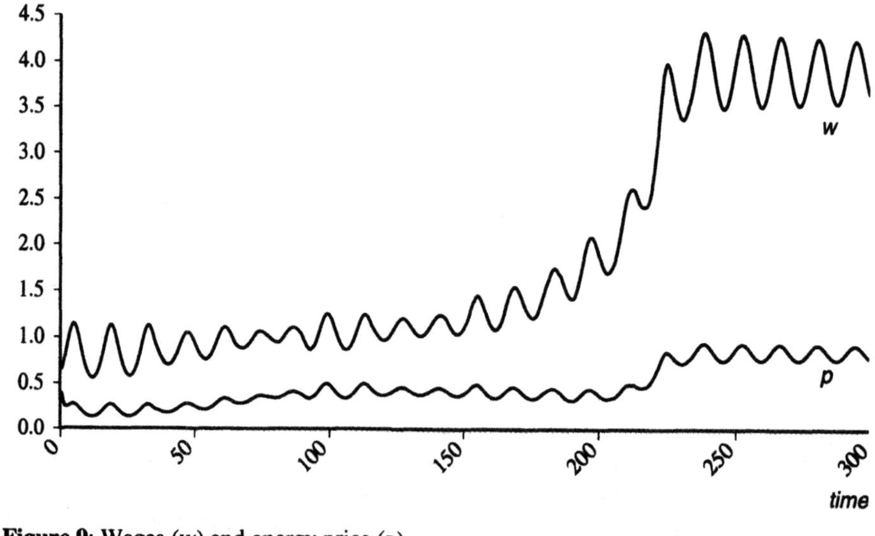

Figure 9: Wages (w) and energy price (p).

There is a lot of disagreement how and if neoclassical models can explain these facts (for a useful discussion, see Scott 1986). However, in the last decade economists inspired by the pioneering work of Schumpeter and Goodwin have achieved considerable progress in understanding the economy as a complex system far from equilibrium. Inventions and innovations continuously destabilize existing economic equilibria. This destabilizing effect of invention and innovation makes an increasing real output per person possible.

We frankly admit that the process of invention and innovation is modelled exogenously. In the next chapter we will offer a possible explanation of the innovation process.

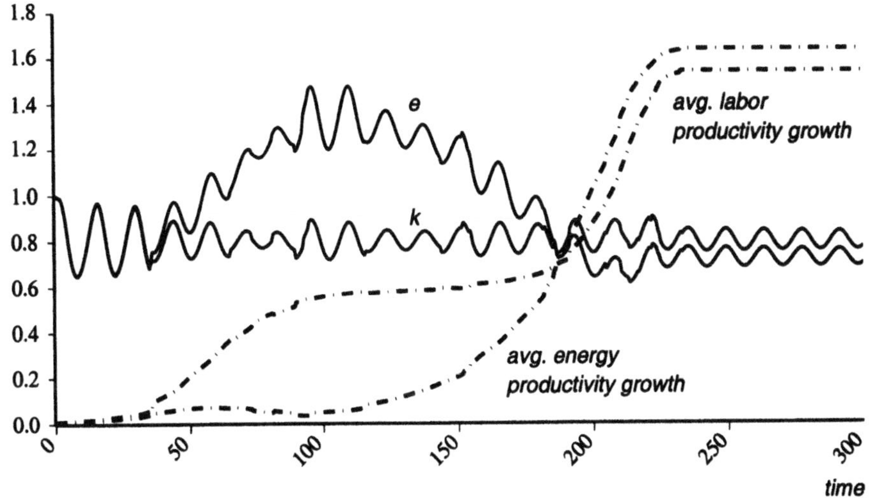

Figure 10: Energy use (e), employment (k), and average productivity growth (weighted according to the technologies' contribution to total output).

16.4 Opinion Formation and Decision Making

16.4.1 Problem

In the previous section, inventions and innovations were amalgamated. This assumption allows us to neglect the role of firms, because every invention leads automatically to an innovation whose success is a product of the diffusion process. However, if innovation should be modeled as a process of its own, then we must clarify who the innovators are and according to which decision procedures they act. We assume that *firms* have to decide when an innovation should take place and which attributes of a technology will promise the greatest success. However, maximizing profits requires actors that are fully informed about future wages and energy prices. Instead, we intend to model actors with bounded rationality. At a first glance, introducing bounded rationality seems like an ad hoc hypothesis. This is not the case under the circumstances of an uncertain future. Therefore, firms need some rules of thumb about how to assess an innovation *ex-ante*. Ex-post the firms are fully aware of whether an innovation has survived the selection pressure of the market.

An appropriate way to model rules of thumb is to endow the firms with suited preferences. In introducing social behavior into our model, we assume that preferences are not fixed, but rather that they are to be determined by an opinion formation process.

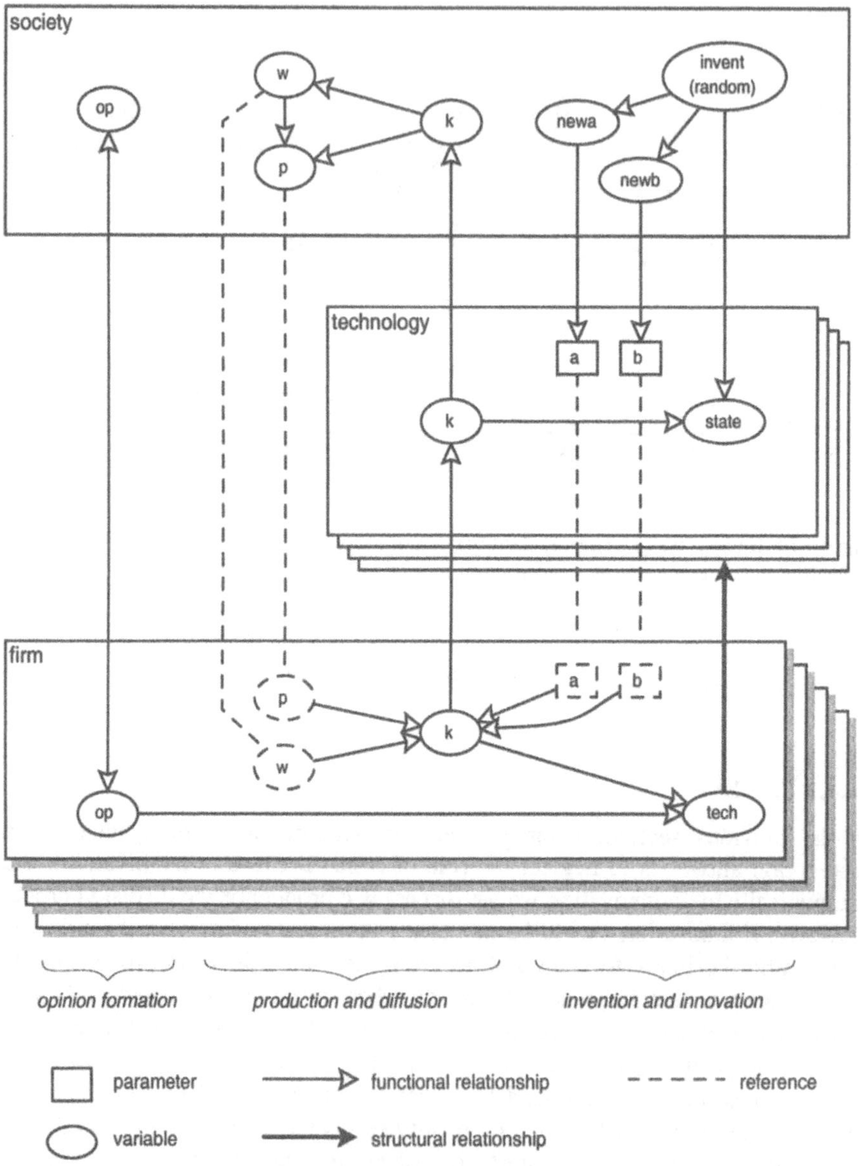

Figure 11: Scheme of the innovation model (three-level model).

16.4.2 Model

Figure 11 shows the architecture of the three-level model. In order to describe opinion formation and innovation we have to extend the invention model by a third type of objects: firms. Once firms are introduced, it is most sensible to let the production process take place in the firms, too. At each moment of time a firm uses exactly one technology for production. A declining profit rate is an indicator of the urgency of innovation. Facing a declining profit rate in a mature market the firms start searching for an innovation. This search can only be successful if the invention process, well known from section 3, has produced enough suited technologies as "candidates".

How should a firm decide which candidate, if any, will perform better in the future? To model firms as actors with bounded rationality, we simply let them rely on their opinion about the relative value of labor productivity compared to energy productivity. A firm favoring energy productivity is said to have a "green" opinion, while a firm with a "blue" opinion favors labor productivity. The opinions of the firms are not constant, rather they can change according to an opinion formation process. At each time step, firms may change their opinion with a certain probability that depends on the overall distribution of opinions in the whole population of firms. Vice versa, this distribution emerges from the aggregation of all individual opinions (see Weidlich/Haag 1983 for details). In this model the social opinion formation process influences the economic process, but note that the opposite is not the case.

16.4.3 Implementation

In the innovation model, the behavior of the society remains roughly the same. Only the macro side of the opinion formation has to be added. Figure 12 shows how the opinions of all firms are aggregated into one macroscopic attribute `op`, that reflects the number of green firms normalized to real numbers between -1.0 (all firms are blue) and 1.0 (all firms are green).

Likewise, only a few changes are necessary to adjust the technology level of our model (see figure 13). The technology-objects have to monitor which firms use them at a moment, so that technology-related simulation output can be produced. Moreover they provide two more attributes `aval` and `bval`, which represent the value that a specific technology has as viewed by a green or a blue firm, respectively. The parameter *`weight`* controls how strong these views differ. So it suffices to set *`weight`* to zero in order to black out the opinion formation process.

```
society :=
{
/* structure */
techs : list of technology;
firms : list of firm;

/* opinions */
op : real := 2 * haselements(firms.op,1) / length(firms) - 1;

/* invention */
... see invention model ...

/* production and diffusion */
... see invention model ...
};
```

Figure 12: Innovation model: object type "society" (macro-level).

```
technology :=
{
/* structure */
soc      : list of society;
allfirms : list of firm;
firms    : list of firm := ... expression to select all
                           firms using this technology ...

/* state */
... see invention model replacing lowerlim by 0 ...

/* invention */
... see invention model ...

/* decision and innovation */
aval : real := power(a, 0.5+weight) * power(b, 0.5-weight);
      /* value of technology for a "blue" firm  */
bval : real := power(a, 0.5-weight) * power(b, 0.5+weight);
      /* value of technology for a "green" firm */

/* production and diffusion */
k : real := pluslist(firms.k_1);
e : real := k*a/b;
};
```

Figure 13: Innovation model: object type "technology" (meso-level).

Last but not least, figure 14 shows the definition of firms that make up the micro-level of the innovation model. Here, we find the expression that determines a firm's opinion. Parameter `pi` indicates a bias for one and against the other opinion, *kappa* denotes how strong the individual opinion is coupled to the overall opinion of the population, and `ny` just serves as a scaling factor.

An innovation takes place if the firm's performance falls short of `lowerlim` and the firm assesses a randomly selected new technology better than the technology currently used. This assessment is done regarding the current opinion of the firm.

```
firm :=
{
/* structure */
soc      : list of society;
alltechs : list of technology;

/* opinions */
/* op = 0 (false): firm prefers high labor productivity */
/* op = 1 (true) : firm prefers high energy productivity */
op : int
   := case op_1 of
      1 :         0 if prob(1,ny * exp(-(pi + kappa*soc.op_1)))
                    else 1;
      default : 1 if prob(1,ny * exp(  pi + kappa*soc.op_1))
                    else 0;
      end;

/* decision and innovation */
qi    : int   := uniform(1,ti_1,soc.nInvented)
                         /* choose any invented technology */
qaval : real := elem(alltechs.aval,qi);
                         /* and get its value for a blue    */
qbval : real := elem(alltechs.bval,qi);
                         /* or a green firm, respectively   */
oaval : real := elem(alltechs.aval,ti_1);
obval : real := elem(alltechs.bval,ti_1);
              /* access values of technology currently in use */
innovate : int
         := (k_1 < lowerlim) &&
            (  (     op  && (qbval > obval))
            || (not(op) && (qaval > oaval)));

ti : int := qi if innovate else ti_1;
                                   /* index of technology to use */

/* production and diffusion */
a : real := elem(alltechs.a,ti);
b : real := elem(alltechs.b,ti);
k : real := k_1 + DT * k_1/c * (1 - soc.w_1/a - soc.p_1/b);
};
```

Figure 14: Innovation model: object type "firm" (micro-level).

16.4.4 Results

As already mentioned, the opinion formation process reduces the complexity and uncertainty about the future for the firms. Figure 15 indicates an opinion formation at the macro-level based on a population of 20 firms and parameters `ny` = 0.02, `pi`

= 0.0, and *kappa* = 0.8. It might be of great interest to analyze simulation results at the level of a single firm. Figure 16 shows the development of a firm acting according (`weight` = 0.2) and not according (`weight` = 0) to its preference, leaving all other parameters constant. The gray areas indicate periods with a preference for energy productive technologies, the white areas those with a preference for labor productive technologies. Obviously, preferences sometimes make a big difference in decision making, with remarkable consequences for performance and employment. The decline forces the firm to search for an new innovation. It also turns out that a firm may have the "wrong" opinion because the profit rate declines dramatically. It must fire employees and its preferences do not allow it to adopt to the wind of change. However, it cannot be concluded from the success of one firm that this could be regarded as a viable indicator for the development of the macroeconomic environment.

Figure 15: Opinion formation process on macro level.

Figure 17 shows that the growth rate of average labor and energy productivity does not change dramatically because of greening investors. At the macroeconomic level the opinion formation process only modulates the economically and technologically driven forces.

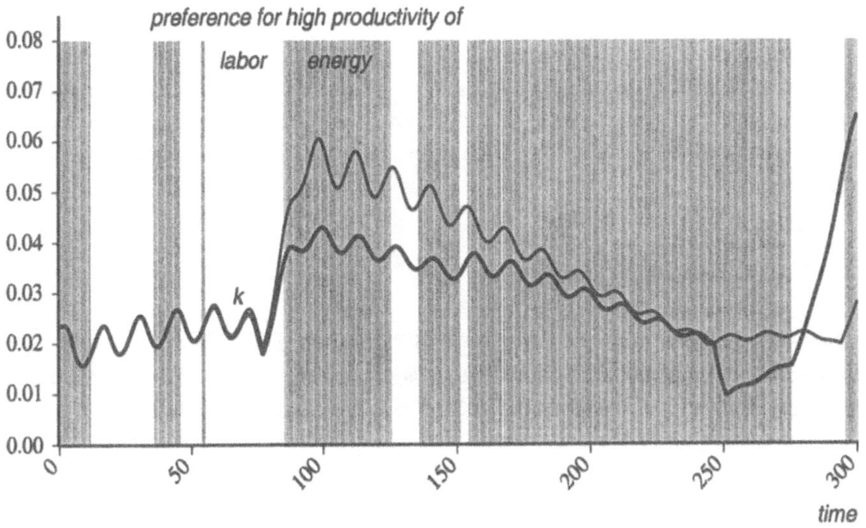

Figure 16: Performance of a firm (measured by k) with and without preferences.

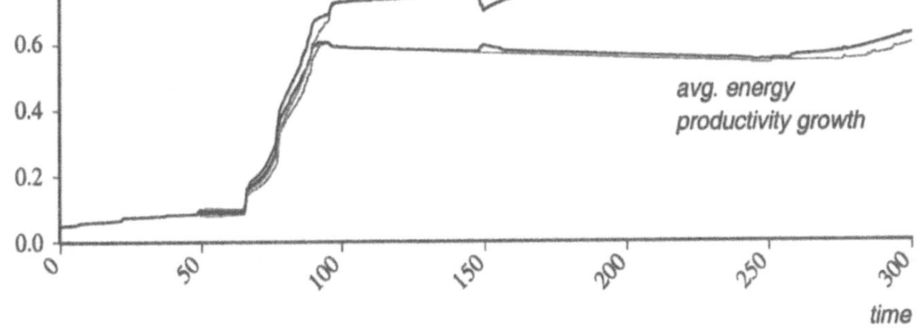

Figure 17: Avrg. growth of energy and labor productivity with/without opinion formation.

One should be wary of concluding that opinion formation, or ideology, does not really matter. The model presented here shows some plausible constellations of parameters. However, we suppose that the results depend on the frequency of inventions occurring in the economy. If inventions occur quite frequently, then the firms will find many options in the pool of technology and therefore the opinion formation process could have a stronger impact. Further research is required to explore whether an increasing invention rate could give opinion a larger importance.

16.5 Concluding Remarks

The models presented here should be understood as a template to enable us to elaborate more sophisticated models to assess some measures of environmental policy (e.g., energy tax, permits and subsidies). We outlined the patterns of growth processes and opinion formation within a disequilibrium model framework. The central idea in our model is that innovation and social conflict induce a permanent state of disequilibrium. Designing a suitable environmental policy needs pluralistic approaches that integrate economic, social and ecological features. The mainstream economy embodied in Computable General Equilibrium Models can be regarded as one reference model with assumptions about rationality and equilibrium which are open to debate. We are convinced that there is a strong need to assess environmental policy in a framework based on bounded rationality and disequilibrium. The models outlined in this paper offer a framework in which such measures can be assessed in a comprehensive and technically sophisticated manner. At this stage the model is highly theoretical and should be further disaggregated to be more realistic. As a next step, the sector of firms should be disaggregated along two dimensions. The first dimension is related to their *rules of behavior* (especially for investment) and the second dimension concerns the *degree of heterogeneity within the population of firms*. Both dimensions will influence the assessment of environmental policy substantially. These aspects deserve a more detailed comment.

The *rules of behavior* in our presented model are quite rudimentary. The firms form no expectations about the future, especially about the development of wages and energy prices. However, expectations about prices could substantially influence their decision for higher energy and labor productivity. Therefore, expectations about prices could be modeled as a feedback loop between the opinion formation process and the selection process of the market. The selection process of the market determines the productivities which also determines wages and energy prices. However, it seems highly plausible that information about future prices will influence investment decisions. Hence, the model could be enriched by opinion formation about price developments which shape the rules for investment decisions. It is inappropriate to conceive of firm behavior in terms of deliberate choices from a broad menu of opportunities for the organization. Even sophisticated problem-solving efforts of an organization always fall into quasi-routine patterns because of a changing environment. In sum, there is a strong request for tools allowing us to model adaptive firms endowed with memory, expectations and goals.

However, it seems quite unrealistic that all firms act according to the same rules of behavior. It seems more plausible that the population of firms can be characterized by a remarkable *degree* of heterogeneity. It seems worthwhile to assess what happens if firms with different rules seek local optima. It is quite easy to devise systems in which each subsystem is optimally adapted to other subsystems around it, but in which the equilibrium is only local (Simon 1996, 46-47). In Darwinian evolution at each incremental step the evolving organism becomes more fit relative to its current environment; but there is no reason for this progress to lead to a global maximum of fitness. Therefore, it may be difficult or impossible to move from one local maximum to another. Hence, from the fact that an economic system is evolving, one cannot conclude that it has reached or is likely to reach a position that bears any resemblance to the equilibria found in the theory of perfect competition. Each firm in the economy is adapting to an environment of other firms evolving simultaneously. The evolution and future of such systems can only be understood from a knowledge of their histories. If the adaptation of business firms and organizations are instances of heuristic search, then we still have to account for the mechanisms that bring the adaptation about. Institutional variation and differentiation occurs much more rapidly and extensively than mutation and selection in the biological world. Nevertheless, the observed inertia of technological and institutional evolution suggests that there are strong stabilizing forces at work (Allen 1988; Hodgson 1993, 257-258). In sum, further research is required to explore what happens within a complex social system with a great degree of heterogeneity.

The last few years have seen a number of remarkable proposals that extend the neoclassical model by emphasizing that technical progress is to a considerable degree endogenous (Nelson 1995, 68). However these neoclassical models are mechanical in the same sense as the old ones. They omit uncertainty and processes of opinion formation as well as surprises in the environment (see also Nelson 1995, 68). Economists who believe in development and evolution apparently do not believe in concepts like global optimization and equilibria. However, there is no obvious consensus in evolutionary economics about how to use the biological analogies of mutation and selection in an appropriate manner.

The multilevel architecture of MIMOSE enabled us to refine the macro model of social conflict and diffusion of technologies to a synergetic one. However, this approach is too limited to describe more complex rules of behavior. In order to model such rules, multi-agent models seem to be more appropriate (Moulin/Caib-Draa 1996). Unfortunately, they do not explicitly support the description of interactions between micro and macro level. We believe that the next step in overcoming these shortcomings should be the development of a tool that supports the integration of multi-agent models on the one hand and different classical approaches on the other (for suggestions and a useful discussion see Brassel et. al. 1997).

With such a framework the outcome of decentralized communication, negotiation, planning and coordination can be analyzed as well as the emergence of a macro level. Therefore it might be interesting to implement some of these multi-agent features in our model, especially to assess certain environmental policy issues. We are aware that this is an ambitious research agenda and that we have taken only small steps in that direction.

References

Allen, P.M. (1975): Darwinian Evolution and a Predator-Prey Ecology. Bulletin of Mathematical Biology 37, 389-405.

Allen, P.M. (1976): Evolution, population dynamics and stability. Proceedings of the National Academy of Sciences USA 73, 655-659.

Allen, P.M. (1988): Evolution, Innovation and Economics. In: Dosi, G., Freeman, C., Nelson, R., Silverberg, G., Soete, L. (eds.): Technical Change and Economic Theory, Pinter Publisher, London, New York, 95-119.

Brassel, K. H., Möhring M., Schumacher E., Troitzsch K.G., (1997): Can Agents Cover all the World? In: Conte, R., Hegselmann, R., Terna, P. (eds.): Simulating Social Phenomena, Springer, Berlin, 41-54.

Edenhofer, O., Jaeger, C. (1998): Power Shifts, The Dynamics of Energy Productivity, Energy Economics 20, 513-537.

Flaschel, P. (1993): Macrodynamics. Peter Lang Verlag, Frankfurt.

Goodwin, R. M. (1989): Essays in Nonlinear Economic Dynamics. Peter Lang Verlag, Frankfurt.

Goodwin, R. M. (1991): Economic Evolution, Chaotic Dynamics and the Marx-Keynes-Schumpeter System. In: Hodgson, G.M., Screpanti, E. (eds.): Rethinking Economics. Markets, Technology and Economic Evolution, Edward Elgar Publishing Limited, Vermont, 138-152.

Hall P. (1994): Innovation, Economics and Evolution. Theorectical Perspectives on Changing Technology in Economic Systems. Harvester Wheatsheaf , New York.

Hodgson, G.M. (1993): Economics and Evolution. Bringing Life Back into Economics. Polity Press, Cambridge.

Karshenas, M., Stoneman, P. (1995): Technological Diffusion. In: Stoneman, P. (ed.): Handbook of the Economics of Innovation and Technological Change, Blackwell, Cambridge, Mass., 265-297.

Metcalfe, St. (1995): The Economic Foundations of Technology Policy: Equilibrium and Evolutionary Perspectives. In: Stoneman, P. (ed.): Handbook of the Economics of Innovation and Technological Change, Blackwell, Cambridge, Mass., 409-512.

Moullin, B., and Caib-Draa, B. (1996): An Overview of Distributed Artificial Intelligence. In: O'Hare, G.M.P., Jennings, N.R. (eds.): Foundations of Distributed Artificial Intelligence, John Wiley et. al., New York, 3-55.
Möhring, M. (1996): Social Science Multilevel Simulation with MIMOSE. In: Troitzsch, K.G., Mueller, U., Gilbert, G.N., Doran, J.E. (eds.): Social Science Microsimulation, Springer, Berlin, 123-137.
Möhring, M., Strotmann, V., Flache, A. (1994): MIMOSE: Eine funktionale Sprache zur Beschreibung und Simulation individuellen Verhaltens in interagierenden Populationen. Universitaet Koblenz-Landau.
Nakicenovic, N. (1987): Technological Substitution and Long Waves in the USA. In: Vasko, T. (ed.): The Long-Wave Debate, Springer, IIASA, Berlin, 76-103.
Nordhaus, W.D. (1994): Managing the Global Commons. The Economics of Climate Change, The MIT Press, Cambridge, Mass.
Nelson, R. (1995): Recent Evolutionary Theorizing About Economic Change. Journal of Economic Literature Vol. XXXIII, 48-90.
Scott, M.F. (1989): A New View of Economic Growth. Oxford University Press, Oxford.
Schumpeter, J. (1934): Theorie der wirtschaftlichen Entwicklung. Eine Untersuchung über Unternehmergewinn, Kapital, Kredit, Zins und Konjunkturzyklus. Duncker und Humblot, Berlin.
Shah, A. and Desai, M. (1981): Growth Cycles with Induced Technical Change. Economic Journal 91, 1006-1010.
Silverberg, G. (1984): Embodied Technical Progress in a Dynamic Economic Model: The Self-Organization Paradigm. In: Goodwin, R.M., Krüger, M., and Vercelli, A. (eds.): Nonlinear Models of Fluctuating Growth, Springer Berlin.
Silverberg, G., Lehnert, D. (1993): Long Waves and 'Evolutionary Chaos' in a simple Schumpeterian Model of Embodied Technical Change. Structural Change and Economic Dynamics Vol 4, No. 1, 9-37.
Simon, H. (1986): On the Behavioral and Rational Foundations of Economic Dynamics. In: Day, R., Eliason, G. (eds.): The Dynamics of Market Economies, North-Holland, Amsterdam, New York, Oxford, 21-41.
Simon, H. (1996): The Science of the Artificial. Third edition. MIT Press, Cambridge, Mass..
Solow, R. (1990): Goodwin's Growth Cycle: Reminiscence and Rumination. In: Velupillai, K. (ed.): Nonlinear and Multisectoral Macrodynamics, Macmillan, London, 31-41.
Troitzsch, K.G. (1997): Social Science Simulation - Origins, Prospects, Purposes. In: Conte, R., Hegselmann, R., Terna, P. (eds): Simulating Social Phenomena, Springer, Berlin, 41-54.
Weidlich, W., Haag, G. (1983): Concepts and Models of Quantitative Sociology. Springer, Berlin.

17. Microsimulating Farm Business Performance

Philip N. Kokic, Ray L. Chambers[1] and Steve Beare[2]

17.1 Introduction

Over the last ten years the Australian Bureau of Agricultural and Resource Economics (ABARE) has developed methodology for forecasting individual level farm business performance which accounts for the dynamic response of farms to macro level price expectations. This paper describes how these methods can be used in microsimulation.

At the core of this methodology is a dynamic supply model (Kokic, Beare, Topp & Tulpulé, 1993). This model predicts an individual farm's response, in terms of the amounts of various products produced, to output price changes according to a profit maximisation criterion and subject to a fixed operating area constraint. The two key stochastic elements in the simulation model are (a) temporal yield variation, which in the Australian context is highly dependent on geographic location, and (b) temporal price variation, which is largely exogenous, since Australia is a price taker, rather than a price setter, in global commodities markets.

In the following section we briefly describe the supply model underpinning the simulations as well as how expectile regression (Newey and Powell, 1987) can be used to disaggregate a farm's individual yield parameters in this model. The spatial smoothing methods in Kokic, Moon, Gooday & Chambers, (1995) are then applied to obtain farm-specific temporal yield distributions which are sampled from using the bootstrap (Efron and Tibshirani, 1993). In Section 3 we describe how these tools were combined to form a microsimulation model of farm incomes. Some conclusions are drawn in Section 4.

17.2 Components of the Microsimulation Model

In this section we briefly describe the main elements of the farm income microsimulation model and their application to AAGIS data. A more detailed descrip-

[1] Department of Social Statistics, University of Southampton, Southampton, UK
[2] Australian Bureau of Agricultural and Resource Economics, Canberra, Australia

tion of the components of the model can be found in Kokic, Chambers and Beare (1997). We begin with an outline of the dynamic supply model which is at the core of the microsimulation process.

17.2.1 Supply Modelling

A farm's profit depends on its level of production and its mix of commodities. Let net farm cash income π be modelled as

$$(1) \quad \pi_{it} = \sum_{j=1}^{m} P_{ijt} Q_{ijt} - C_{it}(Q_{i1t}, \ldots Q_{imt}),$$

where m is the number of commodities, P_{ijt} is the price received for commodity j by farm i in year t, Q_{ijt} is the quantity of commodity j sold by farm i in year t, and C_{it} is the cost function for farm i. Typically unit costs rise more rapidly as output increases (reflecting greater per-unit investment requirements). Thus a cost function of the form

$$C_{it} = b_{i0t} + \sum_{j=1}^{m} b_{ijt} Q_{ijt}^{\mu_{ijt}}$$

for farm i is plausible, where $\mu_{ijt} > 1$.

The area A in which a farm has to operate can also be written as the following function of output quantities:

$$(2) \quad A_i = d_{i0t} + \sum_{j=1}^{m} d_{ijt} Q_{ijt},$$

where d_{ijt}^{-1} is the per unit yield for commodity j on farm i.
Subject to the constraint that the operating area on any farm is fixed, Kokic et. al. (1993) obtained optimum values which maximise the profit function (1). At this optimum total cash costs equals:

$$(3) \quad C_{it} = c_{i0t} + \sum_{j=1}^{m} c_{ijt} Q_{ijt}$$

where

$$(4) \quad c_{ijt} = (P_{ijt} + \lambda_t d_{ijt}) / \mu_{ijt}$$

is the unit cost of producing commodity j, and λ_t is the land and capital investment less depreciation on a per hectare basis. From (1) and (3) it follows that

$$(5) \quad \pi_{it} = \beta_{i0t} + \sum_{j=1}^{m} \beta_{ijt} Q_{ijt}$$

where $\beta_{i0t} = -c_{i0t}$ and $\beta_{ijt} = P_{ijt} - c_{ijt}$, $j > 0$, is the unit net return for commodity j.

Production-price elasticities (derivatives) can also be obtained. These equate to (dropping the subscripts i and t for convenience):

(6) $$\varepsilon_{jj} = \frac{\partial Q_j / \partial P_j}{Q_j / P_j} = (\mu_j - 1)^{-1} P_j (P_j + \lambda d_j)^{-1} (1 - \alpha_j),$$

$$\varepsilon_{jk} = \frac{\partial Q_j / \partial P_k}{Q_j / P_k} = -(\mu_j - 1)^{-1} P_k (P_j + \lambda d_j)^{-1} d_j d_k^{-1} \alpha_k \text{ and } \frac{\partial \lambda}{\partial P_j} = -\alpha_j d_j^{-1}$$

where

$$\alpha_j = d_j^2 Q_j (\mu_j - 1)^{-1} \lambda_j^{-1} \left(\sum_{k=1}^{m} d_k^2 Q_k (\mu_k - 1)^{-1} \lambda_k^{-1} \right)^{-1}.$$

The differential equations (6) give us a method of predicting the production outcome (supply response) of each farm given a forecast price. Furthermore, through (4) the change in unit costs and hence profit can be forecast. In other words, given that farms will always adjust in order to maximise their profit, the behaviour at the micro level is influenced in a specific manner by price changes at the macro level. A method for using finite differencing to solve (6) and hence forecast output and income at the individual farm level is described in Kokic et. al. (1993).

17.2.2 Estimating Farm-Specific Parameters Using Expectile Regression

Since the prices received from each commodity varies from year to year and the cost of production changes according to factors such as water supply, soil quality, management practices, and the type, state and extent of physical structures (such as fences) on the farm, it would be expected that the regression coefficients in (5) vary significantly from one farm to another. Unfortunately variables measuring these effects are not collected in AAGIS, and so they cannot be included in (5). Thus, using ordinary least squares (OLS) regression methods to estimate the parameters in this model will clearly be inadequate if one is interested in the distributional aspects of profit and how these may change if production changes.

The method of expectile regression (Newey and Powell, 1987) can be used to overcome this problem. The expectile normal equations may be written as

$$\sum_i \psi_p\left(y_i - \mathbf{x}_i'\hat{\beta}(p)\right)\mathbf{x}_i = \mathbf{0},$$

where

$$\psi_p(r) = \begin{cases} pr, & \text{if } r > 0, \\ (1-p)r, & \text{otherwise} \end{cases}$$

There will be different expectile regression parameter estimates $\hat{\beta}(p)$ for each value of p.

Expectile regression gives us a method of disaggregating the average behaviour of the overall population of farms (as predicted by OLS regression) into the average behaviour of less and more profitable farms. Let the vector **x** denote the vector of farm output quantities $(Q_1, Q_2, \ldots, Q_m)$. Suppose that for a particular value of p, $\hat{p}_i$ say, the expectile surface $x_i'\hat{\beta}(p_i)$ was to exactly pass through $y_i = \pi_i$. Clearly, if $\hat{p}_i$ was close to one then the ith farm can be seen to be performing relatively efficiently and likewise, if $\hat{p}_i$ is close to zero then the farm would be performing relatively inefficiently. That is, the value $\hat{p}_i$ is a measure of the performance of the ith farm in terms of its profit, its so called technical efficiency (Kokic, Chambers, Breckling and Beare, 1997).

Let $\beta_j(p)$ be the population expectile regression coefficient corresponding to $\hat{\beta}_j(p)$. The model (5) is then effectively replaced by a model of the form

$$\pi_{it} = \beta_0(p_{it}) + \sum_{j=1}^{m} \beta_j(p_{it})Q_{ijt}, \tag{7}$$

where

$$p_{it} = p_i + \eta_{it}$$

and η_{it} is a residual error term. Here p_i is a farm-specific level of performance which depends on unmeasured quantities as described before. The performance measure $\hat{p}_{it}$ obtained by fitting regression expectiles to the survey data are estimates of the random variables p_{it}.

Expectile regression can also be applied to equation (2) to estimate yields. The expectile yield model may be written as

$$A_i = d_0(q_{it}) + \sum_{j=1}^{m} d_j(q_{it})Q_{ijt},$$

where yield performance is measured by $0 \leq q_{it} = q_i + \varepsilon_{it} \leq$. Note that in this case high yield performance corresponds to q_{it} close to zero while low yield performance corresponds to q_{it} close to one.

The q_{it} values in the equation above can be estimated by pooling several year's AAGIS farm survey data together. However, the length of the time series of $\hat{q}_{it}$ for each farm is usually limited to an average of between 5 and 6 observations. This is also the case for the $\hat{p}_{it}$ time series derived from (7).

With AAGIS data, the exact geographic location of each surveyed farm is known. Spatial smoothing techniques can therefore be used to borrow information from neighbouring farms to produce an estimate of the temporal distribution function of the variable of interest for a specific farm (Kokic et. al., 1995). In what follows we denote the estimate of the temporal distribution function for q_{it} by $\hat{F}_i$.

17.3 The Microsimulation Procedure

In this section we describe how to combine the various methods outlined in Section 2 to simulate income. There are several complications involved. In summary these are as follows:

(a) The intended level of production is dependent on expected rather than the realised output prices.
(b) Yield performance is assumed to be statistically independent from one year to the next, and
(c) The level of profitability is dependent on yield.

An effective method for dealing with (a) has not yet been developed. Furthermore only realised, not expected output commodity prices are collected in AAGIS. In order to proceed it has been necessary to assume that the expected and realised output prices are the same. The degree of error introduced by this assumption should however be relatively minor provided the relative change in expected prices is roughly the same as the corresponding relative change in realised prices. In particular, this will be the case when interest is focused on simulating likely alternative income values for farms under current prices.

Assumption (b) is generally not unrealistic as it can be argued that weather conditions from one year to the next at any particular location are essentially statistically independent. If one is unwilling to make such an assumption it may be possible to model the dependencies from one season to the next. However, at least initially, it was not our intention to add this extra degree of complexity to the simulation process.

To deal with (c) the simulation process is broken up into two steps. First income is adjusted for the simulated change in yield and then the yield-adjusted income is forecast on the basis of price changes. The process of simulating a single value of income is described in more detail below. To simplify the presentation the subscript i will be dropped.

The first step is to sample a single value of yield performance, q_{t+1} say, from the estimated temporal distribution function $\hat{F}$. Due to the assumption of independence (b) it is possible to use the simulated value q_{t+1} to determine income in year t+1. This method of sampling from an empirical, or in this case estimated distribution function is often referred to in the statistical literature as the method of bootstrapping, see Efron and Tibshirani (1993). The bootstrap has the advantage that it is a non-parametric method of simulation driven by the actual distribution of the data itself.

After simulating q_{t+1} it is possible to determine yields, $\hat{d}_{j,t+1}^{-1} = \hat{d}_j^{-1}(q_{t+1})$. However, the outputs in year t are based on the realised yields $\hat{d}_{jt}^{-1}$ and so must be adjusted to the simulated value of yield in year t+1. The yield-adjusted outputs are $Q_{jt}^* = Q_{jt}\hat{d}_{jt} / \hat{d}_{j,t+1}$, which continue to satisfy the land area constraint $A = \hat{d}_{0t} + \sum_{j=1}^{m} \hat{d}_{j,t+1} Q_{jt}^*$.

At the same time profit is adjusted for yield by assuming that the profitability measure p_t remains unchanged in the adjustment process, or equivalently the unit net returns $\hat{\beta}_j(p_t)$ remain fixed. This is equivalent to predicting a farm's income (under current output prices) by shifting it along the particular expectile surface that passes through its current value of profit π_t until its output is at the point $(Q^*_{1t},\ldots,Q^*_{mt})$. The process of simulating yield and adjusting income is illustrated in Figure 1 in the case of a single output variable.

Finally, output and unit net returns are projected to year $t+$ using (7) and the differential equations at (6). Denote the resulting simulated value of output for commodity j by $Q_{j,t+1}$ and the corresponding unit net return by $\beta_{j,t+1}$. Profit in year $t+$ is then simulated as

$$\pi_{t+1} = \beta_{0t} + \sum_{j=1}^{m} \beta_{j,t+1} Q_{j,t+1}.$$

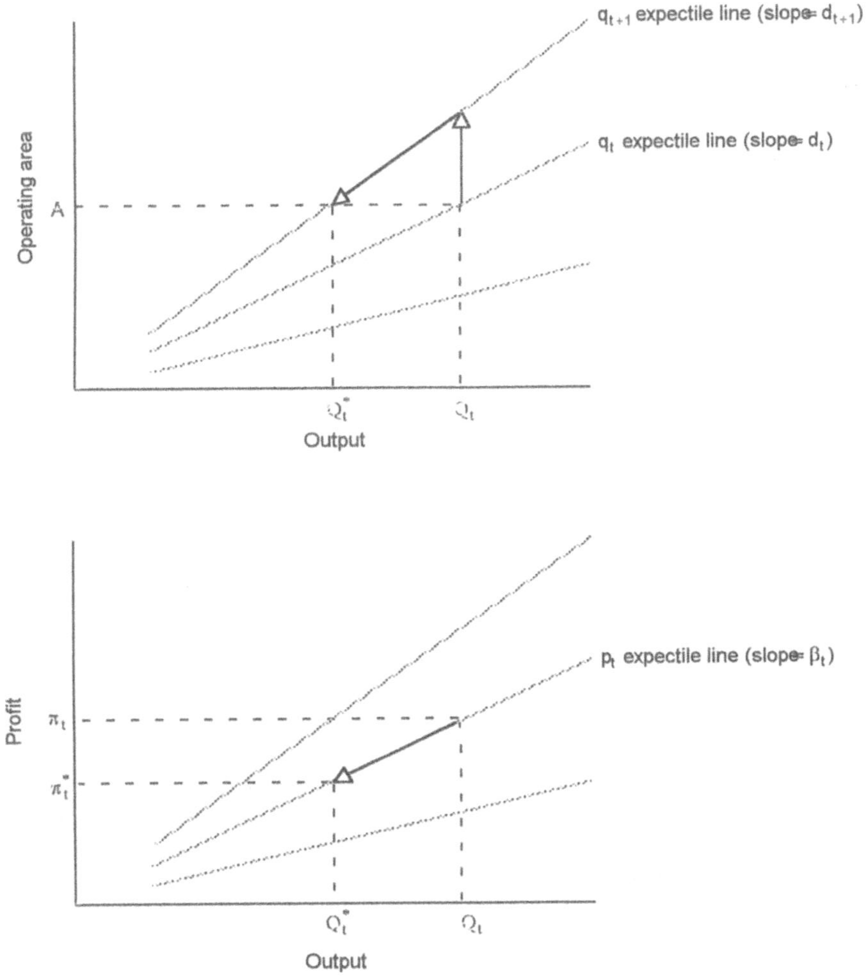

Figure 1: Example of the simulation of a new value of yield

The only difficulty with this approach is determining $\hat{d}_j(q_{t+1})$ as these regression parameters will only be estimated for a finite number of q values. The approach used in Kokic et. al. (1993) was to approximate the function $q \mapsto d_j(q)$ by fitting quadratic splines.

Figure 1 is an example of the simulation of a new value of yield and the adjustment of income prior to projection. In the upper diagram yield is simulated as d_{t+1}^{-1}. Output is then adjusted to Q_t^* by moving along the q_{t+1} expectile line until the land area constraint is met. At the same time the current value of profit π_t is adjusted to π_t^* as illustrated in the lower diagram by shifting along the p_t expectile line.

17.4 Conclusions

Simulation using expectile regression methods offers an alternative, and as can be seen, quite general method of simulating economic microdata. In particular it is not restricted to the assumption of normality, which is often necessary in other approaches. This can be vitally important when considering the tail behaviour of simulated values. Furthermore, expectile regression gives us a method of automatically simulating when the dependent variable changes with each simulation. There is no need for additional modelling of the distribution of the residuals.

References

Efron, B. & Tibshirani, R. J. (1993): An Introduction to the Bootstrap. Chapman & Hall, New York, London.

Kokic, P., Beare, S., Topp, V. & Tulpulé, V. (1993): Australian Broadacre Agriculture. Forecasting Supply at the Farm Level. ABARE Research Report 93.7.

Kokic, P. N., Moon, L., Gooday, J. & Chambers, R. L. (1995): Estimating Temporal Farm Income Distributions Using Spatial Smoothing Techniques, Australian Journal of Statistics 37, 129–143.

Kokic, P., Chambers, R., & Beare S. (1997a): Microsimulating Farm Business Performance. CSDA Technical paper 97.3, Department of Social Statistics, University of Southampton, UK.

Kokic, P., Chambers, R., Breckling, J. & Beare S. (1997b): A Measure of Production Performance, Journal of Business and Economic Statistics 15, 445-451.

Newey, W. K. & Powell, J. L. (1987): Asymmetric Least Squares Estimation and Testing. Econometrica 55, 816–847.

18. Ordering Parameters in the Rule Space of Social Systems

Jürgen Klüver, Jörn Schmidt & Ralph Kier

18.1 Introduction

If there is anything upon which theoretical sociologists might agree then it may be a definition of sociology in terms of social rules: The main task of sociology is the analysis of social rules. Individual actors interact according to specific (social) rules; actors, interactions and rules of interactions constitute social systems, which generate their special dynamics. According to this definition social processes are understood if and only if the rules could be reconstructed by which actors performed their actions. These actions yield specific consequences whether intended or not (Knorr-Cetina 1981); these consequences generate further actions with successive consequences and so on. Therefore the dynamics of social systems is nothing else than the consequences of social actions.

Of course the great sociological theoreticians did not use exactly these terms, with the exceptions perhaps of theoreticians of social systems like Parsons and Luhmann. Yet the firm belief that the concept of social rules is to be placed at the centre of social studies is shared by authors as different as Marx, Weber (e.g. Weber 1907/1982), Parsons, Giddens, Lévi-Strauss or Bourdieu (cf. Bourdieu 1987). It is especially the concept of social rules that can integrate the different traditions of micro and macro sociology (Alexander et al. 1987). Micro sociology emphasises the importance of individual actors and defines society as the result of their actions; macro sociology on the other hand analyses society "as a whole" (Adorno) and looks upon actors as being dependent on the social structures. If one changes the micro sociological focus from actors and actions to the social rules which actors are always following and if social systems are defined by the rules which constitute their dynamics, i.e. the consequences of actions and the successors of actions, then it is immediately possible to unify the different schools of thought in the definition of sociology sketched above. The differences between them become merely those of different concepts for the same thing and of emphasising different methodical procedures. So the *general task* of sociology is the search for *general* features of social rule systems..

Obviously it is impossible to analyse all the very different social systems which history contains and their rule systems. It is equally impossible to model each special social system and to analyse its rule system by investigating the formal model. As there is no sociological classification scheme for social systems sociologists can agree on, one has to look for formal criteria: Instead of analysing each social system in a special model it is possible to investigate *classes of systems.* These classes are characterised by the *ordering or control parameters of their rule systems.* By looking for the parameters of the rule systems one can predict the *general* behaviour (the dynamics) of the system - having regard to the initial states, of course. Therefore the central task of theoretical sociology becomes the investigation of ordering parameters in the rule spaces of social systems; if there are such things as universals or a universal grammar of social action respectively they must be found in such properties of the rule systems of social systems.

The search for general characteristics of social rule systems is, by the way, very similar to the search for grammatical universals in linguistics (Chomsky 1965): As the grammatical universals constitute the universal grammar, universal qualities of social rule systems (may) constitute a universal grammar of social action (see below).

Of course it is an empirical question whether particular social systems can be characterised by specific control parameters. The first task, however, must be to look for *general models of social systems* and to investigate the control parameters which regulate the dynamics and/or the adaptive behaviour of these models. As important work has been done in this respect in the context of Artificial Life (AL) and the investigations about complex adaptive systems following the approach of the Santa Fé Institute, we shall first summarise some results of these investigations.[1]

18.2 Ordering Parameters in Cellular Automata and Boolean Networks

Ordering parameters have been analysed above all rather thoroughly with regard to cellular automata (CA) and Boolean networks (BN). The concentration on CAs and BNs is no restriction of universality because both CAs and BNs are potential universal Turing machines (Berlekamp et al. 1982). According to the Church-Turing Hypothesis this means that every computable function can be computed by a universal Turing machine and therefore by an appropriate CA or BN. For the purposes of simulation, it follows that every system that can be modelled by a formal (computational) system at all can be modelled with a CA or a BN.

CAs and BNs, which are logically equivalent, are discrete systems. A CA consists of a grid of cells, which are in different states. The state of one cell depends on the states of the cells of the so-called neighbourhood and specific rules of transition: If the cells of the neighbourhood of a specific cell are in a certain

[1] These introductory considerations about social rules, social systems and their general properties are discussed in more detail in Kluever 1999.

state, then the state of the cell will pass over to a specified state. A famous example of a CA is the "Game of Life" by Conway (Berlekamp et al. 1982). A cell can be in one of two states - 1 or 0, dead or alive, on or off; the neighbourhood consists of four cells at the sides of the cell (which is represented as a square); this is the so-called von Neumann neighbourhood. One of the transition rules which is especially important goes as follows: If exactly three cells in the neighbourhood of a specific cell are in the state of 1 and if the cell itself is in the state of 0, then the cell passes over to the state of 1.[2]

BNs were first analysed thoroughly by Kauffman (1993). They are artificial networks in which the units - the cells as one can say too - are connected by "logical functions". If one cell is connected with exactly two other cells, then the connecting functions are the well known functions of propositional logics like "and", "or", "if-when", "XOR = either a or b, but not both together" and so on. As G. Boole in the 19th century analysed these logical functions mathematically these functions are also called Boolean functions - hence the name for BNs.

A little example shall illustrate the procedures of BNs. Let us assume a BN with three cells a, b and c. All cells are in one of two possible states 1 or 0, which can be understood as "true" and "false". The logical or Boolean functions of our BN are defined by a "truth table":

a b c	a c b	b c a
1 1 1	1 1 1	1 1 1
1 0 0	1 0 1	1 0 0
0 1 0	0 1 1	0 1 0
0 0 0	0 0 0	0 0 0

This is to be read as follows (we explain only the first table): If a and b are in state 1, then c transforms to the state 1; if a has state 1 and b has state 0, then c transforms to state 0; if a has state 0 and b has state 1, then c goes into state 0; if a and b are in state 0, then c goes into state 0. The rules hold regardless of the state of c before a and b "operate" on c. Obviously the function (a b c) is the "and-function" (logical conjunction) as is the function (b c a); the function (a c b) is the "or-function" (logical disjunction).

A succession of states, i.e. a dynamics of our BN is generated from an initial state at the time *t*:

	a b c
t	1 1 1
t+1	1 1 1
t+2	1 1 1

[2] Informal introductions into CAs are given by Levy (1992) and Poundstone (1987).

In other words, the initial state (1 1 1) is an attractor of period zero. Another dynamics is generated by another initial state:

	a b c
t	0 1 1
t+1	1 1 0
t+2	0 1 1
t+3	1 1 0

In this case, both states are attractors of period one, which means that the systems oscillate between the two states. Our little system has only attractors of period 0 or 1, which is easily verified. Exchanging conjunction and disjunction yields analogous dynamics.

The potential universality of CAs and BNs of course allows dynamics with attractors of arbitrary periods. The logical equivalence of CAs and BNs follows from the simple consideration that the cells of the neighbourhood in a CA are the same as the cells in a BN that operates on another cell. The transition rules of a CA are often expressed in tables like the tables for our BN; Wuensche (1994) even calls BNs generalised CAs.

Whether the dynamics of a CA or BN contains attractors of long periods or only such simple attractors as our little example depends on the control or ordering parameters of their rule systems. It does not depend, as one could assume, on the number of the possible states of the cells or the magnitude of the system. Well known examples of such control parameters are:

(a) The P-parameter (Weissbuch and Derrida) with respect to BNs and the equivalent λ-parameter for CAs (Langton 1992). Both parameters measure the distribution of cell states that result from an application of the transition function(s). The values of the P-parameter are defined in the interval from 0.5 to 1; $P = 1$ means that only one state value is reached by the logical functions; $P = 0.5$ means that all values are evenly generated. For example: In a BN the P-value of the logical disjunction is $P = 0.75$, according to the well known truth table of the disjunction; for the XOR-function we have $P = 0.5$, because the XOR-function distributes the truth-values uniformly. The λ-parameter ranges from 0 to 1. If the rule system of a BN or CA has $P = 1$ (respectively $\lambda = 0$), the dynamics of the CA or BN is characterised by a fixpoint attractor (period zero), $P = 0.5$ generates chaotic behaviour of the system (Kauffman 1993). (The equivalent value of λ is skipped here because it can be expressed only in the formal definition of λ.)

(b) The K-parameter, i.e. the number of "wiring" connections in a BN (the number of variables in a logical function) or the size of the neighbourhood in a CA (Kauffman 1993). $K = 2$ yields dynamics with simple attractors of small periods (a high order of the system). $K > 3$ yields chaotic behaviour.

(c) The number of canalizing functions in a BN or CA: A canalizing (Boolean) function is defined by the property that it is sufficient to know the value of one variable in order to know the result of the function. For example the logical implication "→" is a canalizing function because in the case a → b and a = 0, the whole function gets the value 1, regardless of the value of b (this is the old logical principle "ex falso quodlibet"). XOR is not a canalizing function, as one can see immediately from its truth table: It is always necessary to know the values of both variables in order to know the state value of the dependent variable. The more canalizing functions a BN contains, the more "orderly" is the behaviour, i.e. there are only simple attractors with small periods. (To be more exact: Kauffman (1993) defines this parameter as the ratio of the number of canalizing functions and the number N of the elements of the system.)

(d) The Z-parameter (Wuensche and Lesser 1992) which is similar to the measure of the "temporality" of a CA or BN. It is defined by the probability of determining the value of one cell in a *previous state of* the CA if one knows the actual state and the value of at least one cell in the neighbourhood in the previous state. For example, let us take a Boolean net with K = 2 and investigate three units a, b and c. The logical function f(a,b) = c shall be the logical implication (a→b) = c. In more traditional terms c is the truth value of the logical implication a → b. If c = 1 (true in the actual state), and a = 0 (false) in the previous state, then the value of b in the previous state can only be determined with a probability p = 0.5 (both values of b are possible). So Z = 0.5 for the logical implication. If Z = 1, the behaviour of the BN or CA is chaotic, if Z = 0, there are only point attractors. (In contrast to Wuensche and Lesser we speak of the temporality of the system because here the reversibility of the rule systems become important; for example the reversible Fredkin gates always have Z = 1).

(e) Quite another ordering parameter is given by the b-value in the pay-off-matrix of the iterated prisoner's dilemma (IPD) when modelled with a CA and a matrix of

	C	D
C	1	0
D	b	0

where *b* is the "temptation to defect", i.e. the reward of an aggressive player against a co-operative player (Nowak and May 1993). Nowak and May obtained the results that for b <1.8 and b > 2 the dynamics of the system is characterised by attractors with small periods; the region between 1.8 and 2 gives rise to chaotic trajectories.

(f) Another parameter v has been discovered recently by Jörn Schmidt and Juergen Kluever. It measures a feature of the topology of a BN. V defines a measure of the effects of the units of a BN, that is whether all units act upon the same or a different number of units. v = 0, that is all units act upon the same number of units, defines rule systems with complicated dynamics. By increasing the value of v, i.e. different units act upon different numbers of units, one gets very simple dynamics (for details see Kluever 1999).

The parameters give only statistical insights insofar as there exist exceptions; moreover they can compensate each other: E.g. K = 5 gives chaotic behaviour, but having a lot of canalizing functions (with regard to N) increases the order. These correlations are not well understood because researchers concentrate on their own parameters. Probably the connecting link between the different parameters is the number of canalizing functions as can be seen for example in the case of the Z-parameter: The canalizing function "$\rightarrow$" yields a rather low Z-value. As there are 16 Boolean functions with K = 2 and 14 of them are canalizing functions the high order of BNs or CAs with K = 2 can be easily explained by the fact that each randomly generated BN with K = 2 will contain a lot of canalizing functions. But just now there can be said not much more about these correlations than is expressed by the laconic remarks of Kauffman: "Other features, such as the structure of the wiring diagram among the Boolean variables ... are also control parameters. ... Undoubtedly, further control parameters will be discovered." (1993, 219)

18.3 Meta Parameters in Adaptive Systems

CAs, BNs and systems like these are *self organising systems,* i.e. they generate their dynamics according only to their specific rules and their initial states. They are especially well suited for modelling social self organization, based upon the concept of social rules: that is because they realise a methodical *bottom up approach* (Langton 1988), meaning the modelling of a dynamical system exclusively by local rules of interaction. That is why CAs are often used for modelling social processes (Schelling 1971; Hegselmann 1996; Nowak and Lewenstein 1996; Klüver 1995). Yet despite many advantages CAs and BNs are severely restricted, because they are self organizing systems but not *adaptive.* They do not change their rules or parameters but only their states; rule or parameter changing is only possible by external manipulation - e.g. by a human programmer.

Social systems on the other hand are without doubt self organising systems, but they are also adaptive systems. This means that they are able to change their rules and parameters in reaction to demands of their respective environment. In particular, they can change the ordering parameters of their rule systems. Therefore, an adaptive system is not characterised by one specific value of the control parameters, but by its *variability,* i.e. its specific capability to change its rules. This is a *meta-feature* of adaptive systems.[3]

Logically speaking adaptive systems do not have one set of rules on one logical level but at least two different sets of rules: The one set of rules constitutes the *base system,* i.e. the whole of rule governed interactions and actors of the "real" system. The second set contains the *meta rules of* the system, i.e. the rules by which the base rules are changed. That is why we call adaptive systems *hybrid systems.* Their rule systems are composed of two ore more different rule sets. For example, the learning rule of an artificial neural net is a meta rule insofar as it changes the weight values of the net; these values are themselves simple rules which generate the specific dynamics of the net. As the learning rule depends on the special problem, posed by the environment, the artificial net is an adaptive system with base rules and meta rules. Even more obvious is the example of learning BNs (Wuensche 1994): A kind of learning rule in the fashion of feed back networks changes the Boolean functions, i.e. the wiring connections, and it can change the value of K also.

It is important to recognise that meta rules are not merely logical constructs for building formal hybrid systems like the computer programs described below. Each social system has meta rules as well as rules of interactions, i.e. the base rules. A political system, for example those of western modern societies, contains many rules of interactions which may or may not be written down in form of laws and administrative regulations; the rules of procedures of the parliaments are then nothing else than meta rules which define the way rules of interactions may be changed. Even apparently informal social groups like the play groups of children not only have rather strict rules of interactions but also well defined meta rules which regulate the changing of play rules, social roles in the group and so on (Schwartzman 1978). These examples could easily be multiplied.

If one tries to transfer the concept of ordering parameters to hybrid systems then one has to look for *meta parameters,* i.e. parameters on the meta level of the whole rule system (meaning base rules and meta rules together). Yet it is important to note that meta parameters cannot be characterised by certain dynamics of the system but by *the adaptive success the* system obtains with specific values of the meta parameters. It does not make much sense to speak about "the" dynamics of an adaptive system, when the dynamics is changed permanently by the meta rules. For example, we constructed several hybrid systems consisting of CAs coupled with a Genetic Algorithm (GA). The task of the GA is to change some rules and system parameters according to specific environmental conditions (Klüver 1996). The dynamics of the CAs alone are usually rather simple, meaning that their trajectories contain only attractors of period zero or one. With permanent operations of the GA *on* the CA the whole system exhibits trajectories

---

[3] Farmer (1990) calls systems which are able to change their rules and/or parameters "metadynamical" systems. Being a physicist, he does not use the term, rule changing, but "architecture changing", which is not very well defined.

which are "at the edge of chaos" (Langton 1992) or even beyond that edge. The trajectories of hybrid systems are independent of any initial states - in contrast to "pure" self organising systems - and they are also independent of any initial rule system, provided the changing capacity of the whole system is sufficiently large. Therefore the interesting question with respect to adaptive (hybrid) systems is, are there meta parameters that determine the adaptive success of these systems? After all the adaptability or *variability* of adaptive systems, meaning the rule changing capacity, has the only objective of adapting the system successfully to specific environmental conditions.[4]

Our investigations in this field are still going on and we get still surprised by new results with each new investigation; that is why both the definitions of certain meta parameters and first results are to be taken very tentatively.

Before we start to describe some meta parameters that we investigated, another important difference between the control parameters of self organising systems and meta parameters of adaptive (hybrid) systems must be mentioned:

Control parameters of self organising systems like CAs or BNs are characterised by determining unequivocally the dynamics of the respective system. They are a feature of the rule system and that means that a specific value of the parameter measures the type of trajectory of the system. So the main question with regard to control parameters is, which value of a control parameter measures a specific dynamics? In the terms of classifying CA-dynamics one can also ask which value of a specific control parameter puts the CA into a specific Wolfram class (Wolfram 1986; Langton 1992). The corresponding question with respect to adaptive systems could therefore be, which values of specific meta parameters give the adaptive system a maximal adaptive success and which values minimise the success?

Unfortunately, the question must be a bit more complicated. There is some evidence that the adaptive success of an adaptive system is not necessarily determined by some specific fixed values of the meta parameters - with some exceptions - but by the ability of the system *to change the values of its own meta parameters.* To understand this let us have a look at a rather well known meta parameter, namely the Delta factor in the so called Delta rule of artificial neural nets:

Most of the learning rules in supervised learning of artificial neural nets are "Hebbian learning rules" (Farmer 1990), which means that they are influenced by the famous model of neural learning by Hebb (1949). One of these is the Delta rule, which is usually stated

$$w_{ij}(t+1) = w_{ij}(t) + (s/n * \delta * o),$$

---

[4] Though the main question with regard to adaptive systems is as stated above it cannot be denied that it would be worthwhile to investigate the specific dynamics of hybrid systems also. This is especially the case with hybrid systems - like all real adaptive systems - which do not change their rules permanently but operate some time without rule changing (see below).

whereby w_{ij} (t) is the weight of the connection from neuron j to neuron i at time t, s is a constant "learning rate", n is the number of neurons, o is the output of neuron j at time t and δ is the "error" of the neural net, meaning the difference between the real output and the desired one. The neural net always measures the magnitude of δ and computes its new weight values *depending on the value of d.* Evidently learning can only take place if δ is held as a variable by the system: The nearer the system gets to the desired goal by learning - the solution of a problem - the smaller δ becomes and vice versa.

By generalising this consideration, which is of course well known to all teachers, cognitive psychologists and everybody who has to adapt by learning, the following hypothesis can be stated:

> Meta parameters of adaptive systems may be divided into two classes: class one contains those meta parameters for which specific fixed values determine, *ceteris paribus*, the adaptive success of the system and class two contains those parameters that must be varied by the system in order to realise maximum adaptive success.

It is to be expected that most of the necessary variations of the meta parameters of class two must be similar to d: the further the system is from favourable states the larger must be the values of the meta parameters and vice versa. The system has to adjust its own meta parameters relative to the distance to the attempted goal.

The great biomathematician Fisher was concerned with such a problem sixty years ago when he tried to refute the so-called saltationists. In contrast to those who maintained that nature makes jumps during the biological evolution - *natura facit saltus* - Fisher declared that evolution can only happen by careful adjustments of evolutionary processes. For otherwise evolution could never reach favourable states, i.e. biological species that are able to survive in difficult environments. He illustrated this argument by the problem of adjusting a microscope. If the "right" position of the lens is at 0.5 and if the first setting is 0.6, i.e. the microscope is already moderately well adjusted, then it is reasonable to change the setting by only small factors (the adjuster of course does not know the right value but can evaluate the adjustment by the quality of the picture, which is comparatively good at 0.6). If the adjuster uses a correcting factor of 0.05, then he will reach his goal of an optimal picture and his chance of coming nearer to the desired point of 0.5 with one adjusting step will be fifty-fifty - either he reaches 0.55 or 0.65. If on the other hand the adjuster uses a factor of 0.2 he will never reach his goal because his adjustments will always miss the goal by at least a mistake of 0.1. Fisher had biological reasons for assuming that the microscope is already moderately well adjusted at the beginning. He therefore only needed to consider one side of the adjustment problem (Dawkins 1986). But if the microscope is not well adjusted at the beginning, for example only with an adjusting value of 0.9, then obviously for the same reasons it is sound to start with large values of the correcting factor and to lessen the values the better the picture becomes.[5]

Let us now look at some meta parameters of adaptive systems for which the hypothesis is not as obviously valid as in the cases with δ and the correcting factor when adjusting a microscope (or any instrument).

18.3.1 Parameter h

By distinguishing between base rules and meta rules an obvious criterion for the variability of an adaptive system is the frequency by which the meta rules operate on the base rules. (For the sake of simplicity we refer now only to the cases where adaptive systems contain two logical levels). Let us call this frequency h (for the German word "Häufigkeit").

When looking to social systems, it is clear that they often differ in this aspect. Modern western societies change their rules rather often and so can be characterised by a high h-value; traditional societies such as feudal China or Japan change their rules comparatively seldom.

18.3.2 Parameter r

Frequency of course is not enough for capturing the variability of a system. Equally important is a parameter, which we call "The principle of the sacred cow". By this we mean the "radicality" r of rule changing, measured by the ratio of base rules which *can* be changed during one operation of the meta rules to all rules of the base system. If the number of base rules is n and the number of rules that can be changed at all is k, then

$r = k/n.$

By looking to social systems again, we see that there are always lots of social rules that are not free to be changed. Among these are important traditions, norms based on religion (sacred cows), problem solving strategies that have been successful in the past and so on. In short, generally all those rules are not open to change which belong to the "social identity" of the system. Modern societies are more willing to change most of their rules than traditional societies. A very important constraint for rule changing in most democratic societies is their constitutions (for example, the German "Grundgesetz" forbids the changing of those constitutional rules which refer to Human Rights).

18.3.3 Parameter s

A third meta parameter is s = subtlety. By s we mean a measure for the number of rules changed by *one operation* of the meta system. If only one rule is changed

---

[5] As with δ the solution to the adjustment problem is of course known to everybody in daily life, but probably not consciously. Personally speaking I (Jürgen Klüver) got acquainted consciously with this solution when I was trained in the army to handle a tank gun. The instructing corporal explained this to me very explicitly and very loudly. A very educational institution, the army.

by the meta rules at a time, s is very high; if all rules are changed at once (by the same meta rule!), s is very low. So if n is the number of all rules (open to change) and λ is the number of rules that are changed by one operation of the meta-system, then

$s = 1 - \lambda/n$,

To illustrate this, let us look upon a special CA, modified by a genetic algorithm (GA), which simulates the process of social differentiation, namely the differentiation of a rather homogenous rural society into a class society. We have described this hybrid CA - SOZION - elsewhere (Klüver 1996). The GA operates upon the CA in the manner that certain "mobility parameters" - not to be confused with ordering parameters of the rule systems - are changed. This means that the probability of social mobility through rising or descending into different social classes is raised or lowered according to environmental conditions. This kind of operating on the CA is quite sufficient for the GA to produce satisfactory results. We started experiments on SOZION with rather low s - without knowing about its possible significance - insofar as the GA changed with one operation one whole column of the mobility matrix. This means that all mobility parameters with respect to one social class were changed at once. SOZION was able under these conditions to get rather good results very fast but it could not be brought to improve itself over a specific barrier δ which measures the distance between the "values" of the system and the environmental demands, namely $\delta = 0.08$. [6]

18.3.4 Parameter m

A fourth meta parameter can be called m for magnitude; m measures the quantity of changing *one specific* rule. To give an example, when we combined a CA of the SOZION type with an interactive network (IN, see Stoica in this volume), both coupled with a GA, the GA operated on the CA so that the values of the mobility matrix were changed only rather smoothly, i.e. by raising or lowering by a constant factor of *circa* 0.05 (Klüver et al. 1997). So a mobility parameter is changed as a consequence of a GA-operation, for example from 0.3 to 0.35. The corresponding weight matrix of the IN is changed much more radically: A weight value in the matrix of the IN can be changed by one GA-operation from 0.3 to 0.9.

If m ranges from 0 to 1, then in the case of the hybrid CA, $m = 0.05$, or in the case of the hybrid IN, $m = 1$.

18.3.5 Parameter μ

By measuring only the frequency and other quantitative aspects of rule changing, nothing can be said immediately about the consequences on the dynamics of the

[6] The values of the system and the environmental conditions are both written as a vector; δ is then measured with the Euclidean distance.

base system by changing different rules. It is obvious that base rules can be changed in very different ways and that different changes, i.e. different meta rules, may or may not have consequences in changing the dynamics. To illustrate this, let us have a look again at SOZION: The CA without the GA has a rather high P-value, which means that its rules generate a simple dynamics usually with attractors of period zero. The operations of the GA, regardless of the special values of the meta parameters just mentioned, do not change this P-value which means that the basic dynamics of the CA remains unaltered - in the phases without operation of the GA, of course.

When we introduced additional ways of changing the CA-rules, we got variations of the P-value. For example, when we introduced a meta rule "Change the rule "go from one social class with probability p to the next higher or next lower class" to the rule "go from a social class directly to the class 3 levels higher or lower" ", then the dynamics of the CA changed to trajectories with attractors of period 1 or more. So there are meta rules which do not change the dynamics of the base systems and there are other meta rules which change them. Accordingly, we defined another meta parameter μ with

$\mu_i = 2\,(P_i - P_{i+1})$,

if P_i is the P-value of the base system before the meta rules change the base rules for the i-th time and P_{i+1} is the P-value of the base rules after the meta rules have been applied once for the i-th time. Then μ is the mean value of *all* applications of the meta rules:

$\mu = \Sigma\mu_i /n$,

if n is the number of the meta rules applications during the time period of observation. The hybrid system SOZION has $\mu = 0$.

Some additional remarks are necessary with regard to this meta parameter: We called this meta parameter μ because it operates *on* the λ-parameter (which is logically equivalent to P with the exception that P ranges from 1 to 0.5 and λ from 0 to 1). As μ measures the variation of the dynamics of the base system and as there are many different control parameters there should be a meta parameter for each control parameter. Yet the control parameters are, as we saw, not totally independent of each other. We started with a meta parameter with respect to P or λ. Experiments will be necessary to compare our results with μ to results obtained with meta parameters referring to other control parameters. Nevertheless, as we shall see, it is very likely that one meta parameter like μ is sufficient to know the best adaptive behaviour of hybrid systems with respect to the changing of the base system dynamics.

μ must be defined as a mean value of all operations of the meta rules. The reason for this is rather simple: If specific meta rules are applied to the base rules more than once - the applications then generate a recursive sequence of rule changing - then the effects on the control parameters are generally not the same for each application. If, for example, the meta rule described above, "increase the distance between classes for social rising or descending by 3", then each application of this meta rule changes the dynamics again. One application gives attractors of period 4, the second application yields attractors of period 7 and, as there are only 10 classes in SOZION, the third application gives attractors of period 1! From a generalisation of the famous "theorem of eternal return" by Poincaré it follows that deterministic meta rules must have attractors also, namely attractors of the base rules. An attractor of period zero of the meta rules is a case where the base rules are not changed at all though the meta rules are applied over and over again. Perhaps that is why so often political reforms change practically nothing although everybody believes that many changes will occur.[7]

The meta parameters characterise the *variability* of adaptive systems. If we combine for the sake of convenience the first four meta parameters as a "quantitative" meta parameter k, with

$k = (h,r,m,s)$,

and if we call the variability of a hybrid or meta-dynamical system V and if we take into consideration the fact that hybrid systems may contain more than two logical levels whose number may be called n , then we get

$V = (k,m,n)$.

At the moment, we do not know whether and how it will be possible to express V as a scalar though of course a measure for V may be the length of this vector. Perhaps some creative researcher will find a general formula for V.[8]

Let us now look at first results with regard to the different meta parameters:

(a) Parameter h

In the case of h the results are those expected. Roughly speaking, the greater h, the more successful the adaptation of the system. The relation is not a strictly linear one and of course, these results are only statistically valid as the GA is a stochastic system in both genetic operators. Nevertheless, the trend is unambiguous and confirmed with rather different CAs and also with different INs by Christina Stoica. So it seems obvious that a high value of h is good for an adaptive system, especially in the case of a variable environment. In addition, h is apparently a meta parameter of type one, i.e. a meta parameter where specific values determine more or less adaptive success.

(b) Parameter r

With r things are quite different. Of course a system can be called "adaptive" only, if h and r (like μ and s) are both not zero. Consequently some of our experiments showed that the adaptive success of the system rises with the rise of the r-value from 0 to approximately 0.5. But giving the system higher r-values is

---

[7] Hybrid systems therefore may possess two kinds of attractors: One type is the well known kind of attractors which define the dynamics of the base system; the other type may be called "meta attractors" which define the "dynamics of rule changing".

[8] Ralph Kier will investigate this and other question more thoroughly in his PhD thesis.

often quite counterproductive insofar as the adaptive success decreases with the increase of r. This is especially clear in the case of INs, where the GA often cannot bring the IN to satisfactory values, i.e. approximate the values of the system to the environmental conditions if r = 1. This means of course that all components of the weight matrix are open to change by the GA (we skip the mathematical details of how the system value and the environmental conditions are defined. See Klüver 1996). When r is lowered to about 0.6 or 0.5, the GA is much more successful.

Mathematically speaking the lowering of r is nothing other than the reduction of the "space of freedom" of the system. Therefore, on the one hand, it is understandable that a reduction makes things easier for the GA to find a satisfactory solution out of the vast solution space. On the other side, a fixing of special rules may just block up the ways that the system needs in order to fulfil the demands of the environment. Apparently here an old truism is valid: As many changes as necessary and as few changes as possible. Perhaps this is a case of complexity: If a system has a lot of different possible solutions available - and in complex systems there is always more than one solution to a special problem - than a reduction of r increases the adaptive success. If a system is dependent on only one or few possible solutions, then a severe reduction of r might have unfortunate consequences. This and many other questions are still to be investigated. We are now analysing the properties of h and r by letting several CAs, optimised by GAs, compete against each other and giving them different values of h and r. The initial results show similar trends to the results reported.

Nevertheless, some restrictions must be stated. The reported results with regard to r could not be confirmed for all hybrid CAs, but only for hybrid INs. The reason lies in the fact that the different rules of a CA have different importance for the dynamics of the system. If a CA belongs to the Wolfram class of IV, which means that it has a rather complex dynamics, then some rules are crucial for the dynamics and some are not. For example, the rule of the "Game of Life" cited above is crucial for the fact that the Game of Life belongs to class IV; other rules are not as important. So the changing of such crucial rules of course has much more impact on the behaviour of the system than the changing of not so important rules. In contrast to such CAs, an IN is "symmetric" insofar as all weights are of the same importance for the behaviour of the system. That is why we could experiment with INs regardless to the importance of specific weights.[9]

When analysing r one obviously has to take into consideration this aspect of the different importance of particular rules. In the case of CAs and similar systems this importance can be defined quite naturally by the number of transitions (of cell states) a specific rule generates. According to this, a rule R which generates (principally) n transitions (measured by the number of cell states which can be reached by applying R) is more important than a rule R′ which generates m

[9] The mathematical reason for this is the fact that the weight matrix of an IN is quadratic which is usually not the case with learning networks.

transitions if and only if n > m. It is then possible to classify all rules of a particular CA according to their importance and to analyse r by taking into account the importance of the rules. We hope that the results will confirm the preliminary results reported above.

Our first results indicate that r is also a meta parameter of class one, with regard to the magnitude of the system and the importance of the particular rules of course. If that should be the case, then obviously conservatism may be a rather sound attitude: It does not pay for an adaptive system to change as many rules as possible, even if this could be carried through at all. In many cases it would be much better to change only a fraction of the rules. Yet a lot of more investigations are necessary to confirm these considerations.

(c) Parameter s

s was mainly investigated with SOZION and with hybrid INs, but we have reasons to assume that the results can be generalised. With a high s, meaning only one mobility parameter for one class was changed at a time - for example only the probability of rising from the peasant state into the class of merchants - SOZION took a comparatively long time to reach good values but could improve the results; frequently it reached the "dream value", $\delta = 0$. Rather low s on the other side enabled SOZION to get "adequate" results much faster but usually it could not get better than $\delta = 0.08$ or 0.07, the results mentioned above. If these results can be generalised, s is in itself an optimisation problem for an adaptive system: The lower s, the faster the optimisation process; the higher s, the better the results that can be obtained principally. Whether a system is more interested in quickness than in real optimal results depends of course on the circumstances and the environment: Often time is the important factor because "in the long run we are all dead", as Keynes once remarked very aptly. Yet sometimes "good is not good enough", as a German department store once advertised, so s must be raised at the price of slowing down the optimisation process.

These results indicate that s is a meta parameter of the second kind: An adaptive system that tries to optimise its adaptive process would reach the best results if it kept s as a variable. In the beginning s should be rather small to speed up the adaptive process; the nearer the system comes to the desired states the larger s must be to adjust the optimisation efforts. We try to confirm this assumption with other systems but the trend seems to be unambiguous.

(d) Parameter m

So far m has been analysed only in the case cited above, namely the different behaviours of hybrid CAs and hybrid INs with different values of m. The idea behind these experiments was to let the hybrid system SOZION - CA with GA - "model itself" with an IN, which means that some macro aspects of the CA were captured with an IN (Klüver et al. 1997). The whole program operates this way: First the hybrid CA changes its "mobility matrix" the way sketched above. Then this matrix was taken from an IN as its weight matrix and changed according to the special meta rules of the hybrid IN. Then the changed matrix was fed back to the hybrid CA. In the case of the hybrid CA $\mu = 0.05$, in the case of the hybrid IN m= 1. So the hybrid IN changes the matrix much more drastically than the hybrid CA.

When the hybrid CA gets its mobility matrix from the hybrid IN the CA usually gets much better results in optimising itself to a specific environment than with its "own" matrix, i.e. the matrix which resulted from the more prudent changes. The hybrid CA, as mentioned above, usually got its best results with $\delta = 0.08$; with the mobility matrices changed by the hybrid IN, the hybrid CA often got results with $\delta = 0.001$ and better. So it seems that higher values of ruthlessness, i.e. more changes, yields better results, but things are not so simple. The hybrid IN with high m is often not so good as the hybrid CA with low m. The best results are obtained when the hybrid CA operates with a low m, i.e. in a soft way, on the results it gets from the operations of the IN with a high m. Therefore the best way seems to be to operate with *variable* m and probably it should be varied in the same manner as s and δ. But this assumption has yet to be confirmed, though m rather certainly belongs to class two of the meta parameters.

(e) Parameter μ

In the case of μ we tried some different experimental techniques, namely to let the GA find out "by itself" the optimal values for m. The experiments were run with CAs from the type of the Game of Life, i.e. with two cell states. The results were quite unexpected, yet hindsight, as always, makes them clear:

As mentioned before, adaptation to particular environmental criteria always means the evolution of the system to system states which are - according to some evaluation function - as "near" as possible to some favourable states (see above footnote 6). In the case of CAs this means of course nothing else than the approximation to some distribution of the cell states over the grid. So particular distributions of cell states were given as the goal to which the hybrid system should adapt. The meta parameter μ was held variable, i.e. the GA could change it according to the distance to the goal.

The results were quite unambiguous: In each experimental series the GA first started, after some "warming up phases" due to the stochastic character of the GA with rather large values of m. The nearer the hybrid system came to the target state, the smaller μ was set by the GA. In the end of the runs, when the system had reached some target state or one state sufficiently near to the goal, μ always was set to zero. In other words, the meta parameter μ was used by the system like the δ-factor in the Hebbian learning rules: changes to the control parameter P, which in itself determines the dynamics of the CA without a GA, are handled by the system like the adjustments of instruments. For an adaptive system the dynamics of its base system is just the same as are microscopes or tank guns for human users. In particular, the P-value of the CA which has reached the target states, is practically always exactly the P-value which is characteristic for the distribution of cell states of the target state. It is not only a metaphor when we say that our hybrid systems have discovered the P-parameter for themselves.

Some remarks about these results are necessary:

(1) Obviously the μ-parameter is of the second class, i.e. a meta parameter which is best held variable by the system to obtain optimal adaptive success. This is in itself not trivial because this means that the dynamics of the base system is of only secondary importance for an adaptive system that is able to change these dynamics. Therefore the criterion of adaptive success is indeed the only possibility of measuring meta parameters. The dynamics of the base system is important only for adaptive systems that have small h-values, i.e. which change their base rules only seldom.

(2) With hindsight, as mentioned above, the results are quite understandable: As the target state is defined by a particular distribution of cell states an adaptive system of course does best if it tries to get a P-value for its base system defined by just this distribution. The probability that the hybrid system will reach such a target state is certainly highest with such a P-value. Yet because P, like the other control parameters, is only statistically valid, it was not to be expected *a priori* that the system would reach the solution with the exact P-value, since other solutions are also possible.

(3) When introducing the μ-parameter we mentioned the necessity of defining meta parameters for each known control parameter. Yet in the light of the results obtained with μ, it seems rather likely that this is not the case. If an adaptive system does best if it keeps the changing of the dynamics variable in phase with the changing of the μ-parameter, then this result must hold for the other control parameters also. We can think of no reason why meta parameters defined in regard to the other control parameters should behave differently. Yet with complex and especially adaptive systems one never knows; though we are quite convinced that the results obtained with μ are also valid for other meta parameters, we will do some experiments with these too.

(4) The "discovering" of the control parameter P by our hybrid systems is not the only case where we found behaviours of hybrid systems remarkably similar to "real" systems (e.g. Stoica 1997); often our hybrid systems approached solutions to adaptive problems which are exactly the solutions we find in social reality. In all these cases the systems only got the target and had to find their way for themselves. Sometimes we had the irritating impression that here is something like independent and self-reliant behaviour, though of course all the algorithms that govern the operations of our systems are well known. One need not speculate whether systems like these exhibit the beginnings of "true" Artificial Intelligence, as certainly the independent

solution of problems is one of the marks of true intelligence. [10] Let us rather say that the capability of finding solutions to difficult problems for themselves is one of the most important features of complex adaptive systems in general and social systems in particular. So our results demonstrate that systems like these are very well suited to model the most difficult and important aspects of social complexity; systems without these capabilities are always suited only to capture very restricted aspects of social reality.

(f) *Parameter v*

Another parameter v has been discovered recently by Jörn Schmidt and Jürgen Klüver. This measures a feature of the topology of a BN. v defines a measure of the effects of the units of a BN, that is whether all units act upon the same or a different number of units. v = 0 defines rule systems with complicated dynamics. By increasing the value of v one gets very simple dynamics (for details see Klüver 1999).

18.4 Perspectives and Open Questions

So far, our results raise as many questions as they give answers. Yet they clearly show the way for further investigations. Let me name some of these perspectives:

(i) The combining of different logical levels of rule systems lead into the realms of self reference which are not very well understood. Logical semantics (Tarski 1956) tried to banish such self referential rule systems by declaring that different logical levels should be separated once and for all. Yet social systems are, as all adaptive systems, clearly self referential; therefore rule systems which contain different logical levels of which each is dependent upon the other must be analysed thoroughly. In the previous remarks we only mentioned the changing of base rules by meta rules. Yet the converse is possible also: The meta rules can be changed - for example their h-value - by the proceedings of the base system and therefore by the base rules. The dependency is mutual. We constructed for example a hybrid CA which switches its GA on or off according to the progress the CA makes in relation to specific environmental conditions: the CA determines the h-value of its meta system. So variability is also the capability of varying the meta parameters which, as we saw, may be often the best thing to do for a system. Further investigations have to analyse the consequences. In short, what is needed is a mathematics of self reference and our researches are just a beginning.

(ii) In connection with r it is necessary to mention the importance of memory. We cannot give a logical description of the memory of complex systems in

---

[10] As is well known "classical" AI always had one of its shortcomings exactly at this point: The systems had always just as much "intelligence" as the programmers put into them and practically no independence.

this paper but we must stress that memory is basically nothing else than a reduction of the r-value of an adaptive, i.e. learning system. This can be demonstrated quite clearly by the construction of formal systems like the ART-networks (Grossberg 1987) and learning Boolean nets (Wuensche 1994) on the one side and neuro-physiological results (Edelman 1992) on the other. As memory is an advantage for a system, reductions of r reduce the variability but may increase the competence for problem solving and therefore the adaptive success. This is especially important in the case of a variable environment; our experiments have been carried through so far mostly with constant environments (see e.g. Kauffmann 1993). We have just started some experiments to see whether memory can compensate for optimisation strategies with low efficiency, namely low values of h and r, but wait and see.

Similar reflections must be carried through for the role of the consciousness of an adaptive system. The coupling of a CA with an IN mentioned above can be interpreted very tentatively as the conscious self-modelling of the CA into an IN. As we saw, this resulted in a raising of the μ-value of the whole system, so consciousness may have the effect - and certainly this is one of its functions - to raise or lower the value of some meta parameters. The problems are numerous (see Klüver 1998).

(iii) The similarity of investigating control parameters and meta parameters of complex systems to the search for linguistic universal (Chomsky 1965) was already mentioned in the beginning of this chapter. Indeed, as grammatical universals characterise universal features of all languages, parameters like those dealt with in this paper indicate universal properties of adaptive systems in general and social systems in particular. They are "formal universals" in the sense of Chomsky as the parameters define universal properties of *rule* systems.

Recently in anthropology so called "social universals" have been investigated, namely manners of social behaviour which are to be found in all human societies (Brown 1991). Among these are for example different social roles of the sexes, institutions (rules) of political power and law, rules for integrating the young into society and so on. The universal character of these institutions is explained by Brown and other anthropologists by the universality of the biological human nature. That may be and in this sense these social universals are the parallel to the "substantive universals" in theoretical linguistics. Yet from our experiments it can be seen that the explanation of social universals need not go necessarily to human biology. The behaviour of social systems might be explained more readily by their *mathematical features:* social systems behave the way they do because it is the mathematically simplest way to do it. Of course the importance of human nature must not be denied. Probably human biology plays the role of constraints upon the mathematical freedom the parameters and meta parameters still allow.

(iv) Coming back at last to "real" social systems there are a lot of possibilities for understanding social reality by investigating the control parameters and meta parameters of social systems. To give just one example: The difficulties of understanding modern societies are often mentioned in contemporary theories of society (e.g. Habermas 1981). The difficulties are mostly located in the fact that the dynamics of modern social systems are not to be understood any more by intuition and common sense, in contrast to former societies. Habermas for example postulates explicitly the task of analysing modern social systems by "analytical methods" like the methods of the natural sciences, though he is not able to define these methods.

Both the problem and the methods to deal with the problem can be treated with the concentration upon the objects dealt with in this article. The complicated dynamics of modern societies obviously is a result of both high values of the control parameters and the meta parameters and can therefore be easily understood. As the variability of modern societies is evidently very high in contrast to other societies, so is for example the λ-value. In modern societies the rules of social mobility - the "transition rules" - allow the passing over from one social role to another quite easily. In other words, each social role - each cell state in a CA - can be reached with a rather high probability from any other role - any other cell state. Accordingly the dynamics of modern societies can and must be analyzed by using the methods of investigation described in this article. There are wide new fields of research for empirical and theoretical sociology.

Yet we think there is no other way if we want truly to understand complex systems like the ones we have to deal with in sociology. There has been a lot of efforts in the past to classify social systems from a theoretical point of view; famous examples are the classifications by Marx in the terms of relations of productions or the different levels of social differentiation by social systems theory. Yet all these attempts were rather arbitrary insofar as one could take them or leave them. The formal classifications introduced in this paper with the concepts of control and meta parameters have the advantage that they allow precise definitions and give rise for further investigations. One of these must certainly be the investigation how the classical definitions of different social systems fit into these new categories.

Computer sociology cannot be content to tackle only the logically simple questions. In contrast to the natural sciences social phenomena cannot be simplified artificially but have to be investigated in their logical complexity. That does not mean that it is not possible to analyse simple models of social systems in the computer - on the contrary. It does mean that the models must contain the logical characteristics like self reference which distinguish social systems from natural ones.

There is a lot to do. We have only just started.

References

Alexander, J. C., Giesen, B., Münch, R. & Smelser, N. J. (eds.) (1987): The Micro-Macro-Link. Berkeley, Los Angeles-London

Berlekamp, E., Conway, J. H. & Guy, R. (1982): Winning Ways for Your Mathematical Plays. Academic Press, New York

Bourdieu, P. (1987): Choses Dites. Edition de Minuit, Paris (German edition: Rede und Anwort. Suhrkamp, Frankfurt 1992)

Brown, D. E. (1991): Human Universals. Temple University Press, Philadelphia

Chomsky, N. (1965): Aspects of the Theory of Syntax. MIT Press, Cambridge.

Dawkins, R. (1986): The Blind Watchmaker. Norton, London.

Edelman, G. M. (1992): Bright Air, Brilliant Fire - On the Matter of the Mind. Basic Books, New York.

Farmer, J. D. (1990): A Rosetta Stone for Connectionism. In: Forrest, S. (ed.): Emergent Computation. Physica D 42, 153 - 187. Elsevier (North-Holland), Amsterdam.

Grossberg, S. (1987): The Adaptive Brain. Elsevier, Amsterdam.

Habermas, J. (1981): Theorie des kommunikativen Handelns. Suhrkamp, Frankfurt.

Hebb, D. O. (1949): The Organization of Behaviour. Wiley, New York

Hegselmann, R. (1996): Cellular Automata in the Social Sciences. Perspectives, Restrictions and Artefacts. In: Hegselmann, R., Müller, U. & Troitzsch, K. G. (eds.): Modelling and Simulation in the Social Sciences from the Philosophy of Science Point of View. Kluwer, Dordrecht.

Hegselmann, R., Müller, U. & Troitzsch, K. G. (eds.) (1996): Modelling and Simulation in the Social Sciences from the Philosophy of Science Point of View. Kluwer, Dordrecht.

Kauffman, S. (1993): The Origins of Order. Oxford University Press, Oxford.

Klüver, J. (1995): Soziologie als Computerexperiment. Vieweg, Braunschweig-Wiesbaden.

Klüver, J. (1996): Simulations of Self Organizing Social Systems. In: Faulbaum, F. & Bandilla, W. (eds.): SoftStat 95. Advances in Statistical Software. Lucius, Stuttgart, 425-432.

Klüver, J., Stoica, C. & Schmidt, J. (1997): Simulations of the Self Modelling of Social Systems by a Hybrid System. In: Faulbaum, F. & Bandilla, W. (eds.): SoftStat 97. Advances in Statistical Software. Lucius, Stuttgart, 411-417.

Klüver, J. (1999): The Dynamics and Evolution of Social Systems. New foundations of a mathematical sociology. To be published at Kluwer Academic Publishers, Dordrecht.

Knorr-Cetina, K. (1981): Introduction: The Micro-Sociological Challenge of Macro-Sociology: towards a reconstruction of social theory and methodology. In: Knorr-Cetina, K. & Cicourel, A. V. (eds.): Advances in Social Theory and Methodology. Routledge & Kegan Paul, Boston-London, 1-47.

Langton, C. (1988): Artificial Life - Preface. In: Langton, C. (ed.): Artificial Life. Addison Wesley, Reading MA, pp. I-IX.

Langton, C. (1992): Life at the Edge of Chaos. In: Langton, C., Taylor, C., Farmer, J. D. & Rasmussen, S. (eds.): Artificial Life II. Addison Wesley, Reading MA, 41-91.

Levy, S. (1993): Artificial Life. Penguin Books, London.

Nowak, A. & Lewenstein, M. (1996): Modeling Social Change with Cellular Automata. In: Hegselmann, R., Müller, U. & Troitzsch, K. G. (eds.): Modelling and Simulation in the Social Sciences from the Philosophy of Science Point of View. Kluwer, Dordrecht.

Nowak, M. A. & May, R. M. (1993): The Spatial Dilemma of Evolution. International Journal of Bifurcation and Chaos Vol. 3, 1, 35 – 78.

Poundstone, W. (1987): The Recursive Universe. Oxford University Press, Oxford.

Schelling, T. C. (1971): Dynamic Models of Segregation. Journal of Mathematical Sociology 1, 143 – 186.

Schwartzman, H. (1978): Transformations. The Anthropology of Children's Play. Plenum, New York.

Stoica, C. (1997): Die Vernetzung sozialer Einheiten. PhD-Thesis, Universität Essen.

Tarski, A. (1956): Logics, Semantics, Metamathematics. Oxford University Press, Oxford.

Weber, M., 1907/1982: R. Stammlers Überwindung der materialistischen Geschichts-auffassung. In: Winckelmann, J. (ed.): Gesammelte Aufsätze zur Wissenschaftslehre von Max Weber.

Wolfram, S. (ed.) (1986): Theory and Applications of Cellular Automata. World Scientific, Singapore.

Wuensche, A. & Lesser, M. (1992): The Global Dynamics of Cellular Automata: Attraction Fields of One-Dimensional Cellular Automata. Addison Wesley, Reading MA.

Wuensche, A. (1994): The Ghost in the Machine. In: Langton, C. (ed.): Artificial Life III. Addison Wesley, Reading MA, 465-501.

Biographical Sketches

Wolfgang Balzer

Wolfgang Balzer, born in 1947, became Professor for logic and philosophy at the University of Munich in 1984 where he has taught since then. He is a salient proponent of the 'structural approach' in the philosophy of science, to which he has made several important contributions. During the last decade he has also worked on simulating social institutions. Besides numerous publications on the structure and foundations of science, he has written papers and books on social institutions. Hes is currently working on social simulation programs in Prolog.

Steve Beare

Steve Beare is the Deputy Director of the Australian Bureau of Agricultural and Resource Economics (ABARE), the largest applied economic research agency in Australia. In his current position Dr. Beare is responsible for the development and management of ABARE's economic research and government policy programs, which covers issues relating to minerals, energy, agriculture, forestry, fisheries, trade and the environment. He has been actively involved with the development of research methods for forecasting the financial performance of farm businesses in Australia's agricultural sector with the particular aim of analyzing the distributional impact implied by alternative policy scenarios.

Kai-H. Brassel

Kai-H. Brassel studied computer science and social science at the University of Koblenz, Germany, where he received a masters degree in computer science. His main fields of research are theories of social systems, design and implementation of simulation systems, and object oriented software engineering. Since 1996 he has held a position as academic researcher at the Department of Sociology at the Technical University of Darmstadt.

Ray Chambers

Ray Chambers is Professor of Social Statistics at the University of Southampton, UK. He holds a doctoral degree in Biostatistics from the Johns Hopkins University, Baltimore, USA. Before moving to Southampton in 1995, he was a Senior Lecturer in Statistics at the Australian National University and prior to that, Senior Statistician at the Australian Bureau of Agricultural and Resource Economics. His interest in microsimulation was stimulated by research work initiated at ABARE on methods for forecasting the distribution of farm business performance using micro level sample survey data. His current research interests include methodology for the analysis of data collected in complex surveys and the application of robust and nonparametric methods to statistical data analysis.

Edmund Chattoe

Edmund Chattoe is a Research Fellow in the Department of Sociology at the University of Surrey. He is Associate Director of the Center for Research on Simulation in the Social Sciences (CRESS, http://www.soc.surrey.ac.uk/research/cress.html) and Review Editor of the Journal of Artificial Societies and Social Simulation (JASSS, http://www.soc.surrey.ac.uk/JASSS/JASSS.html). His research interests are the application of Artificial Intelligence techniques to social processes, the implications of cognitivism for decision-making, modelling social evolution and using simulation techniques to integrate theories from different social sciences.

Rosaria Conte

Rosaria Conte is head of the Division "AI, Cognitive and Interaction Modelling" at CNR, Italy. She teaches social psychology at the University of Siena. Her main background is on pragmatics of communication and cognitive science. Her publications are both in the area of Multi-Agent Systems and Agent Theory, and in agent-based simulation. Her research interests include formal computational study of coalition formation, social dependence, norms representation and reasoning, and the role of cognition in the propagation of norms.

Jim Doran

Jim Doran is a Professor in the Department of Computer Science at the University of Essex, with particular interests in multi-agent systems, artificial societies and agent-based modelling. He read mathematics at Oxford, and became a specialist in artificial intelligence studies at the University of Edinburgh. Throughout his career he has also worked on the application of computing and AI techniques in archaeology and anthropology.

Ottmar Edenhofer

Ottmar Edenhofer, trained as an economist at the University of Munich, and is now working as research assistant at the Department of Sociology at the Technical University of Darmstadt. He has published in the field of integrated environmental assessment and evolutionary economics. He is currently finishing a Ph.D. thesis about computer models of social conflict and technological change.

Ilan Fischer

Ilan Fischer is a lecturer at the Department of Behavioral Sciences, Ben Gurion University of the Negev, Israel, and a research fellow at the Department of Psychology, University College London. His research interests include individual and group decision making, confidence and expertise development, and computer simulations of social interactions.

Andreas Flache

Andreas Flache took his master's degree in computer science and social sciences from the University of Koblenz, Germany in 1991. Subsequently, he conducted his doctoral research at the ICS (Interuniversity Center for Social Science Theory and Methodology) at the University of Groningen, The Netherlands. There he took his Ph.D. in 1996. Currently he is postdoc at the Department of Philosophy of the University of Bayreuth, Germany. From July 1999 on, he will carry out a research project at the ICS in Groningen, where he will be research fellow of the Royal Dutch Academy of Sciences. His main research interests concern modelling and computer simulation of social phenomena, social dilemmas and social networks.

Nigel Gilbert

Nigel Gilbert is Professor of Sociology at the University of Surrey, where he is Director of the Center for Research on Simulation in the Social Sciences. He has been associated with the re-emergence on simulation in the social sciences in the 1990s following the publication of Simulating Societies (edited with Jim Doran) in 1994 and is editor of the electronic Journal of Artificial Societies and Social Simulation (JASSS). His main interests are in the use of simulation for understanding science policy and environmental issues, and in the development of sociological theory with the assistance of computational modelling. He has written and edited 13 books and about 110 papers. One of his recent publications is “Simulation for the Social Scientist”, with Klaus G. Troitzsch (open University Press, 1999).

Rainer Hegselmann

Rainer Hegselmann obtained his master's degree in social sciences in 1973 at the University of Bochum, Germany. He took his Ph.D. in philosophy from the University of Essen in 1977. He was professor of philosophy at the University of Bremen 1988-1996. In the academic year 1994/95 he was fellow in residence at The Netherlands Institute For Advanced Study (NIAS). During 1995 he was a guest professor at the Institute For Advanced Study in Vienna, Austria. Since 1996 he has been professor of philosophy at the University of Bayreuth, Germany. His main research interests concern modelling and simulation of social dynamics, moral philosophy, philosophy of science, and the history of early analytical philosophy.

Ralph Kier

Ralph Kier, MA, studied communication science at Essen and wrote his MA thesis about expert systems in the social sciences. At present, he is writing his Ph.D. thesis about complexity in social systems.

Oliver Kirchkamp

Oliver Kirchkamp was born 1962 in Düsseldorf, Germany. He received a diploma in Economics at the University of Bonn in 1992 and then spent a year on a SPES scholarship at DELTA in Paris. He received a Ph.D. in Economics in 1996 at the University of Bonn within the European Doctoral Program of Quantitative Economics an then went to the European University Institute in Florence, Italy, as a Jean Monnet Fellow. Since 1997 he has been head of the experimental laboratory of theSFB 504 at the Universtity of Mannheim. His research interests include learning, evolution, local interaction, experiments, auctions, monetary policy.

Jurgen Klüver

Jurgen Klüver, Prof. Dr. phil., studied philosophy and mathematics at the universities of Kiel and Hamburg. He is now head of the Interdisciplinary Center of Research in Higher Education at the University of Essen. His main research interests are sociology of science, theory of science and the use of computer programs in the social sciences.

Philip Kokic

Philip Kokic is a Senior Advisor in Insiders GmbH, a German company responsible for the development of knowledge based computer systems. He holds a doctorate degree in Mathematical Statistics from the University of Sydney. His current responsibility is in the conception of computer software products adopting innovative mathematical and statistical techniques particularly in the area of banking, finance and insurance. His former position was as a Senior Research Fellow in Social Statistics at the University of Southampton, and prior to that as a Senior Statistician in the Australian Bureau of Agricultural and Resource Economics. In these positions Dr. Kokic worked on projects involved with the microsimulation and forecasting of business performance with a particular emphasis on the analysis of sample survey data.

Bibb Latané

Bibb Latané is Professor of Psychology at Florida Atlantic University and President of Social Science Conferences, Inc, which organizes Nags Head at the Sea Frolic, a subsidized series of scientific meetings, and supports KNITS, a project studying the use of knowledge networking in teaching science. Winner of the 1998 Distinguished Scientist Award from the Society of Experimental Social Psychology, he has taught at Columbia, Ohio State, and UNC.

Mario Marsero

Mario Marsero is a graduate student in Communication Science from the University of Torino, Italy. He is involved in the PSS (Project for Social Simulation) at CNR, Italy.

Michael Möhring

Michael Möhring is a Senior Researcher and Lecturer at the University of Koblenz. His Ph.D. in computer science involved developing a modelling and simulation tool for the Social Sciences (MIMOSE). His main research interests are modelling/simulation methodology and their applications in the social science. He also works on data collection and analysis methods, especially in the fields of online questionaires, statistics and data mining.

Volker Möller

Volker Möller studied mathematics and philosophy at the Universities of Bayreuth and Warwick (UK). He took his master's degree in mathematics at the University of Bayreuth. His master's thesis addressed the controlling of non-linear dynamical systems. From 1996 to 1998 he worked at the Department of Philosophy in Bayreuth, where his main research interest was modelling and simulation of opinion dynamics in a cellular automata framework. Since 1998 he has worked for a software company where he develops and configures systems for the administration of large, heterogeneous computer networks.

Hiroaki Morio

Hiroaki Morio, after Masters degrees in Human Science from Osaka University and Neuroscience from Baylor University, is working on a doctoral degree in social psychology at Florida Atlantic University.

Georg Müller

Georg Müller is a senior assistant at the Department of Social Work and Social Policy of the University of Fribourg (Switzerland) where he teaches social research methodology and statistics. His research interests cover problems of the welfare state, the mathematical modeling of social processes and social indicators research. He is also the author of the book Comparative World Data: A Statistical Handbook for Social Research (Johns Hopkins University Press).

Mario Paolucci

Mario Paolucci is a researcher at the National Institute of Statistics (ISTAT), Italy. His background is in physics. He is involved as a research fellow in the PSS (Project for Social Simulation) at CNR, Italy. His main research interests concern agent-based simulation, theory of complexity and the emergence of social institutions.

Matthew Rockloff

Matthew Rockloff is a senior graduate student at Florida Atlantic University. After obtaining an undergraduate degree in Economics in 1989 from the University of California at Santa Cruz, he spent four years as a corporate financial analyst in the banking industry. He completed a Master of Science degree in Economics at Texas A&M University in 1994, before moving to Florida to pursue a doctorate in Experimental Psychology.

Nicole J. Saam

Nicole J. Saam is Assistant Professor at the Institute of Sociology, Ludwig-Maximilians-Universitaet, Munich, Germany. Her research focuses on modelling and simulation in the social sciences, sociology of organizations, political sociology and Thai studies. Currently she is also working on hybrid learning communities and investigating competence and other requirements for interactions between human actors and artificial agents. Her substantive organizational studies concentrate on theories of management consulting and on organizational reforms in the Thai Buddhist order. She has published numerous papers on simulating sociological theories.

Christof Schatz

Christof Schatz studied Physics (with focus on elementary particle experiments) and Sociological Methods and Modeling in Munich and Geneva. He completed his diploma in Physics in 1994 and worked as scientific assistant at the chair of sociology at the Technical University in Munich. His Ph.D. on the dynamics of social reactions to unemployment situations was completed in 1997. Since 1998 he has worked at Infratest Burke company as scientific manager and programmer of the project AVID96, an empirical based microsimulation, which will be used by the German Ministry of Labor and Social Affairs as a data base for pension system reforms.

Jörn Schmidt

Jörn Schmidt studied chemistry, mathematics and astronomy at the University of Hamburg. He is a research fellow at the Center for Research in Higher Education at Essen. His research fields are sociology of science, theory of science and the use of computer programs in the social sciences.

Christina Stoica

Christina Stoica, Dr. phil., studied educational science at the University of Essen. After completing her MA she studied computer science at the University of Bochum where she received her Ph.D. She is now a research assistant at the Center of Research in Higher Education at the University of Essen.

Ramzi Suleiman

Ramzi Suleiman is Senior Lecturer in the Department of Psychology at the University of Haifa, Research Fellow at the Hary S. Truman Institute for the Advancement of Peace and President of the Galilee Center for Social Research. He is co-founder and co-editor of the Journal of Artificial Societies and Social Simulation (JASSS). His research interests include the study of social identity, social dilemmas, bargaining situations and computer simulation of social phenomena.

Klaus G. Troitzsch

Klaus G. Troitzsch is professor of computer science applications in the social sciences at the Department of Computer Science of the University of Koblenz-Landau in Germany. Originally a political scientist working on political sociology and election studies, for the last twelve years his main interest has been in simulation approaches in the social sciences.

Christoph Urban

Christoph Urban is a Ph.D. student at the Chair for Operations Research and Systems Theory, University of Passau, Germany. His main research interest is in agent-based simulations and reference models.

Subject Index

4GL, 117

Agent, 14, 18, 24, 25, 75, 86, 88, 89, 114, 122, 123, 131, 138, 140, 142, 143, 309
Agent-based modelling, 15
Application, 113, 238, 272, 301
Artificial societies, 17, 313

BDI, 66, 67, 68

Cognition, 14, 15, 88, 91, 94, 95, 97, 98, 99, 106, 107, 108, 148
Complexity, 20, 194, 268, 271, 274, 285
Confidence, 279

Diversity, 131, 133, 134, 148, 196, 205, 210, 213, 271
Dynamics, 14, 26, 27, 44, 60, 89, 148, 159, 162, 177, 195, 211, 269, 271, 274, 272, 273, 371, 372

Elections, 238
Empirical, 178, 214, 285, 301
Evolution, 25, 60, 74, 103, 131, 177, 179, 186, 195, 238, 239, 269, 271, 272, 314, 371, 372
Evolutionary economics, 181
Evolutionary game theory, 181
Expectile regression, 346, 347

Fit, 290, 294

Genetic algorithms, 9, 15, 287

Intelligent agents, 16

Life, 113, 194, 245, 270, 271, 352, 353, 364, 366, 372

Meta parameter, 359
Methodology, 17, 18, 27, 44, 269, 371

Networks, 50, 60, 103, 113, 148, 166, 177, 195, 352
Neural Networks, 48, 50
Nonlinear, 208, 299, 300
Norms, 313
Object, 75, 117, 119, 291
Optimization, 29, 44, 130, 271

PECS-agents, 89, 91, 96
Prolog, 75, 87

Reference models, 88
Representation, 75, 218, 220, 223, 234, 235, 236, 237, 238
Richardson arms race model, 34, 35

Sensitivity analysis, 244, 249
Sets of social rules, 53

Simulation, 3, 4, 6, 14, 16, 24, 26, 44, 59, 60, 61, 65, 74, 75, 83, 86, 113, 114, 115, 117, 130, 131, 148, 177, 204, 206, 207, 208, 210, 214, 215, 221, 246, 269, 270, 271, 272, 296, 300, 301, 302, 304, 313, 350, 371, 372

Social Differentiation, 74, 113
Social network, 128, 176, 197, 318
Social influence, 196, 197, 198, 210, 220, 223, 224, 225, 248

Toolkit, 3, 6, 11

Unemployment, 283, 300

Validation, 269, 274

Author Index

Abelson, R. P., 152, 153, 154, 155, 156, 158, 176, 197, 198, 202, 212, 215
Agah, A., 87, 113,
Agre, P. E., 67, 73, 87, 113
Aiken, M., 279, 299
Albert, A., 265, 269
Alexander, J.C., 351, 371
Alterman, R., 74
Amacher, R. C., 219, 238
Armstrong-Stassen, M., 280, 299
Aronson, E., 216
Aschenbach, G., 279, 299
Asher, H. B., 299
Ashlock, D., 194
Askenazi, M., 130
Attanasio, O. P., 245, 269
Axelrod, J., 275, 278
Axelrod, R., 9, 14, 65, 74,123, 130, 180, 182, 183, 194, 223, 238, 246, 269
Axtell, R., 65, 74, 243, 248, 249, 269

Back, K., 196, 215
Balzer, W., 65, 66, 67, 68, 70, 74
Bandilla, W., 60, 371
Barnett, G. A., 15, 177, 216, 271
Barton, R., 270
Bates, D. M., 276, 299
Beare, S., 344, 345, 347, 350
Bekey, G. A., 113
Belardo, S., 269
Benfer, R. A., 48, 59
Bentham, J., 219, 238
Bentler, P. M., 276, 299
Berlekamp, E., 352, 353, 371
Bernard, P. M., 249, 269
Best, H., 60
Binmore, K., 134, 147, 194, 302, 313
Birch, A. H., 219, 238
Blalock, H. M., 32, 44
Blossfeld, H. P., 276, 299
Boergers, T., 189, 193, 194
Bonacich, P., 177
Bond, A. H., 246, 269
Bonhoeffer, S., 194
Boots, B., 178
Bossel, H., 108, 113, 48, 59
Boster, F., 15, 177, 216, 271
Bourdieu, P., 351, 371
Bourgeois, M., 196, 215, 216
Boyes, W. J., 219, 238
Brams, S., 219, 238
Brandt, R., 131, 147
Brassel, K. H.
Bratley, P., 244, 252, 269, 274, 300
Bratman, M., 131, 134, 148
Breckling, J., 347, 350
Bremer, S. A., 26, 44
Brendel, K., 70, 74
Brent, E. E., 59
Brown, D. E., 369, 371
Bruederl, J., 276, 300
Buckley, W., 56, 59
Burkhart, R., 130
Bush, R. R., 171, 176

Caib-Draa, B.
Caldas, J. C., 85, 113
Carley, K., 8, 14, 16, 65, 74, 177, 249, 269
Carr, E. P., 15
Carrol, G., 270
Carson, C., 74
Cartwright, D., 177
Castelfranchi, C., 113, 131, 133, 134, 137, 148, 302, 305, 306, 307, 312, 31

Cecconi, F., 15
Cerini, A., 15
Chambers, R. L., 344, 345, 347, 350
Chapman, D., 87, 113, 350
Chattoe, E., 243, 262, 265, 269
Cherkasov, I. D., 277, 300
Chomsky, N., 352, 369, 371
Cicourel, A. V., 371
Coelho, H., 85, 113
Cohen, M. D., 269
Cohen, P. R., 66, 74
Cohn, A G., 148
Coleman, J. S., 26, 44, 280, 300
Conte, R., 14, 15, 16, 24, 48, 59, 131, 132, 133, 136, 137, 148, 245, 270, 302, 307, 312, 313
Conway, J. H., 120, 121, 124, 125, 126, 353, 371

Davis, J., 216
Davis, L. D., 14, 269
Davis, R., 71, 74
Dawes, R. M., 166, 176
Dawkins, R., 359, 371
Day, R.
Deif, A. S., 251, 269
Demazeau, Y., 74, 148
Derman, C., 221, 238
Desai, M..
Dieckmann A., 276, 300
Doob, J. L., 157, 176
Doran, J., 3, 4, 14, 15, 16, 17, 20, 24, 44, 74, 85, 113, 130, 134, 217, 245, 269, 270, 272, 301
Doreian, P., 277, 300
Drogoul, A., 87, 113
Duval, R., 128, 130

Earle T. K., 20, 25
Edelman, G. M., 369, 371
Eder, K., 51, 52, 53, 56, 59
Efron, B., 344, 348, 350
Eibl-Eibesfeldt, I., 302, 313
Elliott, E., 265, 270
Ellison, G., 180, 181, 182, 194
Endres-Niggemeyer, B., 60
Epstein M., 65, 74, 269
Eshel, I., 180, 182, 183, 194

Fahrmeir, L., 300
Farmer, J. D., 357, 358, 371, 372
Faulbaum, F., 60, 371
Faust, K., 128, 130
Ferber, J., 87, 113
Ferman, L. A., 299
Festinger, L., 196, 215
Fiacco, A. V., 251, 269
Fikes, R., 75
Finin, T., 9, 14
Fischer, I., 61, 123, 130, 218, 219, 220, 228, 238, 239
Flache, A., 104, 105, 113, 151, 152, 166, 167, 177, 178, 197, 215
Flannery, B. P., 277, 287, 291, 300
Forrest, S., 371
Forrester, J. W., 26, 27, 44, 259, 269
Fox, B., 269, 300
Frederiksen, N., 176, 215
French, J. R. P. jr., 152, 177, 219
Frey, D., 299
Friedman, D., 246, 270
Friedman, J. W., 166, 169, 177
Friedman, M.
Frijda, N. H., 84, 114
Fritzon, R., 9, 14
Furbee, L., 59

Gallant, S. I., 49, 59
Gardner, M., 120, 130
Gasser, L. G., 246, 269
Georgeff, M. P., 66, 75
Gerrard, B., 148
Giddens, A., 53, 59, 351
Giesen, B., 371
Gifi, A., 276, 300
Gilbert, N., 3, 4, 7, 14, 15, 16, 20, 24, 44, 48, 59, 74, 113, 117, 130, 217, 245, 262, 269, 270, 272, 301, 313
Glance, N., 194, 245, 270
Gleser, L. J., 238
Gmytrasiewicz, P .J., 131, 148
Goldman, A. I., 69, 74
Gooday, J., 344, 350
Goodwin, R. M.
Goonatilake, S., 49, 60
Green, L. R., 196, 215
Greene, W. H., 276, 300

Grossberg, S., 369, 371
Gulliken, H., 176
Gumerman, G., 14
Guy, R., 371

Haag, G.
Habermas, 370, 371
Habermas, J.
Haken, H., 277, 300
Hall P., 25, 74, 350
Halpin, B., 48, 60
Hamblin, R. L., 26, 44
Hamerle, A., 277, 300
Hamilton, A. J. J., 219, 238
Hanneman, R. A., 27, 28, 44
Hanssmann, F., 244, 270
Harary, F., 152, 177
Harding, A., 7, 14
Hargreaves-Heap, S., 134, 148
Harrison, R., 243, 248, 270
Harsanyi, J. C., 177, 183, 185, 195
Harton, 196, 212, 215
Harvey, A. C., 246, 270
Haslam, A., 216
Hayduk, L. A., 276, 300
Hazlehurst, B., 7, 14
Hebb, D. O., 358, 371
Hechter, M., 166, 177
Hegselmann, R., 8, 11, 14, 15, 16, 24, 65, 68, 74, 104, 113, 151, 152, 166, 167, 177, 182, 195, 197, 198, 199, 203, 204, 208, 212, 214, 215, 216, 245, 270, 271, 278, 300, 301, 356, 371, 372
Hendry, D. F., 246, 270
Herfurths, M., 60
Herreiner, D., 16
Hey, J. D., 178
Hiebeler, D., 87, 113
Hillier, F., 270
Hodgson, G.M.
Holland, J. H., 9, 14, 115, 124, 129, 130
Holland, J. R., 57, 60
Holland, P. W., 176
Holmstrom-Hintikka, G., 148
Homann, K., 195
Homans, G. C., 303, 313
Horwich, P., 22, 24
Hotelling, H., 180, 195
Huber, P., 74, 254, 256, 270
Huberman, B. A., 194, 245, 270
Huguet, P., 196, 215
Huhns, M. N., 19, 24, 85, 113
Hummon, N., 277, 300
Hutchins, E., 7, 14

Jaccard
Jaccard, J., 276, 300
Jackson, C., 196, 215
Jacobsen, R. B., 26, 44
Jarrett, J. E., 300
Jennings, N., 9, 15, 16, 19, 24, 68, 74, 75
Joereskog, K. G., 300
Johnson, A. W., 20, 25
Johnson, N. L., 222, 238
Jolley, J., 213, 216
Jonker, L. B., 181, 195

Kandori, M., 181, 191, 195
Karshenas, M.
Kauffman, S., 353, 354, 355, 356, 371
Kaylan, A. R., 270
Kelton, W. D., 243, 271
Kelvin, P., 279, 300
Kerr, N. L., 302, 313
Khebbal, S., 49, 60
Kiel, D. L., 265, 270
Kintsch, W., 48, 60
Kirchkamp, O., 179, 180, 182, 185, 193, 195
Kirman, A., 16
Kleijnen, J. P. C., 250, 252, 255, 270, 271
Klüver, J., 48, 49, 51, 57, 59, 60, 351, 356, 357, 361, 364, 365, 368, 369, 371
Knorr-Cetina, K., 351, 371
Köhler, E., 113, 177
Köhler, J., 251, 270
Kohler, T., 4, 14, 15
Kokic, P., 344, 345, 346, 347, 350
Kontopoulos, K. M., 7, 15
Korta, K., 148
Kotz, S., 222, 238
Koza, J. R., 10, 15, 261, 271
Kreifelts, T., 71, 74
Krüger, M.
Kruse, H., 50, 60

L'Herrou, T.
Laird, J. E., 8, 15
Langton, C., 15, 130, 194, 245, 271, 354, 356, 358, 372
Latané, B., 7, 15, 129, 130, 177, 196, 197, 198, 199, 203, 204, 208, 211, 212, 213, 214, 215, 216, 217, 243, 246, 248, 271
Law, A., 52, 243, 271, 272
Legler, D. M., 272
Lehmann, A., 270
Lehnert, D.
Lehrer, K., 157, 177
Leidlmeier, K., 60
Leinfellner, W., 113 , 177
Leinhardt, S., 176
Lemke, F., 260, 271
Lesser, M., 248, 273, 355, 372
Levesque, H. J., 66, 74
Levy, S., 353, 372
Lewenstein, M., 65, 68, 75, 356, 372
Lewis, D., 302, 313
Lieberman, G. J., 270
Liebl, F., 249, 250, 271
Liebrand, W., 65, 74, 148, 177, 239, 301
Lindgreen, K., 195
Lindzey, G., 216
Liu, J., 197, 212, 216
Lomborg, B., 246, 271

Macy, M., 166, 171, 177, 178, 243, 248, 271
Madison, J., 219, 238
Maes, P., 83, 86, 113
Mailath, G. G. , 181, 191, 195
Man Kit Chang, 74
Mangold, R., 60
Manhart, K., 48, 60
Mark, N., 65, 68, 74
May, R. M., 180, 182, 183, 191, 194, 195, 212, 215, 355, 372
Mayes, D. G., 259, 271
Maynard Smith, J., 181, 195
McGarty, C., 216
Mechler, B., 60
Mellars, P., 20, 24
Merz, J., 246, 271, 300
Messick, D., 65, 74, 148, 239
Metcalfe, St.
Metz, W., 243, 272
Meyers, S. D., 249, 272
Michalewicz, Z., 9, 15
Miles, K. F., 272
Mill, J., 219, 238
Miller, J., 56, 60
Miller, J. H., 246, 272
Miller, J. L. L., 26, 44
Minar, N., 117, 130
Mitchell, 213, 216
Mitchell, M.
Mitter, P., 300
Moffat, D., 84, 114
Möhring, M., 12, 15, 117, 130, 243, 251, 272
Möller V., 151, 164, 178, 197, 215
Molnar, P., 118, 130
Moon, L., 344, 350
Mooney, C., 128, 130
Moscovici, S., 196, 216
Moser, P. K., 131, 148
Mosteller, F., 171, 176
Moullin, B.
Müller G., 26, 27, 29, 44, 278
Müller J. P., 9, 15, 68, 74, 75, 114
Müller U., 14, 15, 16, 24, 44, 130, 177, 216, 269, 272, 271, 300, 301, 371, 372
Münch, R., 371
Myers, D. G., 196, 216

Nagl, W., 277, 300
Nauck, B., 300
Nelson, R., 181, 195
Newell, A., 8, 15
Newey, W. K., 344, 346, 350
Nicholson, M., 35, 44
Nordahl, M. G., 180, 182, 183, 195
Nordhaus, W.D.
Norusis, M. J., 5, 15, 36, 44
Norvig, P., 19, 25
Nowak, A., 7, 15, 65, 68, 75, 156, 177, 180, 182, 183, 191, 194, 195, 197, 198, 203, 208, 211, 212, 213, 214, 216, 217, 224, 238, 246, 248, 271, 301, 355, 356, 372

O'Brian, J. J., 272
Ohly, H. P., 48, 60
Okabe A., 164, 178

Olkin, I., 238
Orcutt, G. H., 275, 300
Ören, T. I., 270
Otter, P. W., 301

Paert-Enander, E., 35, 44
Palmer, M., 20, 24
Palmer, R., 272
Paolucci M., 302, 313
Parisi, D., 10, 15
Peitgen, H. O., 216, 300
Penger, O., 60
Pfaffenberger, B., 48, 60
Phillips, J., 35, 44
Pierreval, H., 256, 272
Plaisent, M., 249, 269
Pollack, J. B., 48, 61
Popper, K. R., 28, 44
Poundstone, W., 353, 372
Powell, J. L., 344, 346, 350
Press, W. H., 14, 15, 24, 25, 44, 59, 60, 73, 74, 75, 130, 148, 176, 177, 195, 215, 216, 217, 238, 239, 269, 270, 271, 273, 277, 299, 300, 313, 314, 371, 372
Price, G. R., 91, 11, 181, 195
Prietula, M. J., 14, 65, 74

Quincke, H., 300
Quine, W. v. O., 69, 75
Quinlan, J. R., 256, 272

Rao, A. S., 66, 75
Rasmussen, S., 372
Reiter, L., 243, 248, 272
Retti, J., 60
Reynolds, R. G., 10, 15
Rice, J. P., 271
Richardson, L. F., 27, 34, 35, 36, 39, 42, 44
Ríos Insua D., 251, 272
Rob, R., 113, 181, 191, 195
Rockloff, M. J., 115, 129, 130
Rohwer, G., 299
Rosenbloom, P. S., 15
Roth, A., 166, 178
Roughgarden, J., 271
Rousseau, J. J., 237, 239
Russell, S., 19, 25
Rust, J., 272

Saam, N. J., 243, 248, 265, 267, 272, 275, 301
Sakoda, J. M., 180, 182, 195
Saltelli, A., 255, 270, 272
Samuelson, L., 180, 182, 183, 194
Sandewall, E., 75
Sarin, R., 194
Schachter, S., 196, 215
Schäfer, A, 253, 272
Schaller, M., 196, 217
Schelling, T., 7, 15, 75, 81, 180, 182, 195, 219, 239, 356, 372
Schiffer, R., 69, 75
Schlag, K., 179, 189, 193, 195
Schmidt, B., 88, 96, 108, 114
Schmidt, J., 60, 351, 355, 368, 371
Schrage, L. E., 269, 300
Schumacher E.
Schumpeter, J., 181, 195
Schurmann, 260, 272
Schurmann, J.
Schüssler, R., 75
Schwartzman, H., 357, 372
Scott, M. F.
Screpanti, E.
Selten, R., 183, 185, 195
Shah, A.
Shaked, A., 179, 180, 182, 183, 194
Sheppard, H. L., 299
Shoham, Y., 19, 25, 302, 313, 314
Sichman, J., 133, 136, 148
Simon, H., 26, 44
Singer, H., 277, 300
Singh, M. P., 19, 24, 113
Sloman, A., 95, 96, 114
Smelser, N. J., 371
Smith, R. G., 71, 74
Sobol, I. M., 255, 272
Soerbom, D., 300
Stanley, E. A., 194, 278
Stephens, J. C., 272
Stevens, R. 280, 301
Stewart, I., 265, 272
Stoica, C., 48, 49, 51, 58, 59, 60, 361, 363, 367, 371, 372
Sugihara, K., 178
Suleiman, R., 123, 130, 196, 218, 219, 220, 228, 238, 239
Sundermeyer, K., 83, 84, 86, 114
Swingler, K., 9, 15
Szamrej, J., 197, 198, 204, 217, 238

Szczerbicka, H., 270

Taagepera, R., 35, 44
Tarski, A., 368, 372
Taube, M., 75
Taylor, C., 285, 372
Taylor, M., 169, 178
Taylor, P. D., 181, 195
Tenneholtz, M., 313, 314
Terna, P., 15, 16, 24
Tesfatsion, L., 194
Teukolsky, S. A., 277, 300
Thagard, P., 48, 60
Thaler, R. H., 69, 166, 176
Tibshirani, R. J., 344, 348, 350
Todorov, T., 21, 25
Topp, V., 344, 350
Toussaint, A., 96, 114
Trivers, R. L., 314
Troitzsch, K. G., 3, 9, 14, 15, 16, 24, 44, 48, 61, 74, 130, 177, 216, 271, 272, 275, 301, 371, 372
Tsang, E., 20, 25
Tulpulé, V., 344, 350
Tuomela, R., 148
Tutz, G., 300

Uhrmacher, A., 9, 16, 66, 75
Ullman-Margalit, E., 314

Vallacher, R., 115, 216
Van West, C., 4, 15, 16
Veneziano, V, 148
Vester, F., 247, 248, 257, 272
Vetterling, W. T., 300
Von Maravic, H., 272
Von Martial, F., 74
Vries, P. H., 275, 301

Waal, F. de, 302, 314
Wagner, C., 44, 157, 177
Walker, A., 302, 314
Waltz, D. L., 48, 61
Wan, C. K., 300
Wartenberg, T., 74
Wasserman, S., 128, 130
Watts, D. G., 299
Weber, G., 245, 269, 351, 372
Weibull, J. W., 195, 246, 273
Weinroth, J., 269
Weisbuch, G., 3, 16
Werner, E., 74, 113
Westmeyer, H., 74
Wielemaker, J., 72, 73, 75
Willke, H., 56, 61
Witte, E., 216
Wittmann, J., 252, 273
Wolfram, S., 358, 364, 372
Wooldridge, M., 9, 16, 19, 24, 68, 74, 75, 302, 314
Wright, J. P., 114
Wuensche, A., 248, 273, 354, 355, 357, 369, 372
Wunsch, G., 89, 114

Ye, M., 16
Young, H. P., 181, 191, 195

Zavalloni, M., 196, 216
Zeeman, E., 181, 195
Zeigler, B. P, 3, 16
Zhiang, L., 14
Ziegler, R., 275, 301
Zimbardo, P. G., 93, 114
Zito-Wolf, R., 68, 74